P9-DNC-265

Enzyme Catalysis and Regulation

MOLECULAR BIOLOGY

An International Series of Monographs and Textbooks

Editors: BERNARD HORECKER, NATHAN O. KAPLAN, JULIUS MARMUR, AND HAROLD A. SCHERAGA

A complete list of titles in this series appears at the end of this volume.

ENZYME CATALYSIS AND REGULATION

Gordon G. Hammes
Department of Chemistry
Cornell University
Ithaca, New York

1982

ACADEMIC PRESS
A Subsidiary of Harcourt Brace Jovanovich, Publishers

New York London
Paris San Diego San Francisco São Paulo Sydney Tokyo Toronto

ACADEMIC PRESS, INC.
111 Fifth Avenue, New York, New York 10003

United Kingdom Edition published by
ACADEMIC PRESS, INC. (LONDON) LTD.
24/28 Oval Road, London NW1 7DX

Library of Congress Cataloging in Publication Data

Hammes, Gordon G., Date
Enzyme catalysis and regulation.

(Molecular biology series)
Includes bibliographical references and index.
1. Enzymes. 2. Biological control systems.
I. Title. II. Series.
QP601.H26 574.19'25 82-1597
ISBN 0-12-321962-0 paper AACR2
ISBN 0-12-321960-4 cloth

PRINTED IN THE UNITED STATES OF AMERICA

82 83 84 85 9 8 7 6 5 4 3 2 1

Contents

7. Case Studies of Selected Enzyme Mechanisms

8. Regulation of Enzyme Activity

9. Case Studies of Selected Regulatory Enzymes

10. Multienzyme Complexes

11. Membrane-Bound Enzymes

Preface

This book is based on a course which I have been teaching on and off for 16 years. The course is intended for graduate students and advanced undergraduates. The principle difficulty I have encountered in writing this book (and in giving the course) was deciding what not to include. My desire is to have a book of reasonable size which well-prepared undergraduate and graduate students can use as an introduction to enzyme catalysis and regulation. At the same time, I have tried to make this book sufficiently up to date so as to be a useful reference for research workers. My belief is that if the material in this book is understood, no difficulty should be encountered in reading current literature on enzyme mechanisms. Unfortunately, in order to attain my goal of a reasonable size book, some important topics are necessarily omitted and only a curtailed discussion of others is presented. However, I hope that the scope and excitement of modern research on enzymes is evident.

Some background in biochemistry is assumed so that the first chapter on enzyme structure is relatively brief. The next chapter, which discusses methods of probing enzyme structure, also is not long; a complete discussion would require several additional volumes. Kinetic methods are discussed in some detail, although no attempt is made to provide a compendium of rate laws. Instead the emphasis is on general principles of steady-state and transient kinetics. An overall discussion of enzyme catalysis then attempts to draw together the chemical principles involved. Case studies of a few well-documented enzymes are presented next to illustrate the methods and principles developed earlier. The next two chapters are concerned with the regulation of enzyme activity from a nongenetic viewpoint: this includes a comprehensive discussion of binding isotherms and models for allosterism. Two particular enzymes are utilized as examples of well-studied regulatory enzymes. The last two chapters cover special topics of current interest, namely multienzyme complexes and membrane-bound enzymes. Finally a brief compendium of student-tested problems is provided in the appendix.

I am indebted to many colleagues for their critical comments on portions of the manuscript and for many stimulating discussions. I would especially like to acknowledge G. P. Hess, H. A. Scheraga, P. R. Schimmel, D. A. Usher, and C.-W. Wu. I am indebted to Dr. Richard Feldmann of the National Institutes of Health for the stereo representations of protein structures in Chapter 7. Chapter 8 is based on a review by C.-W. Wu and myself [*Annu. Rev. Biophys. Bioeng.* **3**, 1 (1974)] and Chapter 10 is based on a review appearing in *Biochem. Soc. Symp.* **46** (1981). I would like to thank Dr. Wu, Annual Reviews Inc., and the Biochemical Society for their permission to use portions of the original material. Special thanks are due to my wife, Judy, for her assistance in proofreading and to Joanne Widom for preparation of the index.

The preparation of this manuscript would not have been possible without the very able technical assistance of Connie Wright, Joan Roberts, and Jean Scriber. Since this book is a reflection of my research interests, I would like to acknowledge the financial assistance of the National Institutes of Health and the National Science Foundation, who have supported my research for many years.

Gordon G. Hammes

1

Protein Structure and Dynamics

Since all enzymes are proteins, a logical starting point for the discussion of enzymes is to consider the general features of protein structure. Features of particular interest to enzymology are considered. This topic is treated more fully in some of the references at the end of the chapter.

The *primary* structure of a protein is specified by the order in which the amino acids are linked together through peptide bonds. The most important feature of this structure is the peptide linkage shown in Fig. 1-1. Because of resonance, which gives the N—C bond some double bond character, the peptide bond is planar. Furthermore, the α carbons are always trans. These two features of the peptide bond play a dominant role in determining protein structure. The other covalent linkage of importance is the disulfide bond that joins different parts of the protein chain. Special note also should be made of the imino acids (e.g., proline), which create a very rigid peptide bond. The final structure of a protein is determined by the above factors and optimization of noncovalent interactions involving the peptide backbone and the amino acid side chains. The large variety of amino acid side chains provides the possibility of several different types of noncovalent interactions, namely, van der Waals, pi electron stacking, hydrogen bonding, and electrostatic. In addition, some of these side chains are important in enzymes as acid–base catalysts. The main types of amino acid side chains and their functions are summarized in Table 1-1.

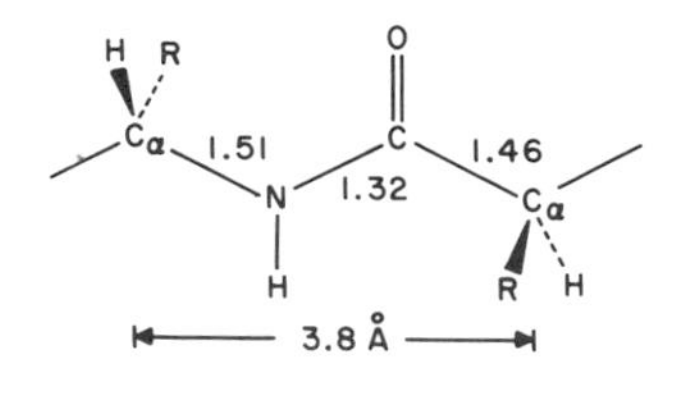

Fig. 1-1. The planar and trans peptide bond with standard bond distances in Ångstroms.

Table 1-1

Amino Acid Side Chains and Their Function

Side chain group	Amino acids	Functions and interactions
Hydrocarbon	Alanine, leucine	van der Waals
Aromatic	Phenylalanine, tyrosine, tryptophan	van der Waals, pi electron stacking
Carboxyl	Aspartate, glutamate	Electrostatic, hydrogen bonding, acid–base catalysis
Amino	Lysine, arginine	Electrostatic, hydrogen bonding, acid–base catalysis
Imidazole	Histidine	Electrostatic, hydrogen bonding, acid–base catalysis, van der Waals, pi electron stacking
Hydroxyl	Serine, threonine, tyrosine	Hydrogen bonding, acid–base catalysis
Amide	Asparagine, glutamine	Hydrogen bonding
Sulfhydryl	Cysteine	Hydrogen bonding, electrostatic, acid–base catalysis

To gain further insight into the nature of protein structure, noncovalent interactions are considered in more detail. Potentially the largest amounts of energy are available from electrostatic interactions. For example, the energy of interaction between two univalent charges is $e^2/\varepsilon r$, where e is the charge of an electron, ε is the dielectric constant, and r is the distance between the charges. In water ($\varepsilon = 80$) for charge separations of a few Ångstroms, the energy is a few kilocalories per mole. The correct dielectric constant to use when considering protein structures is not obvious, but is probably somewhere between that of water and organic solvents ($\varepsilon \sim 3$). Thus the energies involved could be substantial, particularly if the dielectric constant is low and/or clusters of charges are present in the protein. For ions and dipoles, the energy of interaction is $z\mu \cos\theta/\varepsilon r^2$, where z is the charge, μ is the dipole moment, and θ is a dipole orientation angle. In water for a univalent charge, a dipole moment of a few Debyes, and a separation distance of a few Ångstroms, the energy of interaction is less than 1 kcal/mole. However, since the water molecule itself has a substantial dipole moment and is present in large amounts, ion–dipole interactions can be important factors in protein structures.

Whereas the static aspects of electrostatic interactions can be readily formulated, the dynamics are more difficult to ascertain. A few kinetic studies of ion-pair formation have been carried out, and the rate constants appear to be those of diffusion-controlled reactions. The diffusion-controlled rate constants for association and dissociation reactions, k_f and k_d, respectively, can be approximated as (1)

$$k_f = 4\pi D_{AB} af \frac{N_0}{1000} \tag{1-1}$$

$$k_d = \frac{3D_{AB}f}{a^2} e^{[U(a)/kT]} \tag{1-2}$$

$$f = \left(a \int_a^\infty e^{U/kT} \frac{dr}{r^2}\right)^{-1}$$

$$K = k_f/k_d = \frac{4\pi a^3 N_0}{3000} e^{-U(a)/kT} \tag{1-3}$$

In these equations, D_{AB} is the sum of the diffusion constants of the two reactants, a is the distance of closest approach of the reactants, U is the potential energy of interaction between the two reactants, k is Boltzmann's constant, and N_0 is Avogadro's number. For reactions between small molecules with univalent charges of opposite sign, $k_f \sim 10^{10}\ M^{-1}\ \text{sec}^{-1}$ and $k_d \sim 10^{10}\ \text{sec}^{-1}$. As is amplified later for hydrogen-bonding reactions, this implies that the unimolecular rate constants for ion–ion interactions are greater than $10^{10}\ \text{sec}^{-1}$. A model for ion–dipole interactions is the solvation of metal ions by water. For water interactions with alkali metals, the characteristic rate constants for first hydration shell solvation are about $10^9\ \text{sec}^{-1}$. However, for higher valence metals, the water–metal dissociation rate can be considerably slower, and in extreme cases (e.g., Cr^{3+}) can be many hours. These slower rates are not likely to be relevant in proteins.

Another type of electrostatic interaction of great importance in proteins is the hydrogen bond. Some typical hydrogen bonds and lengths are illustrated in Table 1-2. In nonhydrogen-bonding solvents, typical enthalpies of formation are a few kilocalories per hydrogen bond per mole. For example,

Table 1-2

Some Typical Hydrogen Bonds and Bond Lengths

Hydrogen bond	Bond length (Å)
OH · · O	2.7
OH · · N	2.9
NH · · O	3.0
NH · · N	3.1

in $CHCl_3$ the dimerization of 2-pyridone

$$2\ \text{(NH, O)} \rightleftharpoons \text{(NH}\cdot\text{O, O}\cdot\text{HN)} \qquad (1\text{-}4)$$

has an equilibrium constant of 150 M^{-1} and an enthalpy change of -5.9 kcal/mole (*2*). However, in water hydrogen bond formation between solutes is not as favorable because water competes for the hydrogen bonds. Both the standard free energy changes and enthalpy changes for hydrogen bond formation are close to zero in water. In a protein, hydrogen bonds may be shielded from water and, therefore, may be quite stable. This stability is due to the creation of a special structure of the protein. Energetically this means that the stability of the hydrogen bond has been paid for by the energy required for a specific protein conformation. Hydrogen bonding can provide great specificity because of the different possible types of hydrogen bonds and the strong preference for linear bonding.

The dynamics of hydrogen bonding have been studied in a variety of model systems. Some typical data for the dimerization of 2-pyridone in weakly hydrogen-bonding solvents are presented in Table 1-3. In relatively weakly hydrogen-bonding solvents, such as the first four entries in Table 1-3, the association rate is essentially diffusion-controlled, whereas the association rate constants for the last two entries are considerably less than expected for a diffusion-controlled process. This can be understood in terms of the mechanism

$$2\text{P} \underset{k_{-1}}{\overset{k_1}{\rightleftharpoons}} \text{P}\cdots\text{P} \underset{k_{-2}}{\overset{k_2}{\rightleftharpoons}} \text{P—P} \underset{k_{-3}}{\overset{k_3}{\rightleftharpoons}} \text{P=P} \qquad (1\text{-}5)$$

where P represents pyridone, P $\cdot\cdot$ P is a nonhydrogen-bonded dimer that

Table 1-3

Thermodynamic and Kinetic Parameters for the Dimerization of 2-Pyridone

Solvent	ΔG° (kcal/mole)	ΔH° (kcal/mole)	$10^{-9}\,k_f$ (M^{-1} sec^{-1})	$10^{-7}\,k_r$ (sec^{-1})	Reference
$CHCl_3$	-3.0	-5.9	3.3	2.2	*2*
50 wt % dioxane–CCl_4	-2.5	-4.6	2.1	2.9	*3*
Dioxane	-1.6	-1.7	2.1	13.0	*4*
1% Water–dioxane	-1.3	—	1.7	17.0	*4*
CCl_4–dimethyl sulfoxide (1.1 m)	-0.4	—	0.26	14.8	*3*
CCl_4–dimethyl sulfoxide (5.5 m)	0.9	—	0.069	2.7	*3*

forms and dissociates by diffusion-controlled rates, P—P is the dimer with one hydrogen bond formed, and P=P is the complex with two hydrogen bonds formed. If the intermediates are assumed to be in a steady state, the observed rate constants for association and dissociation, k_f and k_r, are

$$k_f = \frac{k_1}{1 + (k_{-1}/k_2)(1 + k_{-2}/k_3)} \tag{1-6}$$

$$k_r = \frac{k_{-3}}{1 + (k_3/k_{-2})(1 + k_2/k_{-1})} \tag{1-7}$$

If the reaction is diffusion-controlled, $k_f \approx k_1$; this is the case when $k_2 > k_{-1}$, i.e., when desolvation of the solute and formation of the first hydrogen bond is faster than diffusion apart of the reactions. Since the value of k_{-1} is about 10^{10} sec^{-1}, the actual rate constant for hydrogen bond formation, k_2, must have a value of 10^{11}–10^{12} sec^{-1}. For the last two entries in Table 1-3, the solvent can form strong hydrogen bonds so the association rate is no longer diffusion-controlled. In these cases, desolvation of the solute is rate determining with a specific rate constant of about 10^8 sec^{-1}. In fact, this rate constant is characteristic of most solvation–desolvation processes involving hydrogen bonds. Thus, the rate constant for the making and breaking of single hydrogen bonds in water is $\geq 10^8$ sec^{-1}. Note that when the dimer formation is diffusion-controlled, $k_r = k_{-1}(k_{-2}/k_2)(k_{-3}/k_3)$. Since k_{-1} is about the same in all cases, k_r is a measure of the thermodynamic stability of the two pyridone–pyridone hydrogen bonds relative to pyridone–solvent interactions.

Hydrophobic interactions are usually rather loosely defined to describe what happens when hydrocarbons are put into water. Actually several types of interactions should be distinguished. Hydrocarbons interact very weakly with each other due to dispersion forces. Also, planar pi electron systems tend to stack on top of each other. Both of these interactions are very short range. When hydrocarbons are put into water, the dominant factor is that hydrocarbons and water do not like to associate. Thus, if a hydrocarbon is solubilized by water, the water tends to form a sheaf around the hydrocarbon in which the water dipoles are strongly oriented through hydrogen bonding. The free energy associated with such interactions can be estimated from measurements of the free energy for the transfer of hydrocarbons from water to a nonpolar solvent. For example, the free energies of transfer for a methylene group and aromatic ring are about -0.7 kcal/mole and -2 kcal/mole, respectively (*5*). If two hydrophobic molecules are present in water, they tend to associate, not because of the strong interactions between the hydrophobic molecules, but because some of the oriented water molecules are released due to the decreased hydrophobic surface area. Schematically this can be represented as

$$\bigcirc + \bigcirc \rightleftharpoons \bigcirc\!\bigcirc + n\mathrm{H_2O}$$

This process has an unfavorable enthalpy change, but a very favorable entropy change. In terms of protein structure, this causes the hydrophobic residues to associate inside the structure—as far as possible from water.

The dynamics of hydrophobic interactions have been studied in model systems by measuring the rates of water solvation and desolvation of hydrophobic molecules. Some typical data are summarized in Table 1-4. The individual rate constants are $\geq 10^8$ $\sec^{-1}$ so that such interactions clearly are very rapid.

While the noncovalent interactions in proteins are generally very weak on an individual basis, hundreds of these interactions exist so that the total energy involved is very large. The hydrophobic interactions can be thought of as the dominant, rather nonspecific, forces leading to the establishment of a protein structure, with hydrogen bonding and electrostatics being especially important in generating specificity. A delicate balance between these noncovalent interactions is achieved in the native protein structure. Since the rate constants for forming and breaking noncovalent interactions are quite large, our discussion thus far would lead us to expect a very floppy and rapidly changing structure. This is not the case because a very important factor has not yet been discussed, namely, *cooperativity*, which coordinates the many weak noncovalent interactions.

The essence of cooperativity for proteins is simple: when a single noncovalent interaction occurs, a second occurs more readily than the first, a third more readily than the second, etc. An example of such a process in a polypeptide is the helix–coil transition. For example, polyglutamic acid exists in an α-helical structure at low pH values in water. The α helix is a

Table 1-4

Time Constants for Solvation–Desolvation of Hydrophobic Molecules

Species	Time constant ($\sec^{-1}$)[a]	Reference
$(\mathrm{PhCH_3})_2\mathrm{NCH_3 \cdot H_2O}$	2.7×10^9	6
Dioxane $(\mathrm{H_2O})_2$	2.8×10^8	7
$(\mathrm{Dioxane})_2(\mathrm{H_2O})_2$	1.0×10^8	7
Glycine or diglycine or triglycine $\cdot\mathrm{H_2O}$	4×10^8	8
Polyethylene glycol	$\sim 10^8$	9

[a] Water dissociation rate constant for first three entries; sum of solvation–desolvation rate constants for the last two entries.

springlike structure with residues three spaces apart on the chain joined by a hydrogen bond between a peptide oxygen and a nitrogen proton. A right-handed α-helical structure is shown schematically in Fig. 1-2a. Note that the hydrogen bonds are approximately parallel to the axis of the helix; 3.6 amino acids are in each turn of the α helix. As the pH is raised, the carboxyl groups ionize and over a very short range in pH the peptide

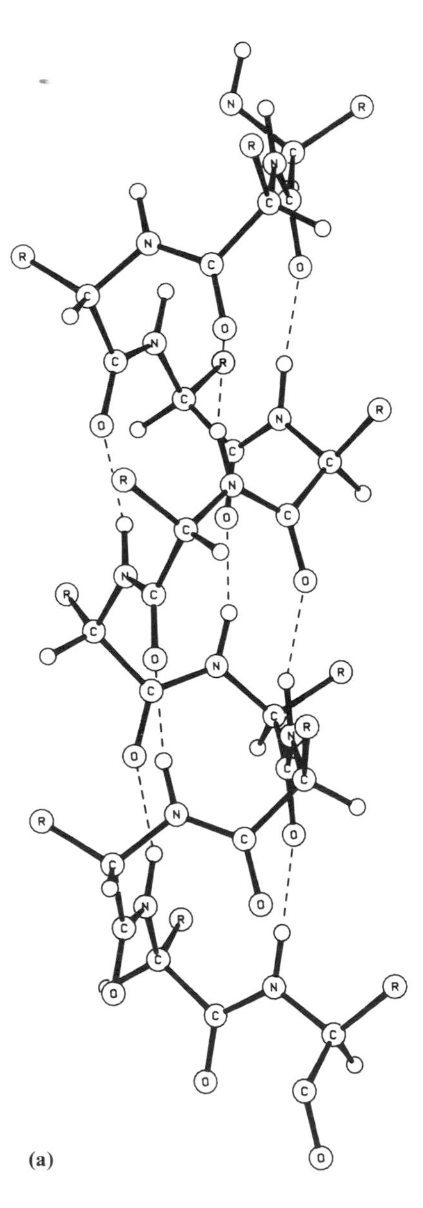

(a)

Fig. 1-2. (a) The right-handed α-helix frequently found in proteins. (b) The anti-parallel β-pleated sheet frequently found in proteins. Hydrogen bonds are designated by dashed lines. (These computer simulations were kindly provided by Marcia Pottle.)

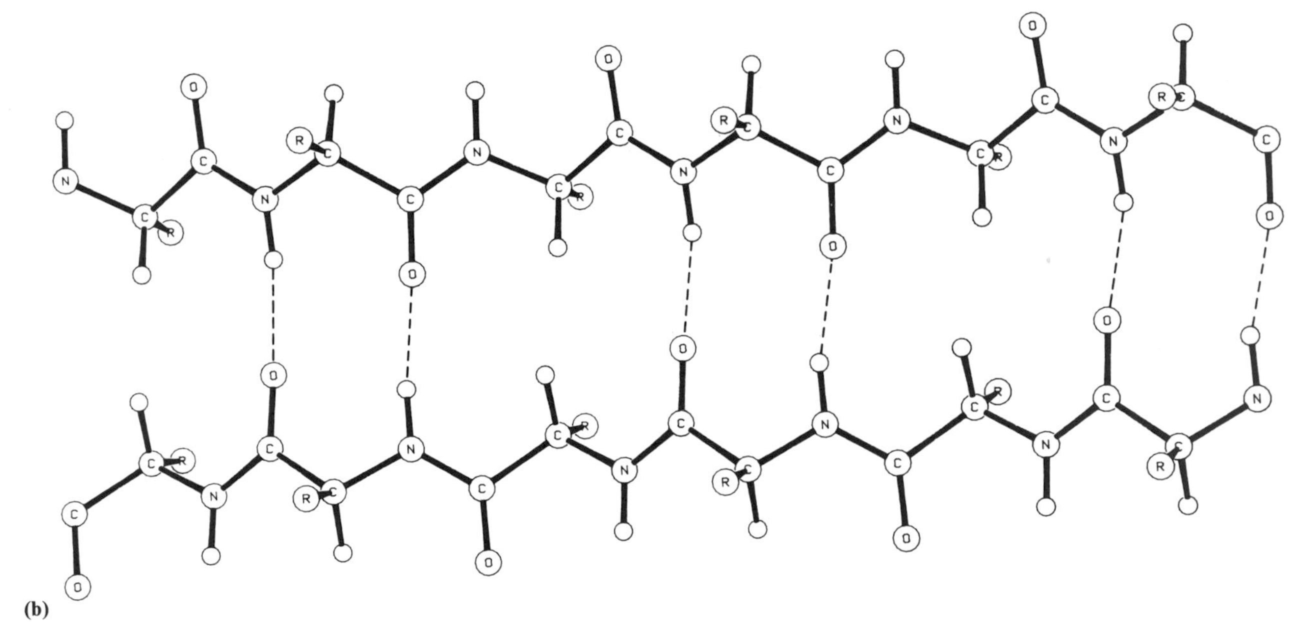

Fig. 1-2. *(Continued)*

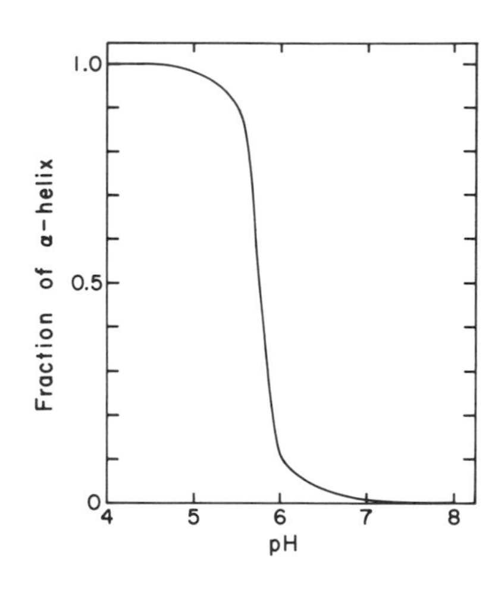

Fig. 1-3. Schematic representation of the helix-coil transition for polyglutamic acid in 0.2 *M* NaCl, dioxane–water (2:1) at 25°C. [Data from P. Doty, *J. Polym. Sci.* **23,** 815 (1957).]

structure becomes a random coil. This behavior is illustrated in Fig. 1-3. The occurrence of this transition over a very narrow range in pH is typical of a cooperative process. For polyglutamic acid the hydrogen bonds between peptide linkages are the dominant interactions involved in the cooperative transition. For proteins similar types of cooperative processes occur, but a larger variety of noncovalent interactions are involved in the cooperativity. The most common manifestation of cooperativity in proteins is denaturation, which occurs over a very narrow range of temperature or pH (e.g., boiling an egg). Elegant theories of cooperative processes have been developed but are not considered here (*10*).

In terms of protein structure, the cooperative nature of the noncovalent structures restricts proteins to a very limited number of structures which are thermodynamically more stable than any others. What does cooperativity do to the dynamics of the noncovalent interactions? For polyglutamic acid, the helix and random coil interconvert at the midpoint of the transition with a specific rate constant of about 10^6 sec^{-1} (*11*). This is considerably smaller than the rate constant for the elementary step of hydrogen bonding ($>10^8$ sec^{-1}), but the interconversion is quite rapid. Thus, the final picture of a protein that emerges is a quite stable structure which can be modulated quite rapidly. In fact, such modulations can occur as rapidly as the simple helix–coil transition but may also occur very slowly. For example, protein denaturation can take hours to occur, and some conformational transitions have been observed that take days.

A local structure in a protein such as the α helix is termed *secondary* structure. Another important secondary structure is the β-pleated sheet, which again is stabilized by hydrogen bonds between peptide linkages. A parallel β-pleated sheet forms when the chains are aligned in the same

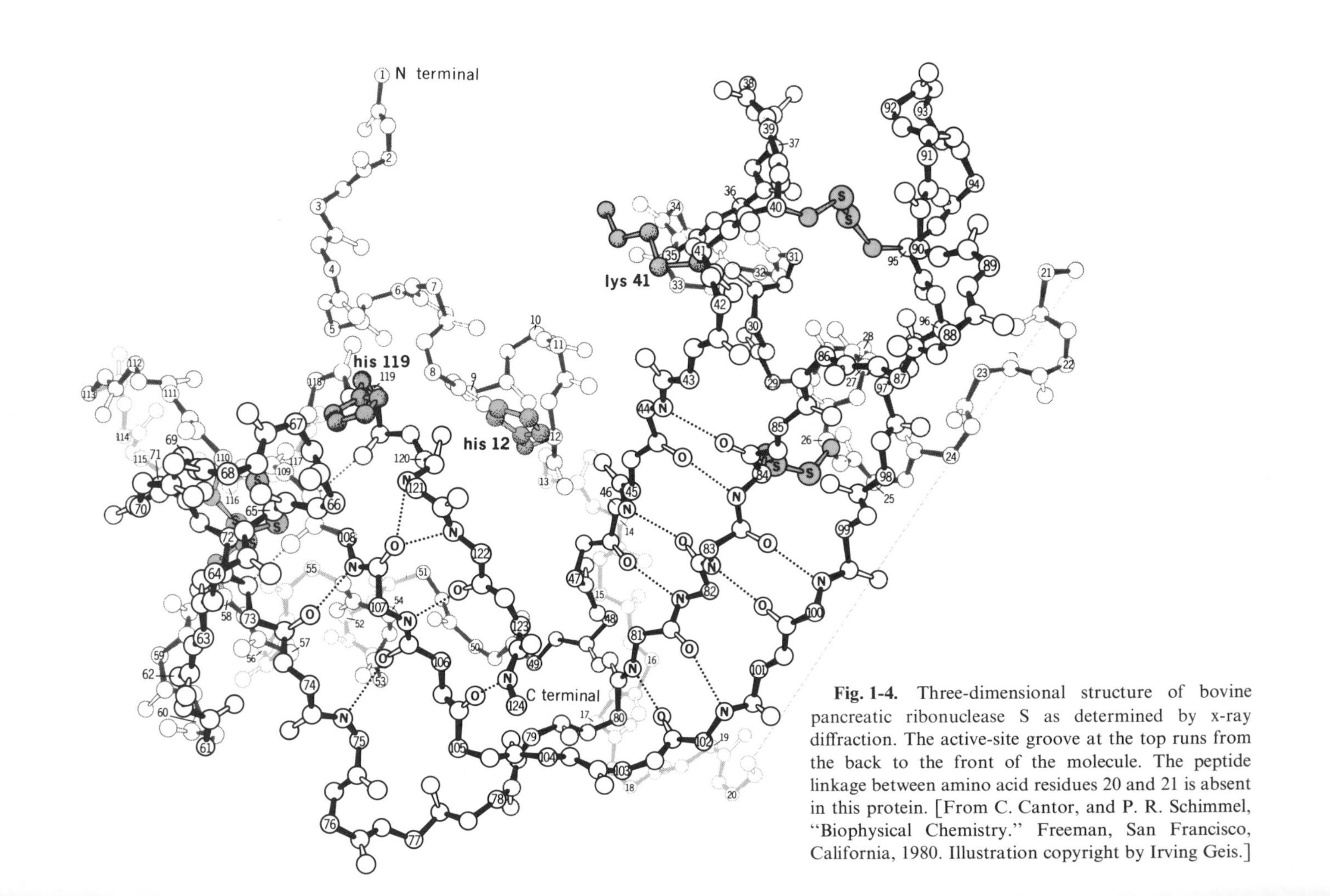

Fig. 1-4. Three-dimensional structure of bovine pancreatic ribonuclease S as determined by x-ray diffraction. The active-site groove at the top runs from the back to the front of the molecule. The peptide linkage between amino acid residues 20 and 21 is absent in this protein. [From C. Cantor, and P. R. Schimmel, "Biophysical Chemistry." Freeman, San Francisco, California, 1980. Illustration copyright by Irving Geis.]

direction; when the chains alternate in direction, an antiparallel β-pleated sheet is formed. An antiparallel β sheet is shown in Fig. 1-2b. The *tertiary* structure of a protein is the complete three-dimensional structure of the polypeptide chain. If multiple polypeptide chains are present, the arrangement of the polypeptide chains with respect to each other is the *quaternary* structure. For example, the three-dimensional structure of the enzyme bovine pancreatic ribonuclease S is illustrated in Fig. 1-4. The groove in the middle of the structure is where the substrates bind.

Many enzymes utilize prosthetic groups (coenzymes) to assist in their catalytic functions. These groups may play a specific catalytic role or a structural role; their interactions with the protein range from relatively loosely bound to covalent incorporation into the protein structure. Prosthetic groups generally are metal ions or rather complex organic molecules. Some important prosthetic groups are shown in Fig. 1-5.

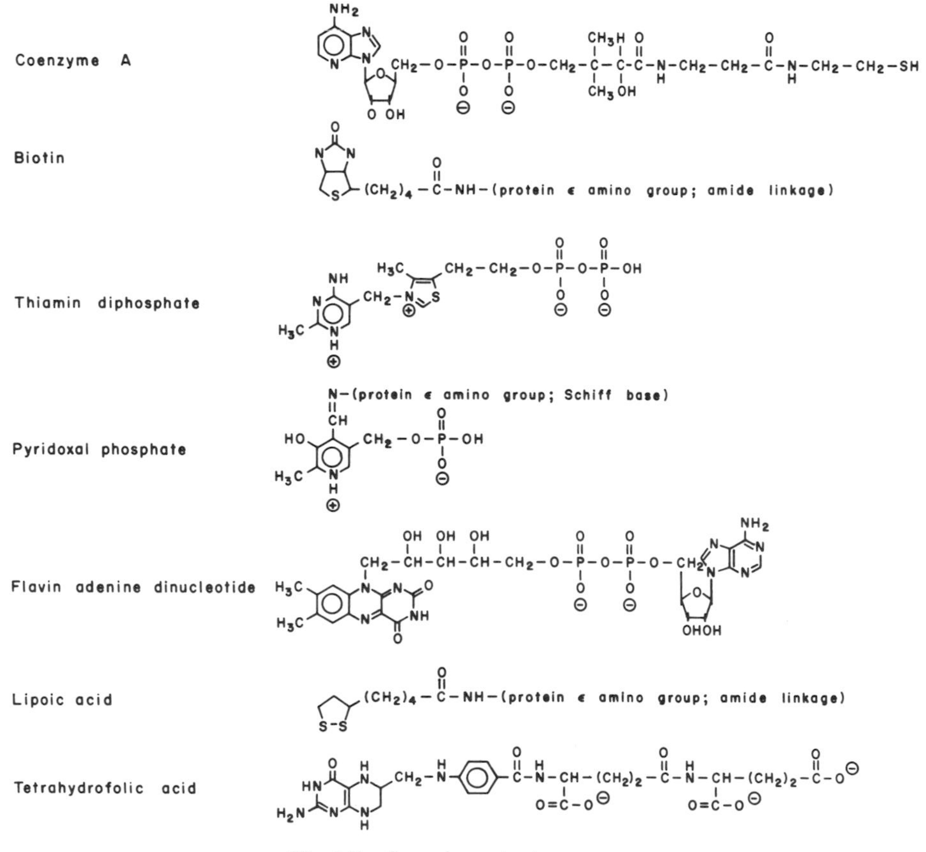

Fig. 1-5. Some important coenzymes.

The binding of substrate to the enzyme is a necessary feature of all enzymatic reactions. The noncovalent interactions involved in the stabilization of the enzyme–substrate complex are the same as those involved in the stabilization of the protein structure. Many interactions are involved, but not a sufficient number for true cooperativity to exist. The strength of the interaction between enzyme and substrate varies considerably: equilibrium dissociation constants generally fall in the micromolar to millimolar range. In all cases studied thus far, very specific binding sites for substrates are found on enzymes; these are called active sites. The site generally is a pocket or groove (cf. Fig. 1-4), with a structure very closely complementary to the substrate: hydrogen bonds form between the substrate and protein backbone or side chains, nonpolar parts of the substrate fit with nonpolar parts of the protein, and favorable interactions occur between electrically charged parts of the molecules. The three-dimensional structure of a dinucleotide binding to the active site of ribonuclease is shown in Fig. 1-6. As is discussed in detail later, the binding of substrates generally triggers a cooperative conformational change of the protein which places the substrate in a hydro-

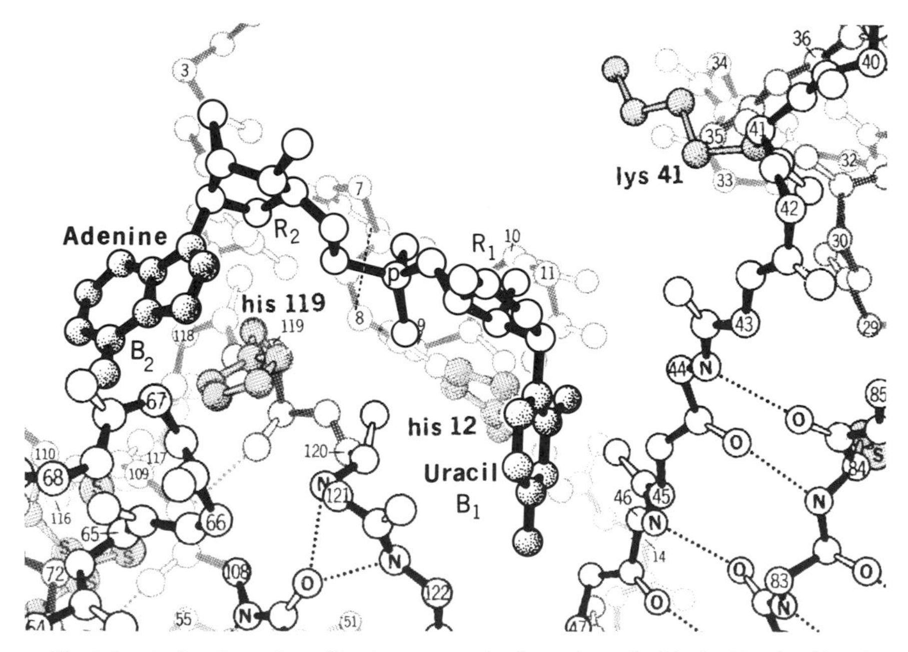

Fig. 1-6. Active site region of bovine pancreatic ribonuclease S with the dinucleotide substrate analog UpcA (uridylyl-3′,5′-adenosine with the 5′ oxygen replaced by a CH_2 group) as determined by x-ray diffraction. B_1 and B_2 are the bases of the UpcA, and R_1 and R_2 are the ribose portions of the UpcA. [From C. Cantor and P. R. Schimmel, "Biophysical Chemistry." Freeman, San Francisco, California, 1980. Illustration copyright by Irving Geis.]

phobic pocket, optimally oriented for catalysis. Thus the noncovalent interactions within the protein are the same as those between enzyme and substrate, and cooperative conformational transitions can be triggered by the addition of enzyme–substrate interactions to the overall structure.

REFERENCES

1. G. G. Hammes, "Principles of Chemical Kinetics." Academic Press, New York, 1978.
2. G. G. Hammes and A. C. Park, *J. Am. Chem. Soc.* **91**, 956 (1969).
3. G. G. Hammes and P. L. Lillford, *J. Am. Chem. Soc.* **92**, 7578 (1970).
4. G. G. Hammes and H. O. Spivey, *J. Am. Chem. Soc.* **88**, 1621 (1966).
5. C. Tanford, *J. Am. Chem. Soc.* **84**, 4240 (1962).
6. E. Grunwald and E. K. Ralph, III, *J. Am. Chem. Soc.* **89**, 4405 (1967).
7. G. G. Hammes and W. Knoche, *J. Chem. Phys.* **45**, 4041 (1966).
8. G. G. Hammes and N. C. Pace, *J. Phys. Chem.* **72**, 2227 (1968).
9. G. G. Hammes and T. B. Lewis, *J. Phys. Chem.* **70**, 1610 (1966).
10. D. Poland and H. A. Scheraga, "Theory of Helix-Coil Transitions in Biopolymers." Academic Press, New York, 1970.
11. A. F. Barksdale and J. E. Stuehr, *J. Am. Chem. Soc.* **94**, 3334 (1972).

GENERAL REFERENCES ON PROTEINS AND ENZYMES

P. D. Boyer, ed., "The Enzymes," 3rd ed. Academic Press, New York [a multivolume treatise].

C. R. Cantor and P. R. Schimmel, "Biophysical Chemistry," Parts I, II, III. Freeman, San Francisco, California, 1980.

A. Cornish-Bowden, "Fundamentals of Enzyme Kinetics." Butterworth, London, 1979.

A. Fersht, "Enzyme Structure and Mechanism." Freeman, San Francisco, California, 1977.

W. P. Jencks, "Catalysis in Chemistry and Enzymology." McGraw-Hill, New York, 1969.

A Meister, ed., "Advances in Enzymology." Wiley (Interscience), New York [a multivolume treatise].

D. V. Roberts, "Enzyme Kinetics." Cambridge Univ. Press, London and New York, 1977.

I. H. Segel, "Enzyme Kinetics." Wiley (Interscience), New York, 1975.

C. Walsh, "Enzymatic Reaction Mechanisms." Freeman, San Francisco, California, 1979.

2

Probes of Enzyme Structure

INTRODUCTION

Many methods of studying enzyme structure exist, and only a few of the more common methods are discussed. For a more complete discussion see, for example, Ref. *1*. Particular emphasis is placed on some of the methods most often used, namely chemical modification, magnetic resonance, and fluorescence.

AMINO ACID SEQUENCING

For a complete knowledge of the structure of a protein, the amino acid sequence is essential. Micro methods and automated instrumentation now permit the determination of the amino acid sequence of virtually any protein that can be obtained reasonably pure in milligram quantities. Alternatively, DNA sequencing is now so easy to do that in many cases the DNA sequence coding for a specific protein is used to deduce the amino acid sequence. Of course, sequencing is still a nontrivial task, especially if the peptide chain is longer than a few hundred amino acid residues. Nevertheless, hundreds of sequences have been determined, and the publication of complete sequences is commonplace. As an example, the complete amino acid sequence of bovine ribonuclease A is shown in Fig. 2-1 (*2*). By convention, the numbering of amino acids always begins at the amino terminus of the polypeptide. Knowledge of the amino acid sequence does not yet permit prediction of the three-dimensional structure, although many theoretical studies are directed toward this goal (cf. *3*).

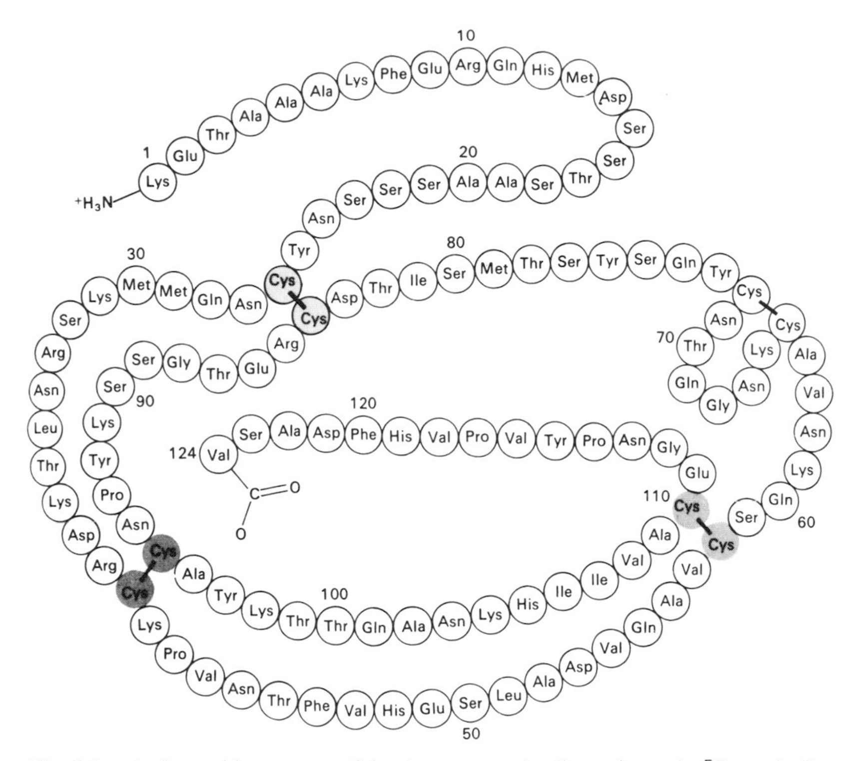

Fig. 2-1. Amino acid sequence of bovine pancreatic ribonuclease A. [From L. Stryer, "Biochemistry," 2nd ed., Copyright © 1981, Freeman, San Francisco, California; original data from C. H. W. Hirs, S. Moore and W. H. Stein, *J. Biol. Chem.* **235**, 633 (1960).]

X-RAY CRYSTALLOGRAPHY

The only certain method for arriving at the three-dimensional structure of a protein is x-ray crystallography. Such structure determinations are still nontrivial: very good crystals and heavy atom derivatives must be obtained, and a formidable amount of data must be collected and analyzed. Nevertheless, the structure of proteins with molecular weights of 50,000 or less can now be regarded as almost routine, and even large structures are actively being pursued. In this regard, the image-reconstruction methods of electron microscopy appear to be very promising additions to x-ray crystallography as primary structural methods (*4*). The resolution of even the very best structures is a few Ångstroms. However, recent x-ray studies at very low temperatures may improve the resolution substantially.

For the enzymologist, a very important question that must be answered before accepting the three-dimensional structure determined with x-ray crystallography is, does the structure represent an active conformation of the enzyme? This question can be most readily answered by carrying out activity measurements on the crystals used for x-ray crystallography. Even in optimal cases, the structure determined generally represents the enzyme locked into a particular conformation. Structural determinations with specific ligands (substrate or inhibitors) bound to the enzyme can be revealing in probing this problem. Low temperature x-ray studies offer special promise since reaction intermediates could be isolated and their structures determined directly. At this point in time, the tertiary structures of tens of enzymes are known. A representation of the ribonuclease S structure has been given in Fig. 1-4.

CHEMICAL MODIFICATION

One of the most commonly used (and misused) methods for studying enzyme structure is chemical modification of specific amino acid residues of the protein (cf. *5*, *6*). An ideal experiment would be to modify one amino acid residue on the enzyme and to determine its effect on the enzymatic activity. Two possible outcomes to this experiment are possible: the enzyme activity is altered or it is unchanged. In the former case, the modified group may be at the active site or quite distant from the active site; since these possibilities cannot be distinguished, the interpretation of the results is ambiguous. In the latter case, the modified group obviously is not involved in the active site, and generally is not of much interest. Thus, even in an ideal experiment, the results of chemical modification cannot be interpreted unambiguously. In real experiments several additional difficulties exist. First, many functional groups of the same type exist on proteins, and all tend to have similar reactivity; therefore, obtaining absolute specificity is difficult. Second, chemical reagents usually are not even specific for group type. Finally, chemical modification usually alters the native structure somewhat, which in turn can alter the enzyme activity. Because of these difficulties, after chemical modification the stoichiometry and heterogeneity of the labeling must be determined, and if possible the specific residues modified. The best analytical method for this purpose obviously is x-ray crystallography since this also would indicate whether a conformational change has occurred. This technique is seldom used simply because the labor is too great. The second choice of analysis is amino acid sequencing, with determination of the specific residue(s) modified. This method is often used, although a considerable expenditure of effort is required. By far the most prevalent methods of determining stoichiometry are spectroscopy (ultraviolet and visible difference spectra and fluorescence) and the use of radioactive labels. These methods, of course, do not give a direct indication of the heterogeneity of labeling or the specific residue(s) modified. In fact, spectroscopic methods permit determination of the stoichiometry only if the spectral properties of the label on the enzyme are precisely known.

A discussion of the many specific reagents that have been utilized to modify enzymes is not profitable for our purposes. However, consideration of some of the factors influencing protein functional group reactivity and reagent group reactivity are now considered. The microenvironment of the protein functional group to be modified is an important factor in the group reactivity. The polarity of the microenvironment can vary considerably; this causes differences in the ionization constants of functional groups which causes differences in reactivities. The polarity of the microenvironment generally has a large effect on tyrosine, cysteine, and carboxyl groups; a smaller effect on amines and imidazole; and not much effect on tryptophan, methionine, and cystine. In a similar vein, internal hydrogen bonding can alter ionization constants. For example, the pK of the carboxyl group in salicylic acid is lowered about one unit through hydrogen bonding to the neighboring hydroxyl, whereas the pK of the hydroxyl group is raised three units through hydrogen bonding to the carboxyl

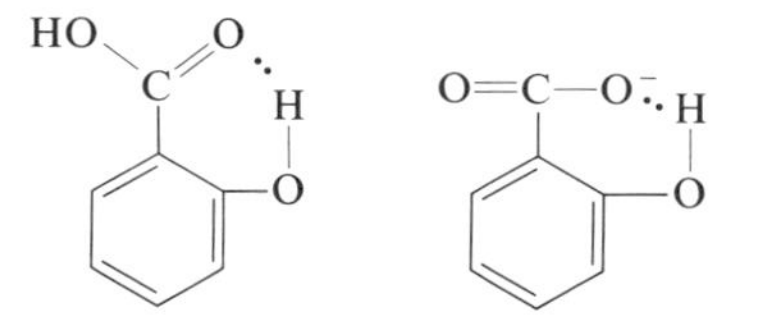

Electrostatic effects also can alter ionization constants: the presence of positive charges encourages formation of proton-poor forms, whereas the presence of negative charges encourages the formation of proton-rich forms. Other factors of importance in functional group reactivity include steric restrictions, charge transfer (such as the overlap of aromatic rings), covalent bond formation (such as with coenzymes), and ligand and metal binding to the protein. In particular, the binding of substrates and inhibitors to the enzyme frequently alters functional group reactivity. Variation in the nucleophilicity of protein functional groups merits special attention as a means of selective group modification. All proton-poor forms are nucleophilic, but the nucleophilicity does not necessarily parallel the ionization constants. However, the reactivity can be varied in similar ways to the ionization constant, i.e., variations in pH and temperature, ligand binding, etc. The order of group reactivity varies with the type of reaction under consideration. For example, the order of reactivity for alkylation is methionine, cysteine, imidazole, amine, phenol, carboxyl, whereas it is cysteine, phenol, imidazole, amine, carboxyl for acylation.

The most important factor that can influence reagent group reactivity is selective adsorption of the reagent on the surface to a restricted number of binding sites. This may be due to rather nonspecific noncovalent interactions, i.e., electrostatic, hydrogen bonding, and hydrophobic, or may be due to specially designed organic molecules that fit into a specific binding site. Electrostatic effects can play a role not only in the binding process but also in the orientation of the reagent. Steric effects can be of major importance. For example, the imidazole on histidine-12 of ribonuclease A is alkylated by α bromo acids: two to four carbon acids work well, five carbon acids work with difficulty, and six carbon acids do not react (*7*). The polarity of the local environment obviously can influence reagent reactivity as well as the protein functional group reactivity, and neighboring groups on enzymes can even serve as catalysts for the chemical modification.

Two types of reagents merit special mention: bifunctional reagents (*8*) and photoaffinity reagents (*9*). Bifunctional reagents have two chemically reactive groups so that they can modify two different groups simultaneously. By using reagents of different lengths information about the spatial proximity of the two modified groups can be obtained. One of the most useful of these reagents is dimethyl suberimidate

$$CH_3O\underset{\underset{NH_2^+}{\|}}{C}(CH_2)_6\underset{\underset{NH_2^+}{\|}}{C}OCH_3$$

because it is fairly specific for amino groups; also the number of methylene groups, and hence its length, can be varied. A related reagent, dimethyl-3,3′-dithiobispropionimidate

$$CH_3O\underset{\underset{NH_2^+}{\|}}{C}(CH_2)_2{-}S{-}S{-}(CH_2)_2\underset{\underset{NH_2^+}{\|}}{C}OCH_3$$

also is used frequently. It has the advantage that the cross-link can be broken after modification by the addition of a reducing agent such as β-mercaptoethanol. In principle, reversible modifications are useful because reversibility is a good indication that the enzyme has not been damaged during modification. Photoaffinity labels are compounds that are stable in the absence of light but become very reactive upon photolysis. Typical reagents are diazo compounds or azides, which give highly reactive carbenes and nitrenes, respectively

$$R\overset{\overset{O}{\|}}{C}{-}CHN_2 \xrightarrow{\text{light}} R\overset{\overset{O}{\|}}{C}{-}\overset{\overset{H}{|}}{C}\!: + N_2$$

$$R{-}N_3 \xrightarrow{\text{light}} R{-}N\!: + N_2$$

The trick is to design a reagent that will bind specifically to the enzyme in the dark; light activation will result in the very reactive species generated reacting with a protein side chain group in the immediate vicinity.

A few examples of studies of the chemical modification of proteins are now considered. Ribonuclease A is a protein of molecular weight 14,000. At pH 5.5–6.0 histidine-12 or histidine-119 can be alkylated readily. At pH 2.8 methionine is also alkylated, and at pH 8.5–10 the lysines also are modified. A detailed analysis of the histidine modification by iodoacetate indicates that either 1-carboxymethyl-histidine-119 or 3-carboxymethyl-histidine-112 is formed, never both, and that both modified enzymes are inactive (*7*). The size and stereochemistry of the acid and the nature of the halogen were varied extensively; some selected results are shown in Table 2-1 where the rate constants for modification are shown for selected reagents. Several points should be noted. First, the reactivity of the enzyme histidines is considerably greater than that of L-histidine. Second, the rate of the reaction is very sensitive to chain length, stereochemistry, and halogen. Third, the partitioning of the modification between histidine-12 and histidine-119 also depends on chain length and stereochemistry. These and other results were interpreted in terms of a very stereorestricted site on the protein, with the two residues being modified about 5 Å apart. Furthermore, the inactivity of the modified enzymes suggested the histidines were at the active site. These conclusions are well substantiated by the x-ray structure. A general problem that merits mention is the difficulty in determining whether a chemical modification completely eliminates the enzymatic activity or the chemically modified species has a very small residual activity. This is because trace amounts of unmodified enzyme are virtually impossible to eliminate, and this means some activity will be observed even if the chemically modified species is completely inactive. The reaction of the cross-linking reagent dimethyl adipimidate

$$CH_3O\underset{\underset{NH_2^+}{\|}}{C}(CH_2)_4\underset{\underset{NH_2^+}{\|}}{C}OCH_3$$

with ribonuclease also has been studied (*10*). Enzymatically active derivatives in which lysine-31 and -37 and -7 and -37 are cross-linked have been

Table 2-1

Rate Constants for the Modification of Ribonuclease A (7)

	$10^4 \times$ Rate constant (M^{-1} sec^{-1})			
Reagent	His-119	His-12	Total	L-Histidine
Iodoacetate	51.1	7.3	58.4	—
Bromoacetate	184.5	20.5	205.	0.086
D-α-Bromopropionate	1.84	4.16	6.00	0.0028
L-α-Bromopropionate	0.66	0.19	0.85	0.0027

isolated. This indicates these lysines are not essential for activity and that lysine -7 and -31 are both near lysine-37.

A classical example of an active site affinity label is tosyl-L-phenylalanine chloromethyl ketone (*11*)

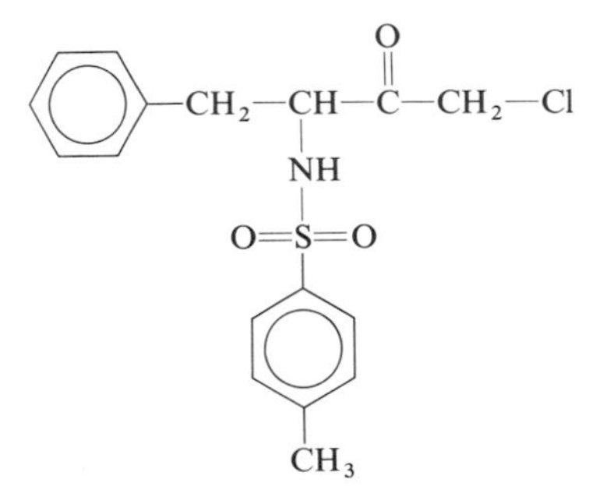

Except for the substitution of the chloromethyl ketone group for methyl and other esters, this is identical to substrates for the enzyme chymotrypsin which catalyzes ester hydrolysis. The chloromethyl ketones are good alkylating groups for sulfhydryls and imidazoles. This reagent was found to inactivate chymotrypsin with a maximum stoichiometry of one label per enzyme molecule, as determined with a radioactive label. In addition, this inactivation was inhibited by substrates and reaction did not occur with denatured enzyme. Amino acid sequencing of the modified protein showed that histidine-57 was alkylated. These results indicate the reagent binds specifically at the catalytic site and modifies a nearby histidine residue.

OPTICAL SPECTROSCOPY

Only a brief description of some of the more important spectroscopic techniques is presented here. Absorption spectroscopy is a useful tool for looking at protein structure because the peptide bond has a strong electronic absorption band in the far ultraviolet (210–220 nm), which is dependent on the conformation of the protein. In addition, the aromatic side chains phenylalanine, tyrosine, and tryptophan have characteristic electronic absorption bands in the near ultraviolet (250–290 nm) that are environment-dependent. Imidazole and disulfides also have absorption bands in the ultraviolet. Thus changes in protein conformation due to variation in pH, binding of ligands, etc. can be monitored by ultraviolet difference spectroscopy. Infrared and Raman spectroscopy, which monitor vibrational transitions, have proved to be less useful because of the complexity of protein spectra, although interesting applications have been made (*12*, *13*). Measurements of the optical rotatory dispersion and/or circular dichroism measure the optical activity of enzymes and have proved to be very useful indicators of the protein conformation since specific structures such as the α helix or β sheet have very characteristic spectra (cf. *1*).

Fluorescence spectroscopy offers uses similar to those described above, and in addition permits the mapping of protein structures through resonance energy transfer measurements (*14, 15*). The excitation of a molecule from its ground singlet electronic state to an electronic excited singlet state is shown schematically in Fig. 2-2. The molecule generally starts in its ground vibrational energy level, but ends up in a variety of vibrational energy levels in the excited electronic energy state. The excitation process occurs in about 10^{-15} sec, and the molecules end up in the ground vibrational energy level of the excited electronic state in about 10^{-12} sec through radiationless transitions (collisions with neighboring molecules and generation of heat). The return to the ground electronic state can occur either through radiationless transitions or through fluorescence which typically occurs in 10^{-8}–10^{-10} sec. As shown in Fig. 2-2, vibrational energy is lost following excitation and fluorescence so that fluorescence always occurs at longer wavelengths than the excitation wavelength. In some cases, the excited singlet state converts to a low lying triplet state (Fig. 2-2). Conversion to the singlet ground state with the emission of light then occurs very slowly since the transition is quantum mechanically forbidden. This is phosphorescence.

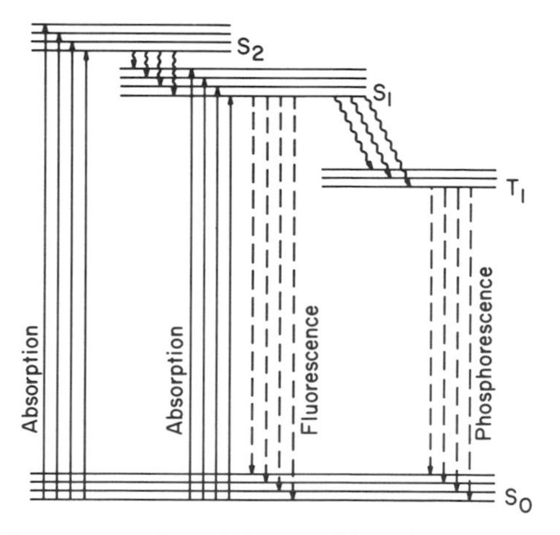

Fig. 2-2. Schematic representation of the transition of a molecule from a ground level singlet state (S_0) to excited singlet (S_1, S_2) and triplet (T_1) states and subsequent fluorescence and phosphorescence. The solid lines represent absorption of light, the dashed lines represent radiative transitions, and the wavy lines represent nonradiative transitions. Some vibrational fine structure is shown. [From G. G. Hammes, "Fluorescence Methods," *in* Protein-Protein Interactions (C. Frieden and L. W. Nichol, eds.). Wiley (Interscience), New York, 1981.]

The return of the excited singlet state molecule, D*, to its ground state can be represented as

$$\begin{aligned} &D^* \xrightarrow{k_f} D + h\nu \quad \text{(emission)} \\ &D^* \xrightarrow{k_{nr}} D \qquad\quad \text{(radiationless)} \end{aligned} \tag{2-1}$$

The decay of the excited state is a first order rate process characterized by the first order rate constant $(k_f + k_{nr})$ where k_f is the rate constant characterizing fluorescence and k_{nr} is the rate constant characterizing all nonradiative processes. The lifetime of the fluorescent molecule, τ_D, is defined as

$$\tau_D = 1/(k_f + k_{nr}) \tag{2-2}$$

The fluorescence lifetime can be determined by applying a very short pulse of exciting light ($\sim$1 nanosec) to the molecule and measuring the decay of emission. Alternatively, the time delay between excitation and emission can be measured. The quantum yield, Q_D, is the fraction of the excited state molecules decaying by fluorescence,

$$Q_D = k_f/(k_f + k_{nr}) \tag{2-3}$$

or alternatively,

$$Q_D = \text{number of quanta emitted/number of quanta absorbed}$$

The quantum yield can be measured directly by steady-state fluorescence measurements. Steady-state fluorescence measurements are typically made by shining an exciting beam of light through the sample and detecting the fluorescence emission at right angles. The fluorescence intensity, I, is

$$I = I_0 Q_D (1 - 10^{-\varepsilon c d}) \tag{2-4}$$

where I_0 is the incident light intensity, ε is the extinction coefficient, c is the concentration, and d is the path length. When $\varepsilon c d \ll 1$ (i.e., for low absorbance)

$$I = I_0 Q_D 2.303(\varepsilon c d) \tag{2-5}$$

and the fluorescence is proportional to the concentration. Fluorescent species are characterized by their excitation and emission spectra. In the former case, the emission is viewed at constant wavelength and the excitation wavelength is varied, whereas in the latter case excitation is at a constant wavelength and the wavelength for viewing emission is varied. With most spectrofluorimeters, only apparent excitation and emission spectra are measured; the true spectra must be obtained by correcting for the wavelength-dependent intensity of the light source and the wavelength-dependent variation of the detector response. The corrections are normally made by comparison with the known corrected spectrum of a standard substance.

In studies of enzymes, the fluorescence of substrates, covalent probes, and amino acid side chains such as tryptophan are useful for monitoring changes in protein structure. In addition, information about the mobility of the fluorescent species can be obtained through polarization measurements. Fluorescent light is polarized because emission occurs in a fixed direction. If a molecule rotates many times during its lifetime, the observed emission polarization is zero; if rotation is restricted during the lifetime of the fluorescing species, polarization is observed. The degree of polarization, P, is defined as

$$P = \frac{I_v - I_h}{I_v + I_h} \tag{2-6}$$

where I_v and I_h are the intensities of the vertically and horizontally polarized components of the emission and the exciting beam either is polarized perpendicular to the vertical and horizontal directions or is unpolarized. For a spherical molecule and perpendicularly polarized excitation

$$\frac{1/P - 1/3}{1/P_0 - 1/3} = 1 + \frac{RT\tau_D}{\eta V} \tag{2-7}$$

where R is the gas constant, T is the absolute temperature, η is the viscosity, and V is the molecular volume. The value of P_0 is obtained by extrapolation of $1/P - 1/3$ versus T/η to $T/\eta = 0$. The rotational diffusion constant for a sphere is $RT/6V\eta$ so the rotational mobility is directly measured by polarization measurements. If the fluorescent probe is rigidly attached to the protein, the rotational properties of the protein are measured. A theory also has been developed for rigid ellipses, but the interpretation of the results becomes quite complex if the fluorescent species has local rotational mobility superimposed on the rotational mobility of the macromolecule.

A unique use of fluorescence is the measurement of molecular distances on macromolecules. One of the mechanisms for nonradiative deactivation of the excited state is the transfer of the energy from fluorescent molecules in the excited state to a second molecule in the ground state. If the second molecule also is fluorescent, the transferred energy can be emitted as a fluorescence characteristic of the second molecule. If the second molecule is not fluorescent, the energy is lost through equilibration with solvent. The conditions for energy transfer to occur are that the fluorescence emission of the fluorescent energy donor overlap the absorption spectrum of the energy acceptor, as shown schematically in Fig. 2-3, and that the energy donor and acceptor are sufficiently close and favorably oriented. A quantitative theory for singlet–singlet energy transfer has been developed (*16*). This theory is valid when the donor and acceptor are not too close ($> \sim 10$ Å).

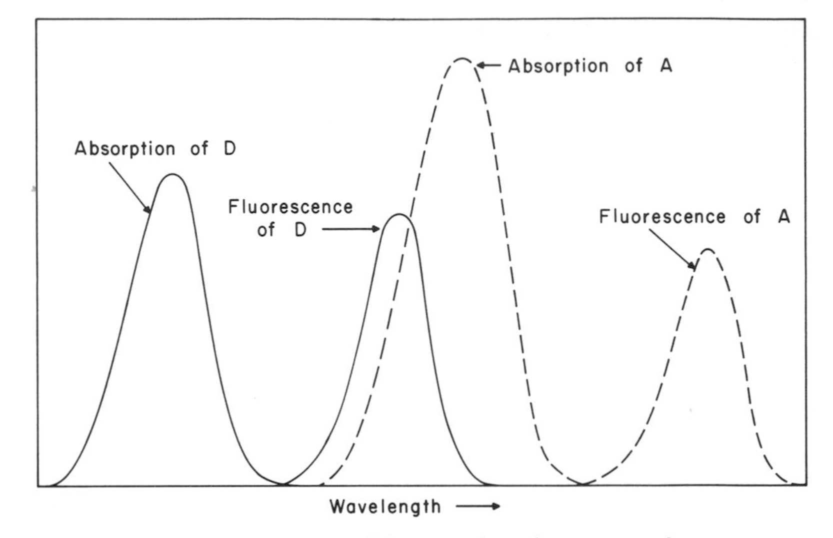

Fig. 2-3. Schematic representation of the spectral overlap necessary for resonance energy transfer between an energy donor *D* and an energy acceptor *A*. *A* need not be fluorescent.

Energy transfer from a donor D to an acceptor A can be represented as

$$\begin{aligned} D^* + A &\xrightarrow{k_t} A^* + D \quad \text{(transfer)} \\ A^* &\longrightarrow A + h\nu' \quad \text{(emission)} \end{aligned} \tag{2-8}$$

or

$$A^* \longrightarrow A \quad \text{(radiationless)}$$

The theory of Förster shows that the rate constant k_t can be written as

$$k_t = (R_\circ/R)^6/\tau_D \tag{2-9}$$

where R is the distance between the energy acceptor and donor, τ_D is the donor excited state lifetime, and $R_\circ$, the distance at which one-half of the fluorescence decay is due to energy transfer, is

$$R_\circ = 9.79 \times 10^3(\kappa^2 J Q_D n^{-4})^{1/6} \text{ Å} \tag{2-10}$$

In the above equation n is the refractive index of the medium, Q_D is the quantum yield of the energy donor, J is the overlap integral (which is a measure of the spectral overlap of the fluorescence emission of the donor and the absorption spectrum of the acceptor), and κ^2 is an orientation factor. The overlap integral is defined as

$$J = \frac{\int \bar{F}(\lambda)\varepsilon(\lambda)\lambda^4 \, d\lambda}{\int \bar{F}(\lambda) d\lambda}$$

where ε is the extinction coefficient of the acceptor, $\bar{F}$ is the corrected fluorescence emission spectrum of the donor, and λ is the wavelength; this integral can be calculated from readily accessible experimental data. The orientation factor arises because the donor and acceptor transition moment vectors can have different orientations and can vary from 0 to 4. The orientation factor cannot be experimentally determined although upper and lower bounds can be calculated from measurements of the fluorescence polarization (*17*). If the donor and acceptor rotate rapidly relative to the fluorescence lifetime, the relative orientations of D and A are random, and $\kappa^2 = 2/3$. This value is often used in calculating $R_\circ$, and normally will not cause an appreciable error in $R_\circ$ because the sixth root of κ^2 appears in Eq. (2-10).

The first order rate constant for fluorescence decay when energy transfer occurs is $(k_f + k_{nr} + k_t)$. The fluorescence lifetime in the presence of an energy acceptor is

$$\tau_{DA} = 1/(k_f + k_{nr} + k_t) \tag{2-11}$$

and the quantum yield is

$$Q_{DA} = k_f/(k_f + k_{nr} + k_t) \tag{2-12}$$

The efficiency of energy transfer, E, is the fraction of the total energy that is transferred

$$\begin{aligned} E &= k_t/(k_f + k_{nr} + k_t) \\ &= 1 - \frac{\tau_{DA}}{\tau_D} = 1 - \frac{Q_{DA}}{Q_D} \end{aligned} \tag{2-13}$$

[Eqs. (2-2), (2-3), (2-11), and (2-12) have been utilized to obtain the final expression for E.] Thus, the efficiency can be readily determined through measurements of the quantum yields or fluorescence lifetimes of the donor in the presence or absence of the acceptor. The efficiency of energy transfer also can be obtained by measuring the induced fluorescence of the acceptor (if it is fluorescent).

The efficiency can be written in terms of $R_\circ$ and R by substitution of Eq. (2-9) into Eq. (2-13) to give

$$E = \frac{(R_\circ/R)^6}{1 + (R_\circ/R)^6} \tag{2-14}$$

For a single donor–acceptor pair, the distance between the donor and acceptor can be calculated from experimentally determined values of E and $R_\circ$. The validity of this theory has been elegantly demonstrated with a series of polyproline oligomers having a donor at one end and an acceptor at the other end (*18*). Since values of $R_\circ$ typically range from 15 to 45 Å, distances from about 10 to 65 Å can be measured with this technique.

The use of energy transfer measurements to obtain structural maps of complex biological structures is a well-established area of research (cf. *15,19*). The primary difficulty usually is placing the energy donor and acceptor specifically at known sites. This involves nontrivial protein chemistry! In addition, the finite size of the probes and the occurrence of multiple donors and acceptors on the same macromolecule can create some difficulties in the establishment of structural maps. Nevertheless, energy transfer measurements are very useful structural tools.

NUCLEAR MAGNETIC RESONANCE

Nuclear magnetic resonance (nmr) has proved to be a very useful tool for studying enzymes, and a number of excellent reviews of the subject are available (*1,20–22*). Only a few specialized topics are considered here. The types of measurements can be divided roughly into five categories.

The magnetic resonance due to the macromolecule can be measured directly; this almost always involves ^{1}H- or ^{13}C-nmr. Because so many similar nuclei are present on a protein, a sharp spectral resolution is difficult. Furthermore, a broadening of resonances occurs due to the relatively slow rotation of the macromolecule; in practice this limits detailed studies to relatively low molecular weight proteins ($< \sim 20{,}000$). Modern high field instruments alleviate these problems a great deal. Very often special environments displace resonances considerably. For example, the nmr spectra of the four imidazole residues in ribonuclease A are shown in Fig. 2-4

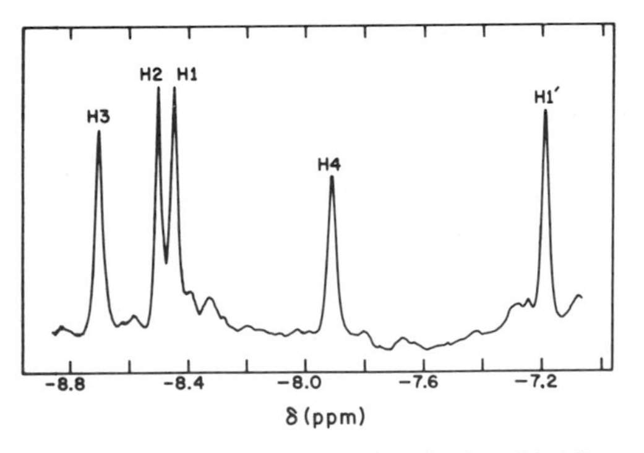

Fig. 2-4. Nuclear magnetic resonance spectrum of the four histidine residues (H1–H4, in ribonuclease A. [Adapted with permission from J. L. Markley, *Biochemistry* **14**, 3546 (1975). Copyright (1975) American Chemical Society.]

(*23*). The effect of varying pH and temperature and of binding ligands to the enzyme can be determined. This can provide information about the ionization constants of individual residues, about the molecular nature of conformational transitions, and about ligand–protein interactions.

The nmr of ligands bound to the macromolecule also can be studied in favorable cases. Direct observation of bound ligand is very difficult, but if a large excess of ligand is used to give a good nmr spectrum and if the ligand goes on and off the protein rapidly, an average of the spectra of free and bound ligand is seen. If equilibration is slow, the lifetime of the bound ligand on the protein can be measured. A special case of this category is considered in more detail later.

The third category of measurements is the nmr of water. This requires the existence of some bound water molecules with magnetic properties different than bulk water. Usually the excess of bulk water molecules obscures this phenomenon, but this is not the case when paramagnetic ions such as Mn^{2+} are present.

Spin labels (covalently attached free radicals) and paramagnetic ions influence neighboring nuclear spins. Sometimes this causes shifts in spectral lines, and sometimes just a difference in line shape. Kinetic and structural information can be obtained by measuring these effects, e.g., the distance between the unpaired electrons and the interacting nuclei. This situation also is considered in more detail later.

Finally the nmr of cations and anions such as ^{23}Na, ^{39}K, and halides can be measured. The sensitivity is quite limited for these cases.

As with any spectral method, in ideal cases the spectra can be interpreted in terms of molecular structure; in less favorable cases the changes observed can be used simply as an empirical monitor of structural changes. The theory of nmr is complex, but magnetic resonance is basically due to the fact that nuclei with nonzero nuclear spins precess around the direction of the magnetic field with a characteristic frequency ω. In a typical experiment, nuclei are put in a kilogauss field and a small oscillating field is applied perpendicular to the kilogauss field; the frequency of oscillation is varied until resonance is observed. This resonance is detected by a receiver coil, usually perpendicular to the magnetic fields. In very simple terms, nmr spectra are characterized by the spin lattice relaxation time, the spin–spin relaxation time, chemical shifts, and spin–spin splitting. The spin lattice relaxation time, T_1, is a measure of the time for nuclei to reach equilibrium with their surroundings after their Boltzmann distribution has been perturbed by the applied field. The relaxation can be regarded as being due to fluctuating magnetic fields of surrounding magnetic dipoles; these fluctuations are characterized by a correlation time τ_c. The spin–spin relaxation time, T_2, is a measure of interactions between neighboring nuclear spins and determines the line width of conventional spectra. The spin–spin relaxation time includes all the relaxation times associated with T_1 plus additional ones due to interacting nuclear spins so that $T_2 \leq T_1$. Chemical shifts are displacements of spectral lines due to varying local magnetic fields that are determined by the electronic environment of the nuclei. For example, the low resolution spectrum of CH_3CHO has two lines due to the two types of protons that are separated by a characteristic chemical shift. Spin–spin splitting is due to the interaction of neighboring nuclear spins. For example, the CHO proton splits the resonances of the CH_3 protons in two, due to the possibility of the CHO proton having two orientations in the magnetic field. Similarly the CH_3 protons have four possible configurations of their nuclear spins ($\uparrow\uparrow\uparrow, \uparrow\downarrow\downarrow, \uparrow\uparrow\downarrow, \downarrow\downarrow\downarrow$) with a weighting of 1:3:3:1 so that the CHO resonance is split into four.

If the nucleus whose resonance is being observed can exist in two environments, the spectrum observed depends on the rate of passage of the nucleus between the two environments. For example, the OH proton of methanol and the protons of water have different chemical shifts. However, a mixture of water and methanol has only a single proton resonance line for these protons with a chemical shift that is a weighted average of those for the pure substances. This is because a rapid exchange of protons occurs between water and methanol; the frequency of the reaction is much greater than the chemical shift of the methanol proton relative to the water protons. During a sweep of the magnetic field through resonance, the protons make many round trips between the two chemical environments. For a nucleus that can exist in two environments, the observation of two resonances is the slow exchange limit and the observation of a single resonance is the fast exchange limit. Intermediate situations also are possible. A schematic representation of the transition between the slow and fast exchange limits is shown in Fig. 2-5. In general, for fast exchange between two environments, the frequency shift, $\Delta\omega$, with respect to an arbitrary standard is

$$\Delta\omega = P_{\mathrm{M}}q\,\Delta\omega_{\mathrm{M}} + (1 - P_{\mathrm{M}}q)\,\Delta\omega_{\mathrm{A}} \tag{2-15}$$

where $P_{\mathrm{M}}q$ is the fraction of nuclei in the M environment, $\Delta\omega_{\mathrm{M}}$ is the frequency shift from the standard to the M environment, and $\Delta\omega_{\mathrm{A}}$ is the frequency shift of the A environment with respect to the standard. This same result applies to both chemical shifts and spin–spin splitting. At the midpoint of the transition between two and one spectral lines, the mean lifetime for exchange, τ_i, is equal to $1/\delta\omega$ where $\delta\omega$ is the frequency shift between the two lines. The mean lifetime for exchange is defined by

$$\frac{1}{\tau_i} = \frac{\text{rate of removal of molecules from } i\text{th state by exchange}}{\text{number of molecules in } i\text{th state}} \tag{2-16}$$

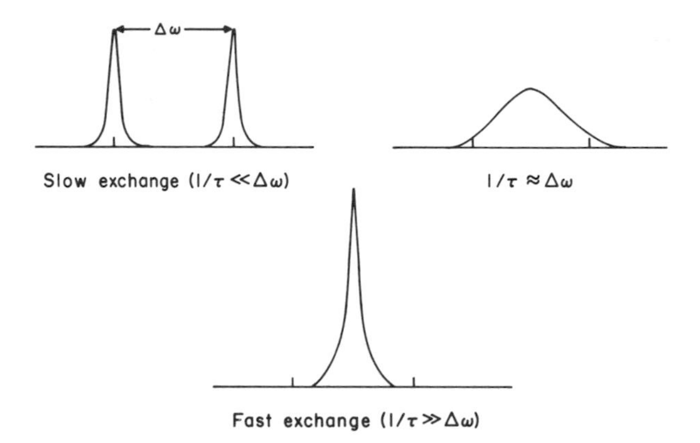

Fig. 2-5. The effect of exchange on the nuclear magnetic resonance spectrum. The exchange occurs between two sites that are equally populated and separated by a chemical shift $\Delta\omega$.

Measurement of τ_i permits determination of the rate constant for exchange between the two environments. The time resolution of this method is determined by $\Delta\omega$; the larger it is, the shorter the time constants which can be measured. Typical exchange lifetimes are $\geq 10^{-2}$ sec although times several orders of magnitude shorter have been measured in favorable cases. As a simple example, consider the exchange of a proton between NH_3 and NH_2^- (*24*)

$$NH_3 + NH_2^- \xrightarrow{k} NH_2^- + NH_3 \tag{2-17}$$

The spectrum of NH_3 has three resonances because N has a nuclear spin with three possible orientations. When NH_2^- is added, the resonances collapse into a single resonance. The coalescence of spectral lines occurs at a concentration of 10^{-7} M NH_2^- and $\delta\omega = 290$ sec^{-1}. In this case, $1/\tau \cong k[NH_2^-]$ where k is the rate constant which can be calculated to be 3×10^9 M^{-1} sec^{-1}. An exact treatment of the effect of chemical exchange on nmr spectra requires a detailed analysis in terms of the general theory of nmr and is not discussed here (cf. *1*).

A very special case of particular interest to enzymologists is now considered, namely, the binding of a ligand to an enzyme where the ligand concentration is much greater than that of the enzyme, and the nmr properties of the bulk ligand are measured. The chemical shift of the ligand in the bulk environment, $\Delta\omega$, with respect to no enzyme being present is

$$\Delta\omega = \frac{(P_M q)\,\Delta\omega_M}{\tau_M^2\left(\dfrac{1}{T_{2M}} + \dfrac{1}{\tau_M}\right)^2 + \tau_M^2\,\Delta\omega_M^2} \tag{2-18}$$

where $\Delta\omega_M$ is the chemical shift between the two environments, τ_M is the mean lifetime in the M (enzyme) environment and T_{2M} is the spin–spin relaxation time in the M environment. The fast exchange limit [Eq. (2-15)] is reached at sufficiently high temperatures so that $1/\tau_M \gg 1/T_{2M}$ and $(\tau_M \Delta\omega_M)^2 \gg 1$. The nmr relaxation times are more useful parameters than the chemical shift, which is typically very small. The relaxation times of the bulk resonances, $T_{1,\text{obs}}$ and $T_{2,\text{obs}}$, are

$$\frac{1}{T_{1,\text{obs}}} = \frac{1}{T_{1,\text{A}}} + \frac{(P_M q)}{T_{1,\text{M}} + \tau_M} \tag{2-19}$$

$$\frac{1}{T_{2,\text{obs}}} = \frac{1}{T_{2,\text{A}}} + \frac{(P_M q)}{\tau_M}\left[\frac{\dfrac{1}{T_{2,\text{M}}}\left(\dfrac{1}{T_{2,\text{M}}} + \dfrac{1}{\tau_M}\right) + \Delta\omega_M^2}{\left(\dfrac{1}{T_{2,\text{M}}} + \dfrac{1}{\tau_M}\right)^2 + \Delta\omega_M^2}\right] \tag{2-20}$$

First consider the spin lattice relaxation time. Two limiting cases are possible: the rate of chemical exchange, $1/\tau_M$, is much greater than $1/T_{1,\text{M}}$ and the rate of chemical exchange is much smaller than $1/T_{1,\text{M}}$. In the former case,

$$\frac{1}{T_{1,\text{obs}}} - \frac{1}{T_{1,A}} = \frac{P_M q}{T_{1,\text{M}}} \tag{2-21}$$

or the spin lattice relaxation time in the M environment determines the observed difference between the measured spin lattice relaxation time in the presence and absence of enzyme; this is the fast exchange limit. In the latter case,

$$\frac{1}{T_{1,\text{obs}}} - \frac{1}{T_{1,\text{A}}} = \frac{P_M q}{\tau_M} \tag{2-22}$$

or the rate of chemical exchange is measured; this is the slow exchange limit. In practice, these two limits are achieved by making use of the fact that τ_M has a considerably larger temperature coefficient than $T_{1,\text{M}}$. Therefore, at sufficiently high temperatures the fast exchange limit is observed, whereas at sufficiently low temperatures the slow exchange limit is reached. A schematic representation of $1/T_{1,\text{obs}} - 1/T_{1,\text{A}}$ versus the reciprocal absolute temperature is shown in Fig. 2-6. The situation for the spin–spin relaxation time is identical if $\Delta\omega_M = 0$. If $\Delta\omega_M \neq 0$, the definitions of slow and fast exchange require that $\Delta\omega_M^2 \gg 1/T_{2,\text{M}}^2$, $1/\tau_M^2$ and $(T_{2,\text{M}}\tau_M)^{-1} \gg \Delta\omega_M^2$, respectively, in addition to the usual conditions. Also, an intermediate case exists in which the spin–spin relaxation time is dependent on $\Delta\omega_M$ and, therefore,

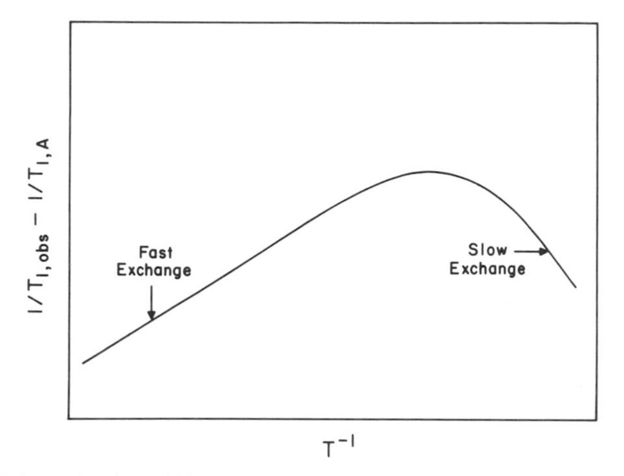

Fig. 2-6. Schematic plot of $1/T_{1,\mathrm{obs}} - 1/T_{1,\mathrm{A}}$ versus the reciprocal absolute temperature.

is frequency-dependent. Although the spin lattice relaxation time is more difficult to measure than the spin–spin relaxation time because pulsed magnetic fields must be used, the spin lattice relaxation time usually can be interpreted with considerably more ease.

Thus far the absence of paramagnetic species has been assumed. However, paramagnetic ions can provide a very useful perturbation of the nmr relaxation times. Paramagnetic species alter the correlation times that modulate the oscillating fields which ultimately determine the relaxation times. Two correlation times can be distinguished: one modulates the field through dipolar interactions, whereas the other, the scalar correlation time, modulates it through chemical bonds. The dipolar correlation time, τ_c, is

$$\frac{1}{\tau_c} = \frac{1}{\tau_{\mathrm{R}}} + \frac{1}{\tau_{\mathrm{S}}} + \frac{1}{\tau_{\mathrm{M}}} \tag{2-23}$$

where τ_{R} is the relaxation time for rotation and τ_{S} is the electron spin relaxation time for the paramagnetic species. For example, for aquo complexes of paramagnetic ions such as Mn^{2+}, $\tau_{\mathrm{R}} \sim 10^{-11}$ sec, $\tau_{\mathrm{M}} \gg 10^{-11}$ sec, and $\tau_{\mathrm{S}} \geq \tau_{\mathrm{R}}$ in most situations. The scalar correlation time, τ_e, is

$$\frac{1}{\tau_e} = \frac{1}{\tau_{\mathrm{S}}} + \frac{1}{\tau_{\mathrm{M}}} \tag{2-24}$$

For paramagnetic metal ions, P_{M} is the ratio of the concentration of paramagnetic ion to the concentration of ligand and q is the number of ligands in the coordination sphere of the paramagnetic ion. The contribution of the paramagnetic species to the spin lattice relaxation time contains two terms, one due to dipolar coupling and the other to scalar coupling. For virtually all cases of interest, the scalar term is negligible so that the paramagnetic contribution can be approximated as

$$\frac{1}{T_{1,\mathrm{M}}} = \frac{2}{15} \frac{\mathrm{S}(\mathrm{S}+1)\gamma^2 g^2 \beta^2}{r^6} \left(\frac{3\tau_c}{1+\omega^2\tau_c^2} \right) \tag{2-25}$$

where S is the electron spin quantum number, γ is the magnetogyric ratio, g is the nuclear g factor, β is the Bohr magneton, ω is the nuclear spin resonance frequency, and r is the distance between the nucleus whose resonance is observed and the paramagnetic species. The only two unknown parameters in this equation are r and τ_c. Since τ_c can be determined (see below), determination of the paramagnetic contribution to $T_{1,\mathrm{M}}$ is another method of mapping distances on the protein. The paramagnetic contribution to the spin–spin relaxation time is considerably more complex because the scalar term cannot be neglected

$$\frac{1}{T_{2,\mathrm{M}}} \approx \frac{1}{15} \frac{\mathrm{S}(\mathrm{S}+1)\gamma^2 g^2 \beta^2}{r^6} \left(4\tau_c + \frac{3\tau_c}{1+\omega^2\tau_c^2} \right) + \frac{\mathrm{S}(\mathrm{S}+1)\mathrm{A}^2}{3} \tau_e \tag{2-26}$$

In this equation A is a scalar coupling term. As might be expected, the spin-lattice relaxation time rather than the spin–spin relaxation time is normally used to determine distances on proteins.

In principle, two types of parameters can be obtained from measurements of the nmr relaxation times, τ_{M} (slow exchange), which gives information about the kinetics of ligand binding, and $T_{1,\mathrm{M}}$ or $T_{2,\mathrm{M}}$ (fast exchange), which give information about the distance between the paramagnetic species and the ligand on the protein (providing τ_c can be determined). Two variables are available to sort things out, the temperature and the frequency of the spectrometer. The slow exchange region ($\tau_{\mathrm{M}} \gg T_{1,\mathrm{M}}, T_{2,\mathrm{M}}$) has three important characteristics.

1. $1/T_{1,\mathrm{obs}} - 1/T_{1,\mathrm{A}} = 1/T_{2,\mathrm{obs}} - 1/T_{2,\mathrm{A}} = P_{\mathrm{M}}q/\tau_{\mathrm{M}}$;
2. the relaxation times are frequency independent;
3. $1/T_{1,\mathrm{obs}} - 1/T_{1,\mathrm{A}}$ and $1/T_{2,\mathrm{obs}} - 1/T_{2,\mathrm{A}}$ increase with increasing temperature since τ_{M} decreases with increasing temperature.

Unfortunately $1/T_{1,\mathrm{obs}} - 1/T_{1,\mathrm{A}}$ also may increase with increasing temperature in the fast exchange region. One possibility is that fast exchange occurs, but τ_c is determined by τ_{S}, which can have a positive temperature coefficient.

This possibility can sometimes be eliminated by an independent determination of τ_S with electron paramagnetic resonance. A second possibility is that fast exchange occurs, but $(\omega\tau_c)^2 > 1$ and τ_c becomes shorter with increasing temperature. This requires $1/T_{1,obs} - 1/T_{1,A}$ to be frequency dependent, which is not true in the slow exchange limit. Also in neither of these cases should $1/T_{1,obs} - 1/T_{1,A} = 1/T_{2,obs} - 1/T_{2,A}$.

If $1/T_{1,obs} - 1/T_{1,A}$ decreases as the temperature increases, then fast exchange must be occurring and $T_{1,M}$ can be measured. To calculate r, the correlation time must be determined. This can be done by measuring the frequency dependence of the spin lattice relaxation time as long as $(\omega\tau_c)^2$ is comparable to or greater than unity. If this is not the case, τ_c must be estimated through independent knowledge of τ_R, τ_S, and τ_M. For example, for small complexes of aqueous Mn^{2+}, $1/\tau_c \sim 1/\tau_R \sim 3 \times 10^{10}$ sec^{-1}. Since r depends approximately on $\tau_c^{1/6}$, a precise knowledge of τ_c is not necessary for reasonably precise distance determinations. In some cases, a knowledge of both the spin lattice and spin-spin relaxation times also permits determination of τ_c.

As examples of the kinetic information about enzyme systems that can be obtained from nmr, the dissociation rate constants, $1/\tau_M$, for a few enzyme–ligand systems are presented in Table 2-2. Some results of determinations of distances between paramagnetic centers and nuclei by nmr are given in Table 2-3. The results from simple systems, where direct comparison can be made with x-ray diffraction measurements, are presented as well as a few results from enzyme systems.

Table 2-2

Selected Examples of Kinetic Parameters of Protein–Ligand Interactions Determined by nmr

Protein	Paramagnetic center	Ligand	$\log(1/\tau_M)$ sec^{-1}	Reference
Pyruvate kinase	Mn^{2+}	H_2O	7.9	*25*
		FPO_3^{2-}	4.5	*25*
Pyruvate carboxylase	Mn^{2+}	H_2O	6.2	*26*
		Pyruvate	4.3	*27*
		α-Ketobutyrate	3.9	*27*
Aldolase	Mn^{2+}	H_2O	5.8	*28*
		Fructose diphosphate	4.7	*28*
Alcohol dehydrogenase	ADP-R. (R· = free radical)	H_2O	4.2	*29*
		Ethanol	2.8	*30*
		Acetaldehyde	2.9	*30*
		Isobutyramide	2.3	*30*

Table 2-3

Selected Examples of Ligand–Mn^{2+} Distances Determined by nmr

Enzyme	Ligand		Distance		Reference
			nmr[a]	x-ray Diffraction	
—	F^-	Mn—F	2.1	2.08–2.15	*25, 31*
—	Imidazole	Mn ··· HC(2)	≥3.1	3.27	*21, 32*
		Mn ··· HC(5)	3.4	3.24	
Pyruvate kinase	FPO_3^{2-}	Mn ··· F	3.0–5.8		*25*
Pyruvate carboxylase	Pyruvate	Mn ··· CH_3	3.5–6.6		*27*
	α-Ketobutyrate	Mn ··· CH_3	3.5–6.6		
		Mn ··· CH_2	3.5–6.6		
Creatine kinase + ADP	Creatine	Mn ··· CH_2	8.7		*33*
		Mn ··· CH_3	9.4		

[a] The range in distances is due to the uncertainty in τ_c.

A useful empirical method has been developed for measuring the binding of a paramagnetic ion or free radical to a macromolecule (*21*). An enhancement factor, ε, is defined as

$$\varepsilon = \frac{1/T^*_{1,\mathrm{obs}} - 1/T^*_{1(0)}}{1/T_{1,\mathrm{obs}} - 1/T_{1(0)}} \tag{2-27}$$

Here * indicates the presence of the macromolecule and (0) indicates the absence of paramagnetic ion. An analogous enhancement can be defined in terms of T_2. In the enhanced system, τ_R is increased so that τ_c is determined by τ_S or τ_M. The net result is a decrease in $T_{1,M}$ and an increase in ε. This approach has been used to study the interactions of enzymes with Mn^{2+} and substrates, with T_1 of water being monitored. Three enhancements were measured: ε_b, Mn^{2+} plus enzyme; ε_a, Mn^{2+} plus substrate; and ε_T, Mn^{2+} plus enzyme and substrate. The enzymes studied fall into three classes. In one class, ε_b and ε_a are less than ε_T. This means the substrate is necessary to bring Mn^{2+} into the macromolecular environment; this implies a structure such as enzyme–substrate–metal. A metal or enzyme bridge, rather than a substrate bridge, cannot be excluded in some unusual cases. In another class, $\varepsilon_b > \varepsilon_T$; this implies structures such as

enzyme–metal–substrate or enzyme⟨metal | substrate⟩ (enzyme bonded to both metal and substrate, with metal bonded to substrate)

In the third class, $\varepsilon_b = \varepsilon_T$, which suggests the structure substrate–enzyme–metal. This approach also is useful for titrations of the enzyme with paramagnetic species and determination of enzyme–ligand binding constants.

SPIN LABELS

Spin labels are free radicals that are usually covalently incorporated into the enzyme structure. The most common are nitroxide free radicals such as

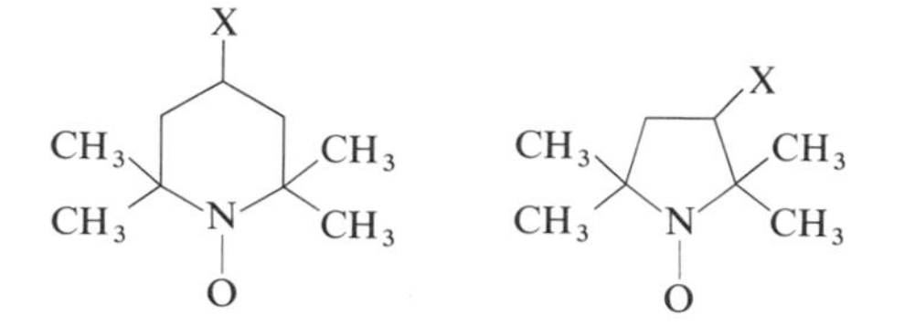

Here X is often —CH_2I, which readily reacts with protein sulfhydryl groups. Spin labels are useful empirical monitors of enzyme conformation since their electron paramagnetic resonance (epr) spectra are environment-dependent. The epr spectra also give information about the probe mobility and polarity. The distance of the probe from specific nuclei can, of course, be determined by nmr measurements as described above. The epr spectrum of a nitroxide radical has three resonances because of the three possible orientations of the nuclear spins of nitrogen in the magnetic field. The spectrum depends on the orientation of the unpaired electron spin with respect to the magnetic field. This is shown in Fig. 2-7 where the spectra of a nitroxide radical are shown for several differently oriented crystals and a rigid glass. The z axis is parallel to the nitrogen 2p orbital associated with the unpaired electron. The epr spectra are usually presented as first derivatives of the

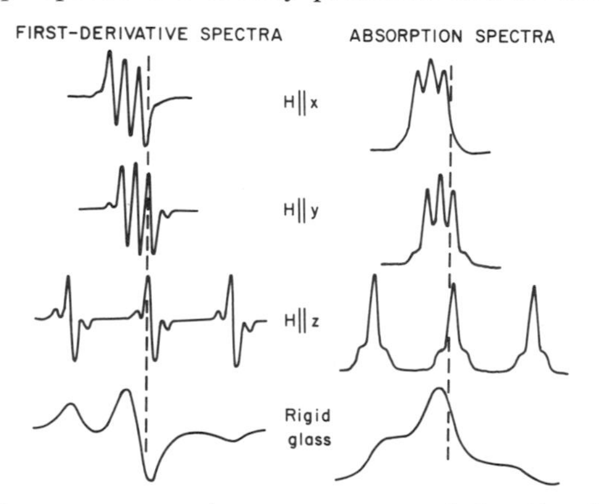

Fig. 2-7. The electron paramagnetic resonance spectra of nitroxide spin labels in a rigid matrix. The label was oriented relative to the laboratory magnetic field, H, as indicated where the z axis is parallel to the nitrogen 2p oribital of the unpaired electron and the x axis is parallel to the N–O bond. The dashed line is a reference sample. [Adapted with permission from O. H. Griffith, L. J. Libertini, and G. R. Birrell, *J. Phys. Chem.* **75**, 3417 (1971). Copyright (1971) American Chemical Society.]

absorption spectra simply because this is experimentally convenient. Note that quite different spectral line shapes and positions are found for different orientations. If the spin label is not rigidly oriented, the spectral lines become sharper, and the spectra of different possible orientations are averaged. The mobility of the spin label, that is its rotational relaxation time, can be calculated from an analysis of the spectral line shapes. In order for averaging to occur, rotation must be rapid relative to the frequency separation of lines, which is typically about 10^9 sec^{-1}. Changes in mobility due to ligand binding, variations in pH, etc. are indicative of protein conformational changes. The epr spectra also are altered by the polarity of the local environment.

The calculation of distances from the spin label to nuclei using nmr poses several theoretical difficulties that have not yet been considered: the tumbling motion may not be isotropic, the g value is not isotropic, and the multiplet nature of epr spectra must be taken into account. The distances between bound saccharides, histidine-15, and tryptophan-123 on lysozyme have been determined (34). A spin label was attached to the N(3) of the imidazole on histidine-15 and the nmr of nuclei on a bound saccharide was monitored. In addition, a spin-labeled polysaccharide or a noncovalent spin label was bound to the enzyme, and the nmr of imidazole or bound polysaccharide was measured. The distances determined are summarized in Fig. 2-8; included in this figure are measurements of distances estimated from x-ray crystallography.

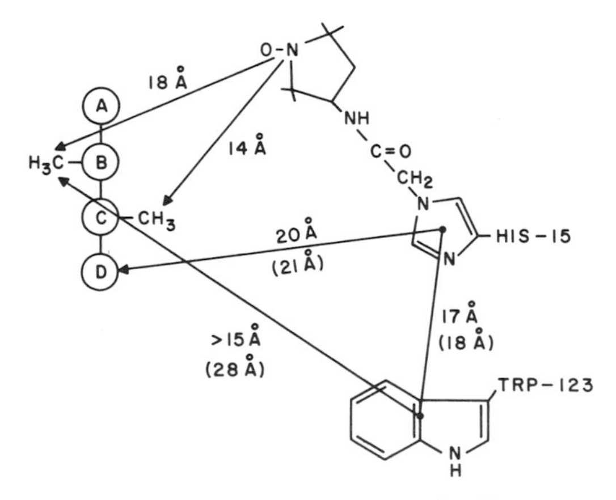

Fig. 2-8. Distances between groups on lysozyme determined by nuclear magnetic resonance with a spin label attached to the imidazole on histidine-15 or noncovalently bound near tryptophan-123 or at D. Monomeric units of bound oligomeric N-acetylglucosamine are represented by A, B, C, D. The numbers in parentheses have been obtained by x-ray crystallography. [Adapted with permission from R. W. Wien, J. D. Morrisett, and H. M. McConnell, *Biochemistry* **11**, 3707 (1972). Copyright (1972) American Chemical Society.]

A more extensive discussion of spin labels can be found in Refs. *1*, *22*, and *35*.

REFERENCES

1. C. R. Cantor and P. R. Schimmel, "Biophysical Chemistry." Parts I–III. Freeman, San Francisco, California 1980.
2. D. G. Smyth, W. H. Stein, and S. Moore, *J. Biol. Chem.* **238**, 227 (1973).
3. C. B. Anfinson and H. A. Scheraga, *Adv. Prot. Chem.* **29**, 205 (1975).
4. R. W. Crowther and A. Klug, *Ann. Rev. Biochem.* **45**, 161 (1975).
5. E. Shaw, *in* "The Enzymes" (P. Boyer, ed.), 3rd ed. Vol. 1, p. 91. Academic Press, New York, 1970.
6. L. A. Cohen, *in* "The Enzymes" (P. Boyer, ed.), 3rd ed. Vol. 1, p. 148. Academic Press, New York, 1970.
7. R. L. Heinrikson, W. H. Stein, A. M. Crestfield, and S. Moore, *J. Biol. Chem.* **240**, 2921 (1965).
8. F. Wold, *Methods Enzymol.* **25**, 623 (1972).
9. V. Chowdhry and F. H. Westheimer, *Ann. Rev. Biochem.* **48**, 293 (1979).
10. F. C. Hartman and F. Wold, *Biochemistry* **6**, 2439 (1967).
11. G. Schoellmann and E. Shaw, *Biochemistry* **2**, 252 (1963).
12. H. E. van Wart and H. A. Scheraga, *Methods Enzymol.* **49**, 67 (1978).
13. R. D. B. Fraser and E. Suzuki, *in* "Physical Principles and Techniques of Protein Chemistry" (S. J. Leach, ed.), Vol. B, p. 273. Academic Press, New York. 1977.
14. R. C. Chen and H. Edelhoch, eds., "Biochemical Fluorescence Concepts," Vols. 1 and 2. Dekker, New York, 1975 and 1976.
15. G. G. Hammes, *in* "Protein-Protein Interactions" (L. W. Nichol and C. Frieden, eds.), p. 257. Wiley (Interscience), New York, 1981.
16. Th. Förster, *in* "Modern Quantum Chemistry," Part III (O. Sinanoglu, ed.), p. 93. Academic Press, New York, 1965.
17. R. E. Dale, J. Eisinger, and W. E., Blumberg, *Biophys. J.* **159**, 577 (1976).
18. L. Stryer and R. P. Haugland, *Proc. Natl. Acad. Sci. U.S.A.* **58**, 719 (1967).
19. L. Stryer, *Ann. Rev. Biochem.* **47**, 819 (1978).
20. J. J. Villafranca and F. M. Raushel, *Ann. Rev. Biophys. Bioeng.* **9**, 363 (1980).
21. M. Cohn and A. S. Mildvan, *Adv. Enzymol.* **33**, 1 (1970).
22. R. A. Dwek, "Nuclear Magnetic Resonance in Biochemistry: Applications to Enzyme Chemistry." Oxford Univ. Press (Clarendon) London and New York, 1973.
23. J. L. Markley, *Biochemistry* **14**, 3546 (1975).
24. R. A. Ogg, *Discuss. Faraday Soc.* **17**, 215 (1954).
25. A. S. Mildvan, J. S. Leigh, Jr., and M. Cohn, *Biochemistry* **6**, 1805 (1967).
26. M. C. Scrutton and A. S. Mildvan, *Biochemistry* **7**, 1490 (1968).
27. A. S. Mildvan and M. C. Scrutton, *Biochemistry* **6**, 2978 (1967).
28. A. S. Mildvan, R. O. Kobes, and W. J. Rutter, *Biochemistry* **10**, 1191 (1971).
29. A. S. Mildvan and H. Weiner, *Biochemistry* **8**, 552 (1969).
30. A. S. Mildvan and H. Weiner, *J. Biol. Chem.* **244**, 2465 (1969).
31. M. Griffel and J. W. Stout, *J. Am. Chem. Soc.* **72**, 4351 (1950).
32. M. Cohn, *Q. Rev. Biophys.* **3**, 61 (1970).
33. A. C. McLaughlin, J. S. Leigh, and M. Cohn, *J. Biol. Chem.* **251**, 2777 (1976).
34. R. W. Wien, J. D. Morrisett, and H. M. McConnell, *Biochemistry* **11**, 3707 (1972).
35. L. J. Berliner, ed., "Spin Labeling I and II." Academic Press, New York, 1976 and 1978.

3

Steady-State Enzyme Kinetics

INTRODUCTION

The study of enzyme kinetics under steady-state conditions is by far the most common way of investigating enzyme mechanisms. Because enzymes are such very efficient catalysts, very low concentrations suffice to make a reaction proceed at a measurable rate: concentrations of 10^{-8}–10^{-10} M are typical. Substrate concentrations are usually greater than 10^{-6} M. When the total enzyme concentration is much less than the total substrate concentrations, all of the enzyme species can be assumed to be in a steady state after a very short induction period. This vastly simplifies the rate laws for the enzymatic reactions. In this chapter, a few important prototype reactions are considered in detail. Many good discussions of steady-state enzyme kinetics are available (*1–5*).

SINGLE SUBSTRATE–SINGLE PRODUCT

The simplest possible enzyme mechanism is the conversion of a single substrate to a single product. For example, the enzyme fumarase catalyzes the interconversion of fumarate and L-malate

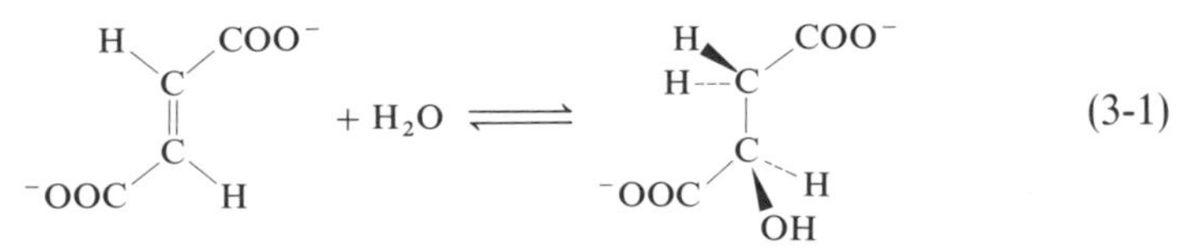

If the initial velocity (rate), v, of this reaction is studied as a function of substrate concentration, [S], starting with either only fumarate or only L-

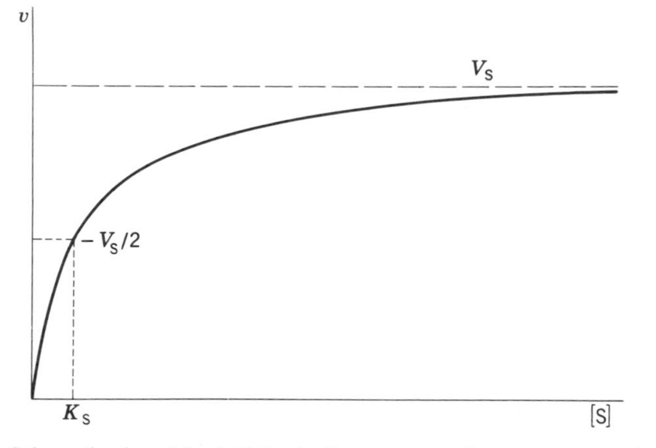

Fig. 3-1. Schematic plot of the initial velocity, v, versus substrate concentration, S, for an enzyme reaction involving one substrate.

malate, the results obtained in Fig. 3-1 are obtained at constant enzyme concentration. The maximum initial velocity reached is called the maximum velocity, V_S, and is directly proportional to the total enzyme concentration $[E_0]$. The ratio $V_S/[E_0]$ is often called the turnover number or k_{cat} and is a direct measure of the catalytic efficiency of the enzyme. The substrate concentration at which the initial velocity reaches one-half the maximum velocity is called the Michaelis constant, K_S. A simple mechanism consistent with these findings is the Michaelis–Menten mechanism

$$E + S \underset{k_{-1}}{\overset{k_1}{\rightleftharpoons}} X \underset{k_{-2}}{\overset{k_2}{\rightleftharpoons}} E + P \tag{3-2}$$

where S and P are the substrate and product, E is the free enzyme, and X is a reaction intermediate. Since X is present in a steady state

$$-\frac{d[X]}{dt} = (k_{-1} + k_2)[X] - k_1[E][S] - k_{-2}[E][P] = 0 \tag{3-3}$$

Conservation of mass and the restriction that $[E_0] \ll [S_0]$ requires that

$$\begin{aligned} [E_0] &= [E] + [X] \\ [S_0] &= [S] + [P] \end{aligned} \tag{3-4}$$

The disappearance of substrate and appearance of product is given by

$$-\frac{d[S]}{dt} = \frac{d[P]}{dt} = k_1[E][S] - k_{-1}[X]$$

Combining the above equations, we obtain

$$-\frac{d[\mathrm{S}]}{dt} = \frac{d[\mathrm{P}]}{dt} = \frac{(k_1 k_2[\mathrm{S}] - k_{-1}k_{-2}[\mathrm{P}])[\mathrm{E}_0]}{k_1[\mathrm{S}] + k_{-2}[\mathrm{P}] + k_{-1} + k_2} \tag{3-5}$$

This equation can be written as

$$-\frac{d[\mathrm{S}]}{dt} = \frac{(V_\mathrm{S}/K_\mathrm{S})[\mathrm{S}] - (V_\mathrm{P}/K_\mathrm{P})[\mathrm{P}]}{1 + [\mathrm{S}]/K_\mathrm{S} + [\mathrm{P}]/K_\mathrm{P}} \tag{3-6}$$

with

$$\begin{aligned} V_\mathrm{S} &= k_2[\mathrm{E}_0] & V_\mathrm{P} &= k_{-1}[\mathrm{E}_0] \\ K_\mathrm{S} &= (k_{-1} + k_2)/k_1 & K_\mathrm{P} &= (k_{-1} + k_2)/k_{-2} \end{aligned} \tag{3-7}$$

For initial velocity measurements, $[\mathrm{P}] = 0$ so that

$$v = -\frac{d[\mathrm{S}]}{dt} = \frac{V_\mathrm{S}}{1 + K_\mathrm{S}/[\mathrm{S}]} \tag{3-8}$$

This equation has exactly the dependence on substrate concentration shown in Fig. 3-1: As $\mathrm{S} \to \infty$, $v \to V_\mathrm{S}$ and when $[\mathrm{S}] = K_\mathrm{S}$, $v = V_\mathrm{S}/2$. The Michaelis constant is not an equilibrium constant, but a steady-state constant measuring the ratio $[\mathrm{E}][\mathrm{S}]/[\mathrm{X}]$ in the steady state. An alternative derivation of Eq. (3-8) is to note that $[\mathrm{E}_0] = [\mathrm{E}] + [\mathrm{X}] = [\mathrm{X}](1 + K_\mathrm{S}/[\mathrm{S}])$ and $v = k_2[\mathrm{X}]$.

A number of different ways of plotting the initial velocity data are used. Probably the most common is to plot $1/v$ versus $1/[\mathrm{S}]$ since

$$\frac{1}{v} = \frac{1}{V_\mathrm{S}} + \frac{K_\mathrm{S}}{V_\mathrm{S}}\frac{1}{[\mathrm{S}]} \tag{3-9}$$

This plot, the Lineweaver–Burke or double reciprocal plot, gives a strong weighting to results at low substrate concentrations and should not be used without proper weighting statistics. Equation (3-9) can be rearranged to give

$$\frac{[\mathrm{S}]}{v} = \frac{K_\mathrm{S}}{V_\mathrm{S}} + \frac{[\mathrm{S}]}{V_\mathrm{S}} \tag{3-10}$$

and a plot of $[\mathrm{S}]/v$ versus $[\mathrm{S}]$ gives a straight line with a more even weighting of the data. A third method of plotting the data is the Eadie–Hofstee plot. Multiplication of Eq. (3-9) by vV_S gives

$$v = V_\mathrm{S} - vK_\mathrm{S}/[\mathrm{S}] \tag{3-11}$$

A plot of v versus $v/[\mathrm{S}]$ is linear, but since v is the variable with the greatest error, its appearance in both the ordinate and abscissa can cause appreciable deviations from linearity. These three plots are illustrated in Fig. 3-2. For

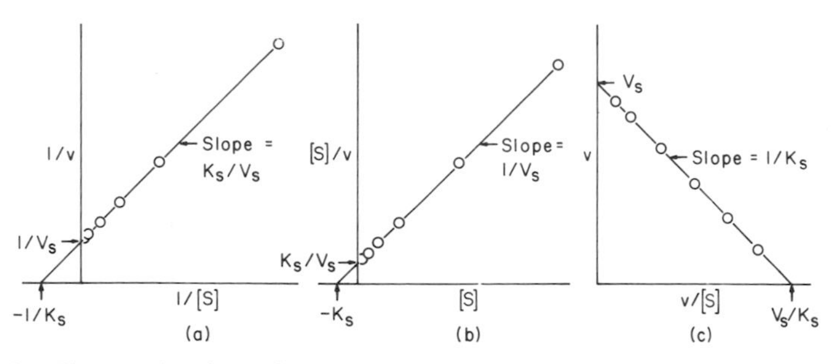

Fig. 3-2. Plots used to determine V_S and K_S from initial velocity data for a single substrate reaction. (a) $1/v$ versus $1/[S]$; (b) $[S]/v$ versus $[S]$; (c) v versus $v/[S]$. Here v is the initial velocity and $[S]$ is the substrate concentration.

those with access to a computer and a program for carrying out a nonlinear least squares analysis, the kinetic parameters are best obtained by a direct fit of the data to Eq. 3-8. Still another method for obtaining the four steady-state parameters is to integrate Eq. (3-6) directly by numerical analysis of the time course of the reaction on a computer. This method has not yet obtained wide usage. Although four steady-state parameters can be obtained, they are not all independent. At equilibrium, $-d[S]/dt = d[P]/dt = 0$ and Eqs. (3-5) and (3-6) give the Haldane relationship

$$K_{eq} = \frac{[P_{eq}]}{[S_{eq}]} = \frac{k_1 k_2}{k_{-1} k_{-2}} = \frac{V_S}{K_S}\frac{K_P}{V_P} \tag{3-12}$$

where K_{eq} is the equilibrium constant for the overall reaction and $[P_{eq}]$ and $[S_{eq}]$ are the equilibrium concentrations of product and substrate. If all four steady-state parameters are known, the four rate constants characterizing the mechanism [Eq. (3-2)] can be calculated.

Unfortunately the single intermediate Michaelis–Menten mechanism is not unique in predicting the rate law of Eq. (3-6). In fact, the form of this rate equation is the same regardless of how many intermediates are put into the mechanism. For example, if two intermediates are assumed

$$E + S \underset{k_{-1}}{\overset{k_1}{\rightleftharpoons}} X_1 \underset{k_{-2}}{\overset{k_2}{\rightleftharpoons}} X_2 \underset{k_{-3}}{\overset{k_3}{\rightleftharpoons}} E + P \tag{3-13}$$

the rate equation is given by Eq. (3-6) with

$$K_S = \frac{k_2 k_3 + k_{-1} k_3 + k_{-1} k_{-2}}{k_1(k_{-2} + k_2 + k_3)} \qquad K_P = \frac{k_{-1} k_{-2} + k_{-1} k_3 + k_2 k_3}{k_{-3}(k_{-2} + k_2 + k_{-1})}$$

$$V_S = \frac{k_2 k_3 [E_0]}{k_3 + k_{-2} + k_2} \qquad V_P = \frac{k_{-1} k_{-2} [E_0]}{k_{-1} + k_{-2} + k_2}$$

While the Haldane relationship between the four steady-state parameters still is valid, obviously all six rate constants cannot be calculated. Analysis of the general case where an arbitrary number of reaction intermediates exists, however, shows that lower bounds to all of the rate constants in the mechanism can be determined from the four steady-state parameters (*6*). For a mechanism with n intermediates the lower bounds are

$$\begin{aligned} k_{i+1} &\geq V_S/[E_0] \quad (i \neq 0) & k_1 &\geq (V_S + V_P)/[E_0]K_S \\ k_{-i} &\geq V_P/[E_0] \quad (i \neq n+1) & k_{-(n+1)} &\geq (V_S + V_P)/[E_0]K_P \end{aligned} \tag{3-14}$$

The equalities can be seen to be valid for the case $n = 1$, and the inequalities can be easily derived for $n = 2$. At low substrate concentrations ($[S] \ll K_S$), $v = (V_S/K_S)[S]$. The apparent second order rate constant $V_S/([E_0]K_S) = k_{cat}/K_S$ is sometimes referred to as a measure of the catalytic efficiency of the enzyme since it combines both V_S and K_S. The analysis presented here indicates that this is a lower bound to the rate constant for the combination of enzyme and substrate. Steady-state kinetic experiments, in fact, are unable to give direct information about the nature and number of reaction intermediates. This finding is not too surprising since only the substrate and product, not the intermediates, are observed in steady-state studies.

KING–ALTMAN METHOD

A very useful general method of deriving the steady-state rate equations for enzyme reactions has been presented by King and Altman (7). It is unique in that the rate law can be written down without having to solve simultaneous equations. The proof underlying this method involves the theory of determinants and is not presented here. The method is applicable to any catalytic process in which the concentration of substrate(s) is much greater than that of catalyst. The mechanisms previously discussed are now reconsidered in the light of the method of King and Altman. First the mechanisms are written in terms of the enzyme species (I and II).

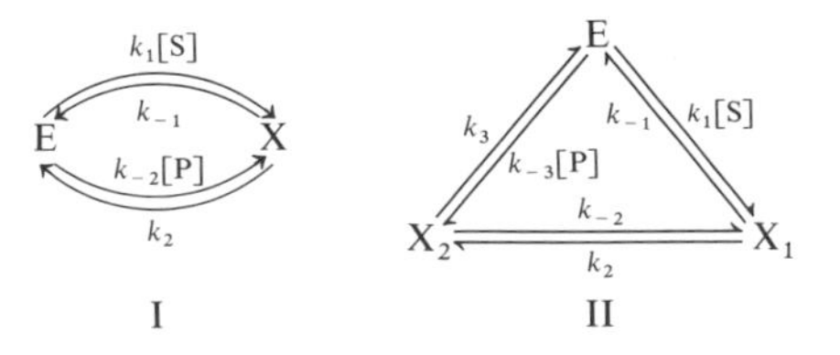

A schematic way of calculating the ratio of any enzyme species to that of the total enzyme concentration was derived by King and Altman. Consider all possible paths leading directly to a given species within the given reaction mechanism; block out in succession one individual step in the pathway and omit the reverse pathway. The results for the species E in the two mechanisms being considered are shown in III and IV.

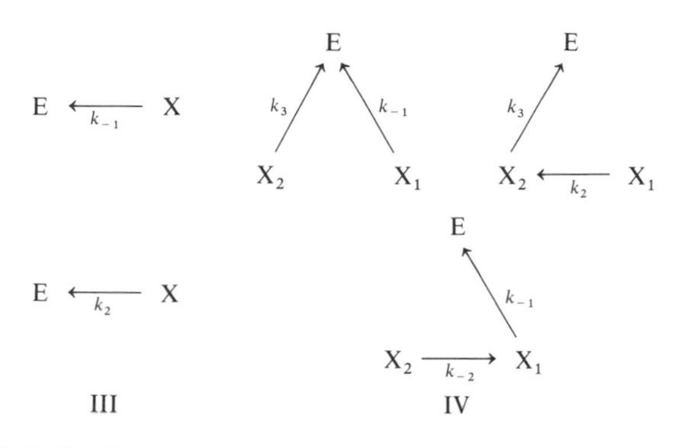

The ratio $[E]/[E_0]$ is equal to the sum of the product of the rate constants involved in each of the possible paths divided by a term D, which will be defined shortly. Thus

$$\text{I.}\quad \frac{[E]}{[E_0]} = \frac{k_{-1} + k_2}{D}$$

$$\text{II.}\quad \frac{[E]}{[E_0]} = \frac{k_{-1}k_3 + k_2k_3 + k_{-1}k_{-2}}{D}$$

In general, each term of the numerator involves rate constants (and concentrations) associated with reaction steps which individually or in sequence lead to the species in question. If the number of enzyme species is n, $n - 1$ rate constants are found in each term in the numerator and are associated with $n - 1$ different enzyme-containing species. All the possible combinations of $n - 1$ rate constants which conform to this requirement are present as numerator terms. In a similar manner, an expression can be obtained for other enzyme species

$$\text{I.}\quad \frac{[X]}{[E_0]} = \frac{k_1[S] + k_{-2}[P]}{D}$$

$$\text{II.}\quad \frac{[X_1]}{[E_0]} = \frac{k_1k_3[S] + k_1k_{-2}[S] + k_{-2}k_{-3}[P]}{D}$$

$$\frac{[X_2]}{[E_0]} = \frac{k_1k_2[S] + k_2k_{-3}[P] + k_{-1}k_{-3}[P]}{D}$$

The denominator term for a given mechanism is simply the sum of all of the numerator terms; thus

$$\text{I. } D = k_{-1} + k_2 + k_1[\text{S}] + k_{-2}[\text{P}]$$

$$\begin{aligned}\text{II. } D = {} & k_{-1}k_3 + k_2k_3 + k_{-1}k_{-2} + k_1k_3[\text{S}] + k_1k_{-2}[\text{S}] \\ & + k_{-2}k_{-3}[\text{P}] + k_1k_2[\text{S}] + k_2k_{-3}[\text{P}] + k_{-1}k_{-3}[\text{P}]\end{aligned}$$

The rate law for each of the mechanisms is

$$\text{I.} \quad -\frac{d[\text{S}]}{dt} = k_1[\text{E}][\text{S}] - k_{-1}[\text{X}] = [\text{E}_0]\left(\frac{k_1[\text{S}][\text{E}]}{[\text{E}_0]} - \frac{k_{-1}[\text{X}]}{[\text{E}_0]}\right)$$

$$\text{II.} \quad -\frac{d[\text{S}]}{dt} = k_1[\text{E}][\text{S}] - k_{-1}[\text{X}_1] = [\text{E}_0]\left(\frac{k_1[\text{S}][\text{E}]}{[\text{E}_0]} - \frac{k_{-1}[\text{X}_1]}{[\text{E}_0]}\right)$$

If appropriate substitutions are now made, the same rate equation as Eq. (3-6) is obtained.

An additional simplifying rule is that all parallel steps in a mechanism can be replaced by a single step with the effective rate constants being the sum of the rate constants for all paths in a given direction. Thus the one intermediate Michaelis–Menten mechanism can be written as

$$\text{E} \underset{k_{-1}+k_2}{\overset{k_1[\text{S}]+k_{-2}[\text{P}]}{\rightleftharpoons}} \text{X}$$

The numerator for the ratio $[\text{E}]/[\text{E}_0]$ is simply the rate constant for the backward reaction, whereas the numerator for the ratio $[\text{X}]/[\text{E}_0]$ is the rate constant for the forward reaction.

This schematic method is applicable to all mechanisms, including multisubstrate ones, and its use saves considerable time and effort for all but the very simplest mechanisms.

INHIBITION OF ENZYME CATALYSIS

An important method for investigating substrate specificity and the structure of the active site is to study the effect on the rate of catalysis of substances that are structurally similar to the substrate. In general, the rate is decreased by such substances, and this phenomenon is called inhibition. Several of the kinetically distinct types of inhibition that are observed are now considered. Probably the most commonly studied type of inhibition occurs when the inhibitor reacts with the same binding site on the free enzyme as the substrate. Because the substrate and inhibitor compete for the same binding site, this is called *competitive* inhibition. This can be accommodated into the simple Michaelis–Menten mechanism [Eq. (3-2)] by addition of the equilibrium

$$E + I \rightleftharpoons EI \tag{3-15}$$

where I is the inhibitor which binds to the enzyme with an equilibrium dissociation constant of

$$K_I = \frac{[E][I]}{[EI]}$$

Straightforward application of the steady-state approximation leads to

$$[E_0] = [E] + [X] + [EI] = [X]\left\{1 + \frac{K_S}{[S]}(1 + [I]/K_I)\right\}$$

and

$$v = k_2[X] = \frac{V_S}{1 + (K_S/[S])(1 + [I]/K_I)} \tag{3-16}$$

[In terms of the King–Altman method, dead end equilibria such as Eq. (3-15) can be taken into account by noting that

$$\frac{[EI]}{[E_0]} = \frac{[I]}{K_I}\frac{[E]}{[E_0]} = \frac{[I]}{K_I}\left(\frac{k_{-1} + k_2}{D}\right)$$

where now $D = (k_{-1} + k_2)(1 + [I]/K_I) + k_1(S)$ (assuming $(P) = 0$).] Several different methods of plotting the data can be used to determine K_I. Rearrangement of Eq. (3-16) gives

$$\frac{1}{v} = \frac{1}{V_S} + \frac{K_S}{V_S[S]}(1 + [I]/K_I)$$

If $1/v$ is plotted versus [I] at varying substrate concentrations, a series of straight lines is obtained which intersect at $K_I = -[I]$. If the usual double reciprocal plot of $1/v$ versus $1/[S]$ is made at varying inhibitor concentrations, the lines have a slope of $(K_S/V_S)(1 + [I]/K_I)$, and a secondary plot of the slope versus [I] gives K_I. Alternatively, Eq. (3-16) can be written as

$$\frac{[S]}{v} = \frac{[S]}{V_S} + \frac{K_S}{V_S}(1 + [I]/K_I)$$

A plot of $[S]/v$ versus [I] at varying substrate concentrations is a series of parallel lines with a slope of $(K_S/V_S)/K_I$ and an intercept of $[S]/V_S + K_S V_S$, whereas a plot of $[S]/v$ versus [S] at varying inhibitor concentrations is a series of parallel lines with a slope of $1/V_S$ and an intercept of $(K_S/V_S)(1 + [I]/K_I)$. These four types of plots are illustrated in Fig. 3-3.

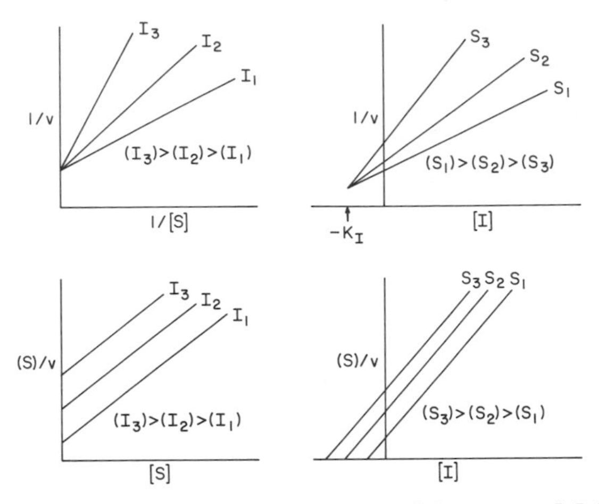

Fig. 3-3. Schematic representations of plots of $1/v$ and $[S]/v$ versus $1/[S]$, $[S]$ or $[I]$ for the case of competitive inhibition. The subscripts designate different constant concentrations of the substrate, S, and inhibitor, I; v is the initial steady-state velocity.

The effect of a competitive inhibitor is to alter the Michaelis constant. If both the maximum velocity and Michaelis constant are altered to the same extent by the inhibitor, the inhibition is *uncompetitive*. Mechanistically this can arise if the inhibitor binds appreciably only with the reaction intermediate X

$$X + I \rightleftharpoons XI, \quad K_I = [X][I]/[EI]$$

The rate law for this case is

$$v = \frac{V_S}{1 + K_S/[S] + [I]/K_I} = \frac{V_S/[1 + [I]/K_I)}{1 + \left[\frac{K_S}{1 + [I]/K_I}\right]\frac{1}{[S]}} \tag{3-17}$$

Again this equation can be rearranged into two forms for convenient determination of the kinetic parameters

$$\frac{1}{v} = \frac{1}{V_S}(1 + [I]/K_I) + \frac{K_S}{V_S}\frac{1}{[S]}$$

$$\frac{[S]}{v} = \frac{[S]}{V_S}\left(1 + \frac{[I]}{K_I}\right) + \frac{K_S}{V_S}$$

For the first of these equations, a plot of $1/v$ versus $1/[S]$ at varying inhibitor concentrations is a series of parallel lines and a plot of $1/v$ versus $[I]$ at varying substrate concentrations also is a series of parallel lines. For the second equation, a plot of $[S]/v$ versus $[S]$ at varying inhibitor concentra-

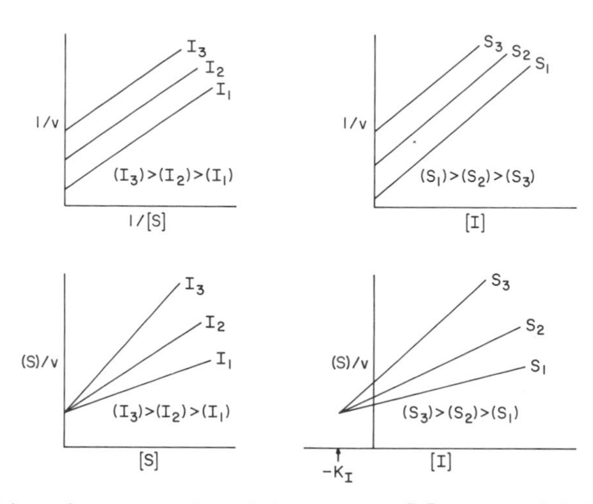

Fig. 3-4. Schematic representations of plots of $1/v$ and $[S]/v$ versus $1/[S]$, $[S]$ or $[I]$ for the case of uncompetitive inhibition. The subscripts designate different constant concentrations of the substrate, S, and inhibitor, I; v is the initial steady-state velocity.

tions is a series of lines intersecting on the ordinate at $[S]/v = K_S/V_S$, whereas a plot of $[S]/v$ versus $[I]$ at varying substrate concentrations is a series of lines intersecting at $K_I = -[I]$. These plots are shown in Fig. 3-4. This type of inhibition is not often observed.

When only the maximum velocity of the reaction is altered by the inhibitor, the inhibition is *noncompetitive*. Mechanistically this occurs when the inhibitor binds to both the free enzyme and the intermediate with the same equilibrium dissociation constant. In this case

$$v = \frac{V_S/(1 + [I]/K_I)}{1 + K_S/[S]} \qquad (3\text{-}18)$$

This equation can be rearranged as usual to give

$$\frac{1}{v} = \frac{1 + [I]/K_I}{V_S} + \frac{K_S}{V_S}\frac{(1 + [I]/K_I)}{[S]}$$

and

$$\frac{[S]}{v} = \frac{(1 + [I]/K_I)}{V_S}[S] + \frac{K_S}{V_S}(1 + [I]/K_I)$$

The four plots that can be generated from these equations are shown in Fig. 3-5. Note that these three types of inhibition can be readily distinguished from the various plots. If the inhibition constants for the binding of inhibitor to the free enzyme and the intermediate are different, both the apparent

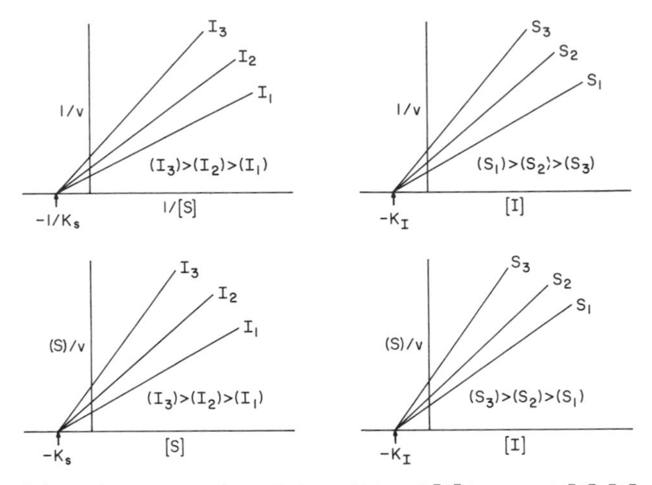

Fig. 3-5. Schematic representations of plots of $1/v$ and $[S]/v$ versus $1/[S]$, $[S]$ or $[I]$ for the case of noncompetitive inhibition. The subscripts designate different constant concentrations of the substrate, S, and inhibitor, I; v is the initial steady-state velocity.

maximum velocity and Michaelis constant are altered by the inhibitor. The initial velocity in this case is

$$v = \frac{V_S/(1 + [I]/K_I')}{1 + \dfrac{K_S(1 + [I]/K_I)}{[S](1 + [I]/K_I')}}$$

where K_I and K_I' are the inhibition constants for binding to the free enzyme and intermediate, respectively. This usually is called *mixed inhibition*, and the three cases previously discussed are limits of this more general mechanism.

Obviously more complex mechanisms of inhibition can be developed, but they are not considered here.

MULTIPLE SUBSTRATES

Many enzyme reactions have two substrates and two products, and the rate laws for such reactions now are considered as prototypes for multiple substrate reactions. Two examples are the phosphorylation of creatine catalyzed by creatine kinase

$$\text{ATP} + \text{creatine} \rightleftharpoons \text{ADP} + \text{creatine phosphate}$$

and the transamination reaction catalyzed by asparate aminotransferase

$$\text{Aspartate} + \text{ketoglutarate} \rightleftharpoons \text{oxalacetate} + \text{glutamate}$$

For the overall reaction

$$A + B \rightleftharpoons C + D \tag{3-19}$$

three types of limiting mechanisms are possible. First the substrates can be taken up in a defined order and the products released in a defined order with both substrates (and both products) required to be simultaneously on the enzyme. This is called a *ternary complex* mechanism with a *compulsory pathway* and can be represented as

$$\begin{aligned} E + A &\underset{k_{-1}}{\overset{k_1}{\rightleftharpoons}} EA \\ EA + B &\underset{k_{-2}}{\overset{k_2}{\rightleftharpoons}} X_1 \underset{k_{-3}}{\overset{k_3}{\rightleftharpoons}} ED + C \\ ED &\underset{k_{-4}}{\overset{k_4}{\rightleftharpoons}} E + D \end{aligned} \tag{3-20}$$

A second possible mechanism is one in which the first substrate transfers a chemical moiety to the enzyme and the first product is released. The second substrate then binds to the enzyme and the chemical moiety is transferred to it to produce the second product. This is called a *binary complex* or a *shuttle* or a *ping pong* mechanism and can be written as

$$\begin{aligned} A + E &\underset{k_{-1}}{\overset{k_1}{\rightleftharpoons}} X_1 \underset{k_{-2}}{\overset{k_2}{\rightleftharpoons}} C + X_2 \\ X_2 + B &\underset{k_{-3}}{\overset{k_3}{\rightleftharpoons}} X_3 \underset{k_{-4}}{\overset{k_4}{\rightleftharpoons}} D + E \end{aligned} \tag{3-21}$$

These two mechanisms can be readily distinguished by steady-state kinetic studies since they produce different rate laws. The expressions for the initial velocity can be obtained easily with the King–Altman method. For the mechanism in Eq. (3-20)

$$\frac{[E_0]}{v} = \phi_1 + \frac{\phi_2}{[A]} + \frac{\phi_3}{[B]} + \frac{\phi_4}{[A][B]} \tag{3-22}$$

with

$$\phi_1 = \frac{k_3 + k_4}{k_3 k_4}, \ \phi_2 = \frac{1}{k_1}, \ \phi_3 = \frac{k_{-2} + k_3}{k_2 k_3}, \ \phi_4 = \frac{k_{-1}}{k_1}\phi_3$$

and for the mechanism in Eq. (3-21)

$$\frac{[E_0]}{v} = \phi_1' + \frac{\phi_2'}{[A]} + \frac{\phi_3'}{[B]} \tag{3-23}$$

with

$$\phi_1' = \frac{k_2 k_4}{k_2 + k_4}, \ \phi_2' = \frac{k_{-1} + k_2}{k_1 k_2}, \ \phi_3' = \frac{k_{-3} + k_4}{k_3 k_4}$$

Both rate laws predict that a plot of $[E_0]/v$ versus $1/[A]$ should be linear, but for the first mechanism the lines obtained at different concentrations of B should be intersecting, whereas for the second mechanism they should be parallel. A schematic representation of these plots is presented in Fig. 3-6. All of the kinetic parameters can be obtained by use of plots similar to those discussed for simpler reactions. In both cases, A and B appear symmetrically in the rate law so that it is not possible to say which substrate combines with the enzyme first.

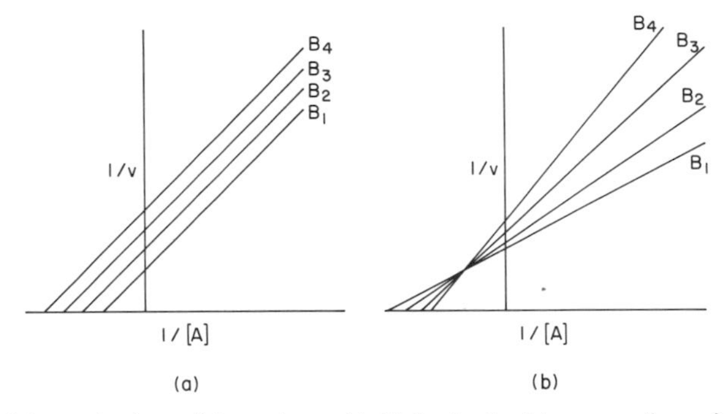

Fig. 3-6. Schematic plots of the reciprocal initial velocity, $1/v$, versus the reciprocal concentration of substrate A for a two substrate reaction. (a) Successive binary complex formation between enzyme and substrates; (b) ternary complex formation between both substrates and enzyme. The concentrations of B are such that $B_1 > B_2 > B_3 > B_4$.

A third limiting mechanism is one in which a ternary complex forms, but the substrates equilibrate with the enzyme independently and rapidly relative to the rate determining step. This *random, rapid equilibrium* mechanism can be represented as

$$\begin{aligned}
&A + E \overset{K_1}{\rightleftharpoons} EA \\
&B + E \overset{K_2}{\rightleftharpoons} EB \\
&EA + B \overset{K_2}{\rightleftharpoons} X_1 \overset{k}{\rightleftharpoons} X_2 \rightleftharpoons EC + D \text{ (or } ED + C) \\
&EB + A \overset{K_1}{\rightleftharpoons} X_1 \overset{k}{\rightleftharpoons} X_2 \rightleftharpoons EC + D \text{ (or } ED + C) \\
&C + E \overset{K_3}{\rightleftharpoons} EC \\
&D + E \overset{K_4}{\rightleftharpoons} ED
\end{aligned} \tag{3-24}$$

where the K_is are equilibrium constants. The rate law in this case is identical to Eq. (3-22) with

$$\phi_1 = \frac{1}{k}, \ \phi_2 = \frac{1}{K_1 k}, \ \phi_3 = \frac{1}{K_2 k}, \ \phi_4 = \frac{1}{K_1 K_2 k}$$

Thus, a distinction between the first and third mechanisms cannot be made from simply measuring the initial velocity at various substrate concentrations. The third mechanism requires that $\phi_4 = \phi_2\phi_3/\phi_1$, but the experimental precision rarely is sufficient to use this as a criterion of mechanism. The two mechanisms can be distinguished, however, by measuring the effect of product inhibition on the initial rate. Experiments can be carried out with varying concentrations of A, B, and C in the absence of D and with varying concentrations of A, B, and D in the absence of C. The steady-state initial velocities for the first mechanism in these two cases is

$$\frac{[\mathrm{E_0}]}{v} = \phi_1(1 + \theta_2[\mathrm{C}]) + \frac{\phi^2}{[\mathrm{A}]} + \frac{\phi_3}{[\mathrm{B}]}(1 + \theta_3[\mathrm{C}]) + \frac{\phi_4}{[\mathrm{A}][\mathrm{B}]}(1 + \theta_3[\mathrm{C}]) \tag{3-25}$$

$$\frac{[\mathrm{E_0}]}{v} = \phi_1 + \frac{\phi_2}{[\mathrm{A}]}(1 + \theta_1[\mathrm{D}]) + \frac{\phi_3}{[\mathrm{B}]} + \frac{\phi_4}{[\mathrm{A}][\mathrm{B}]}(1 + \theta_1[\mathrm{D}]) \tag{3-26}$$

In these equations the θ_i are constants that can be expressed in terms of rate constants. Since the substrates do not appear symmetrically in these equations, it is possible to determine which substrate binds first, which binds second, which product is released first, and which is released last. [Note that C is noncompetitive with respect to A in Eq. (3-25) and D is competitive with respect to A in Eq. (3-26).] The corresponding initial velocity equations for the third mechanism are

$$\frac{[\mathrm{E_0}]}{v} = \phi_1 + \frac{\phi_2}{[\mathrm{A}]} + \frac{\phi_3}{[\mathrm{B}]} + \frac{\phi_4}{[\mathrm{A}][\mathrm{B}]}(1 + K_3[\mathrm{C}]) \tag{3-27}$$

$$\frac{[\mathrm{E_0}]}{v} = \phi_1 + \frac{\phi_2}{[\mathrm{A}]} + \frac{\phi_3}{[\mathrm{B}]} + \frac{\phi_4}{[\mathrm{A}][\mathrm{B}]}(1 + K_4[\mathrm{D}]) \tag{3-28}$$

The forms of the rate laws for these two mechanisms now are clearly different, and an experimental distinction is possible. In practice, this distinction is not always clear-cut because the rate laws are sufficiently complex and the experimental error sufficiently large to cause some ambiguity. In addition, mechanistic steps not in the original schemes may occur when three substrates are present simultaneously.

A more general form of the ternary complex mechanism can be written as

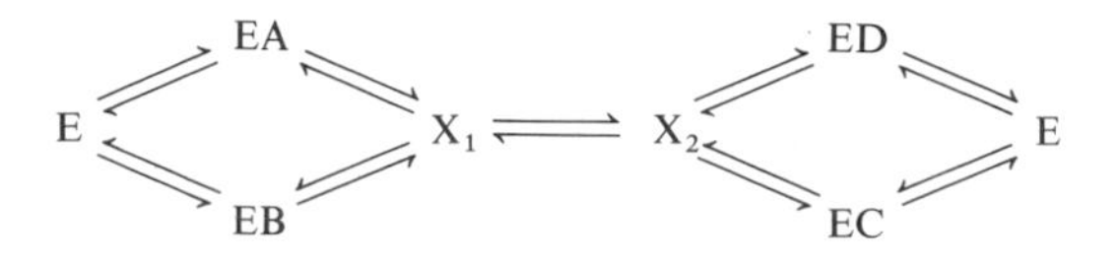

However, a complete steady-state treatment of this mechanism leads to a rate law that is too complex for practical use. As might be expected from the earlier discussion of a single substrate–single product mechanism, the forms of the rate laws for the two substrate–two product mechanisms are independent of the number of reaction intermediates. Also lower bounds to the rate constants can be calculated from the steady-state parameters (*8*).

The three mechanisms just discussed often are written in terms of the diagrams in Fig. 3-7. Also a nomenclature has been suggested in which the type of mechanism is classified as ordered, ping pong, or random followed by the number of substrates and products (*4*). Thus the mechanisms are ordered bi bi, ping pong bi bi, and random bi bi. This special terminology is not utilized in this volume.

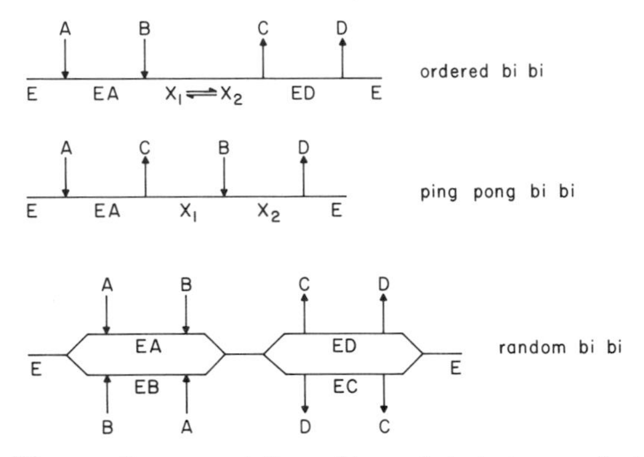

Fig. 3-7. Diagramatic representations of two substrate–two product reactions.

Reactions with more than two substrates and two products can be analyzed using the same techniques as the two substrate–two product reactions. Obviously, the kinetic equations are very complex.

ISOTOPE EXCHANGE AT EQUILIBRIUM

An interesting method for determining information about the reaction pathway for multiple substrate reactions is the measurement of isotope exchange at equilibrium (*9, 10*). With this technique, a mixture of the enzyme and substrates is allowed to come to equilibrium with the total enzyme concentration being much smaller than the total substrate concentration. A small radioactive concentration of one of the substrates is then added to the system, and the rate of appearance of radioactivity in one of the products is measured. Derivation of the rate equations is complex, and only an abbreviated treatment is given here. Fortunately, the principles of this experiment can be discussed without knowledge of all of the detailed kinetic equations. Consider as an example the introduction of radioactive A (designated as A*) which appears as C* in a two substrate–two product compulsory pathway mechanism. The mechanism can be written as

$$E + A^* \underset{k_{-1}}{\overset{k_1}{\rightleftharpoons}} EA^*$$

$$EA^* + B \underset{k_{-2}}{\overset{k_2}{\rightleftharpoons}} X^* \underset{k_{-3}}{\overset{k_3}{\rightleftharpoons}} ED + C^*$$

$$ED \underset{k_{-4}}{\overset{k_4}{\rightleftharpoons}} E + D$$

If EA* and X* are assumed to be in a steady state, the initial velocity ($[C^*] = 0$) is

$$v^* = k_3[X^*] = \frac{k_1 k_2 k_3 [E][A^*][B]}{k_{-1}(k_{-2} + k_3) + k_2 k_3 [B]}$$

Moreover, the concentration of E is determined by the equilibrium concentrations of unlabeled substrate and can be found with the King–Altman method to be

$$[E] = \frac{[E_0]}{1 + \frac{k_1}{k_{-1}}[A] + \frac{k_{-4}}{k_4}[D] + \frac{k_1 k_2 [A][B]}{k_{-1} k_{-2}}}$$

The concentration of C does not occur explicitly because the ratio of concentrations of products to reactants must obey the equilibrium expression $(k_1 k_2 k_3 k_4)/(k_{-1} k_{-2} k_{-3} k_{-4}) = [C][D]/[A][B]$. These two expressions can be combined to give

$$v^* = R_{AC}[A^*]/[A]$$

where R_{AC}, which is defined by the above equations, is the rate of conversion of A to C at equilibrium and is a function of the substrate concentrations and the rate constants and is proportional to the total enzyme concentration. Since the rate of conversion of A to C is equal to the rate of conversion of C to A at equilibrium, the net rate equation for the exchange reaction can be written as

$$\frac{d[C^*]}{dt} = R_{AC}([A^*]/[A] - [C^*]/[C])$$

Furthermore $[C^*] + [A^*] = [C^*_\infty] + [A^*_\infty]$ and $[C^*_\infty]/[A^*_\infty] = [C]/[A]$ where the subscript infinity denotes the equilibrium concentration, so that

$$\frac{d[C^*]}{dt} = R_{AC} \frac{[A] + [C]}{[A][C]} ([C^*_\infty] - [C^*])$$

Finally this equation can be integrated to give

$$\ln(1 - F) = -R_{AC}\frac{[A] + [C]}{[A][C]}t \tag{3-29}$$

where F is the fractional equilibration, $[C^*]/[C^*_\infty]$. This important result states that the rate of isotope exchange at equilibrium always is first order regardless of the mechanism. Mechanistic information is contained in R_{AC}, and further analysis shows that $R_{BC} > R_{AC}$, $R_{BD} > R_{AD}$, $R_{BC} > R_{BD}$ and $R_{AC} > R_{AD}$. This result is intuitively expected; the rate is slower when more intermediates must be passed through. Thus, isotope exchange experiments at equilibrium can determine the order of addition and release of substrates without a complex kinetic analysis. Information about the substrate and product binding constants also can be obtained by studying the rates at varying substrate and product concentrations. Furthermore, ordered and random, rapid equilibrium mechanisms can be distinguished since the above rate inequalities do not prevail for the latter mechanism. The disadvantages of this method are the complexity of the rate law at equilibrium, which may even be different from that predicted because of the variety of possible enzyme–substrate–product complexes, and the side reactions which may occur during the relatively long incubations due to enzyme impurities.

pH DEPENDENCE OF ENZYME CATALYSIS

Reaction rates in enzyme systems are often extremely sensitive to variation of the pH. In general, enzymes exhibit maximum catalytic activity at a definite pH. The optimum pH is generally in the vicinity of pH 7 (± 1), although exceptions are well known. The dependence of the kinetic parameters of the fumarase reaction on pH is illustrated in Figs. 3-8 and 3-9. Several different possible effects of pH on the reaction must be distinguished. In the first place, many substrates may have ionizable groups and only one of the ionized forms of the substrate may be acted upon by the enzyme. Since substrate ionization constants can be determined quite easily and precisely, it is generally (but not always) fairly easy to determine which form of the substrate undergoes reaction. For example, in the case of fumarase the doubly charged acid anions are the reactive species. In some cases, difficulties may arise from a close coupling of substrate ionizations with enzyme ionizations.

Two reasons are most often given for the effect on the enzyme of varying pH. Changes in pH can produce substantial structure changes in the enzyme, and this partial denaturation, i.e., loss of native structure, can produce a

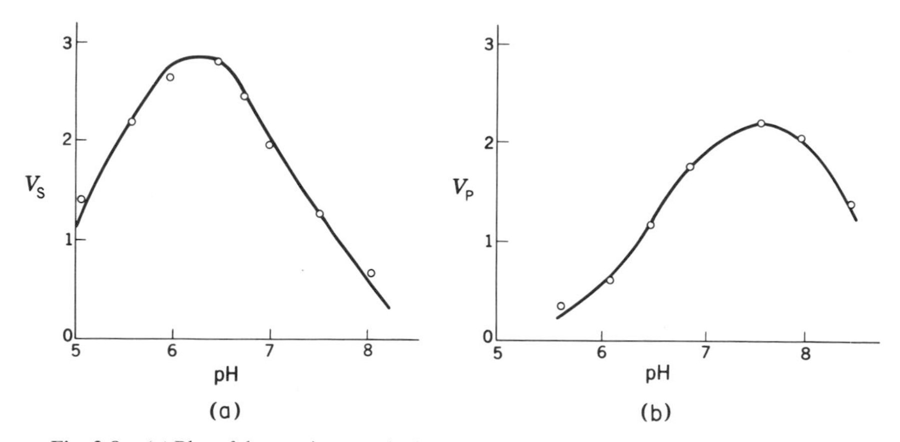

Fig. 3-8. (a) Plot of the maximum velocity for fumarate as a function of pH in 0.01 *M* acetate buffer at 25°; the solid line is a theoretical curve. (b) Plot of the maximum velocity for *l*-malate as a function of pH in 0.01 *M* acetate buffer at 25°; the solid line is a theoretical curve. [Adapted from C. Frieden and R. A. Alberty, *J. Biol. Chem.* **212**, 859 (1955).]

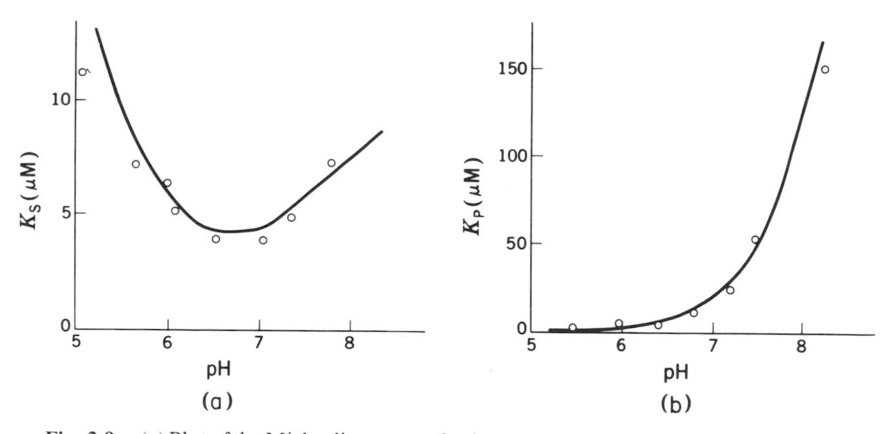

Fig. 3-9. (a) Plot of the Michaelis constant for fumarate as a function of pH in 0.01 *M* acetate buffer at 25°; the solid line is a theoretical curve. (b) Plot of the Michaelis constant for *l*-malate as a function of pH in 0.01 *M* acetate buffer at 25°; the solid line is a theoretical curve. [Adapted from C. Frieden, and R. A. Alberty, *J. Biol. Chem.* **212**, 859 (1955).]

sharp decline in the specific activity of the enzyme. Usually such effects occur only at extreme pHs and can be recognized by a general lack of reproducibility of the experimental results and variation of the rate with past history of the enzyme, e.g., the length of time the enzyme has been exposed to the extreme pHs. The type of variation of rate parameters with pH illustrated for fumarase is usually explained as being due to ionizations at the active site of the enzyme, with only one of the ionized forms of the enzyme being catalytically active. This behavior can be incorporated into the simple one-intermediate mechanism by Eq. (3-30)

$$\begin{array}{ccccc} EH_2 & & XH_2 & & EH_2 \\ K_a \upharpoonleft\downharpoonright & & K_{xa} \upharpoonleft\downharpoonright & & \upharpoonleft\downharpoonright \\ EH + S & \rightleftharpoons & XH & \rightleftharpoons & EH + P \\ K_b \upharpoonleft\downharpoonright & & K_{xb} \upharpoonleft\downharpoonright & & \upharpoonleft\downharpoonright \\ E & & X & & E \end{array} \tag{3-30}$$

Here hydrogen ions have been omitted for the sake of simplicity and the K_is define ionization constants in the usual fashion; e.g.,

$$K_a = \frac{[EH][H^+]}{[EH_2]}$$

Since protolytic reactions are usually very fast (especially in the presence of buffers, where proton transfer reactions occur readily), all the protolytic steps can be assumed to be in equilibrium throughout the course of the reaction. The derivation of the rate equation is exactly as before except that the enzyme-conservation equation is given by

$$[E_0] = [EH] + [E] + [EH_2] + [XH] + [X] + [XH_2] \tag{3-31}$$

or

$$[E_0] = [EH]\left(1 + \frac{[H^+]}{K_a} + \frac{K_b}{[H^+]}\right) + [XH]\left(1 + \frac{[H^+]}{K_{xa}} + \frac{K_{xb}}{[H^+]}\right) \tag{3-32}$$

and

$$\frac{d}{dt}([X] + [XH] + [XH_2]) = 0$$

$$= -(k_{-1} + k_2)[XH] + k_1[EH][S] + k_{-2}[EH][P] \tag{3-33}$$

The resulting rate equation is readily seen to have the same form as Eq. (3-6) but now

$$K_S = K'_S \frac{1+[H^+]/K_a+K_b/[H^+]}{1+[H^+]/K_{xa}+K_{xb}/[H^+]}, \qquad K_P = K'_P \frac{1+[H^+]/K_a+K_b/[H^+]}{1+[H^+]/K_{xa}+K_{xb}/[H^+]}$$

$$V_S = V'_S \frac{1}{1+[H^+]/K_{xa}+K_{xb}/[H^+]}, \qquad V_P = V'_P \frac{1}{1+[H^+]/K_{xa}+K_{xb}/[H^+]} \tag{3-34}$$

where now the primed quantities are the same functions of the rate constants as given by Eq. (3-7). Note that the ratio $V_S/K_S[E_0]$ (and $V_P/K_P[E_0]$) is dependent only on the ionization constants of the free enzyme. This function exhibits a maximum at a definite pH, which should be the same for both the forward and reverse reactions. This type of behavior is illustrated in Fig. 3-10 for the fumarase reaction. Both pK_a and pK_b can be determined and have been found to be 6.2 and 6.8 at 25°C for the case illustrated (*11*).

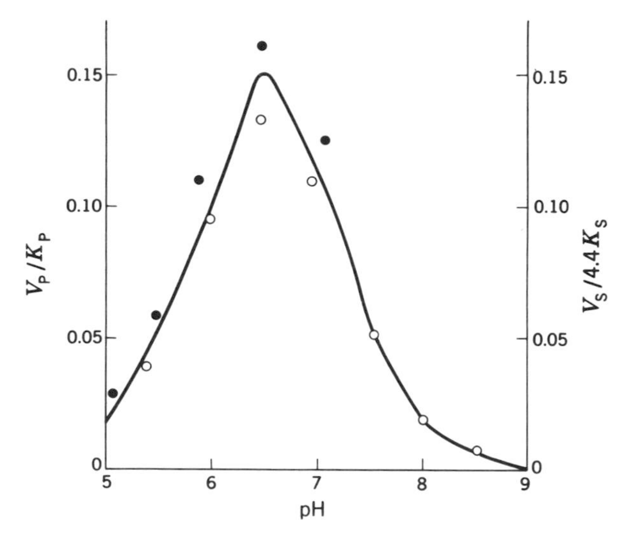

Fig. 3-10. Plot of $V_S/4.4K_S$ (●) and V_P/K_P (○) versus pH in 0.01 *M* acetate buffer at 25°. The factor 4.4 is the equilibrium constant for the overall reaction and normalizes the ordinates. The solid line is the theoretical curve calculated according to the equations given in the text. [Adapted from C. Frieden, and R. A. Alberty, *J. Biol. Chem.* **212**, 859 (1955).]

The mechanism assumed also predicts that V_S and V_P have identical dependencies on pH which is not found for the fumarase reaction. This indicates at least one more reaction intermediate must be added to the mechanism. A mechanism like that in Eq. (3-30), but with *n* reaction intermediates has been analyzed (*6*). The results obtained indicate that the kinetic parameters depend on pH exactly as for the one intermediate mechanism. However, the ionization constants of the intermediate are complex averages involving rate constants and the ionization constants of all of the intermediates. The ratios V_S/K_S and V_P/K_P still depend only on the ionization constants of the free enzyme. The pH dependence of these parameters obviously offers the best possibility for an unequivocal interpretation. In contrast, the apparent ionization constants of the reaction intermediates cannot be reliably interpreted since the number of reaction intermediates is unknown. A further complication can arise if parallel pathways occur, that is if steps such as $E + S \rightleftharpoons X$ and $EH_2 + S \rightleftharpoons XH_2$ are included in the mechanism. Exactly the same rate law would be obtained, but the interpretation of the pH dependence of the kinetic parameters would be altered. Even the pK values determined from the pH dependence of V_S/K_S and V_P/K_P could be complex functions of rate and ionization constants. Finally, in some cases the protolytic reactions may not occur rapidly relative to other steps in the reaction mechanism. This situation requires a more complex analysis (*5*).

The pH dependence of the kinetic parameters of multisubstrate reactions can be derived in an analogous fashion. As with the single substrate–single product mechanism, a ratio of kinetic parameters always exists that depends only on the ionization constants of the free enzyme (neglecting parallel pathways). As before, only the pK values of the free enzyme can be interpreted with reasonable confidence. However, in general it should be realized that the pK values determined from kinetics may not correspond to true ionization constants.

The best method for obtaining pK values from kinetic data is direct fitting of the data to theoretical equations. However, a convenient graphical method for estimating pK values from V_S (or V_P) and V_S/K_S (or V_P/K_P) is to plot log V_S or log V_S/K_S versus pH (*12*). For example, consider the equation

$$\log V_S = \log V'_S - \log(1 + [H^+]/K_{xa} + K_{xb}/[H^+])$$

At very low pH values, $\log V_S = \log V'_S - \log([H^+]/K_{xa})$ so that the plot has

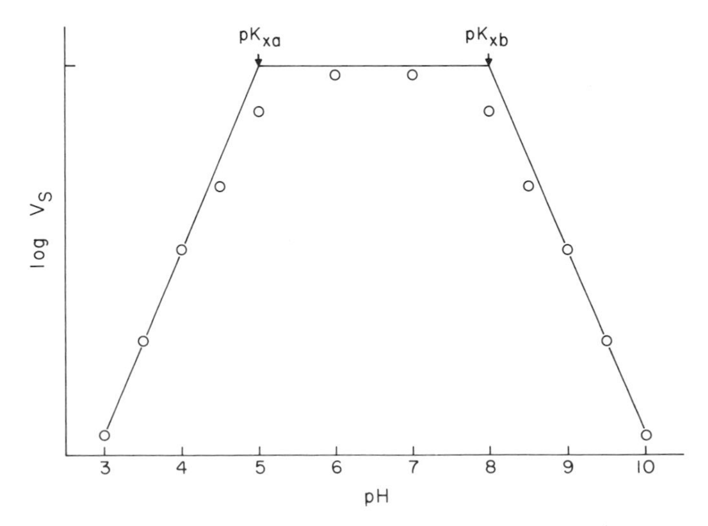

Fig. 3-11. A plot of log V_S versus pH. Here $\log V_S = \log V'_S - \log[1 + (H^+)/K_{xa} + K_{xb}/(H^+)]$; p$K_{xa}$ = 5.0 and pK_{xb} = 8.0. The lines are the limiting behavior of this function and can be used to estimate the pK values.

a slope of $+1$. At very high pH values, $\log V_S = \log V'_S - \log(K_{xb}/[H^+])$ and the plot has a slope of -1. If the pK values are well separated (several units or more), an intermediate region exists where $\log V_S = \log V'_S$. Examination of these equations shows that the intersection points of these straight line segments occurs at pH = pK_{xa} and pH = pK_{xb}. A schematic drawing of a plot of $\log V_S$ versus pH is shown in Fig. 3-11. A similar plot can be made for the Michaelis constants, but this requires all four ionization constants to be well separated. The logarithmic plots are most useful for preliminary analyses of the data and/or for cases where the number of ionizable groups determining the kinetic parameters is small.

TEMPERATURE DEPENDENCE OF ENZYME CATALYSIS

The temperature dependence of enzyme catalysis is very difficult to interpret in mechanistic terms. The reaction rate will precipitously drop at a sufficiently high temperature due to thermal denaturation of the enzyme. Below this temperature, plots of $\log V_S$, $\log K_S$, $\log(V_S/K_S)$, etc. versus the reciprocal temperature more often than not are straight lines. However, the interpretation of the apparent activation energy is not clear. Included in this activation energy could be the activation energy of the rate constant for the rate determining step, enthalpy changes for ionization, and other equilibria in the mechanism, or activation energies for thermal conformational changes. If straight lines are not observed, curvature could be due to heat capacity effects, a change in the rate controlling step, or thermally induced conformational changes of the enzyme.

Limited mechanistic information can be obtained from a study of the temperature dependence of the steady-state kinetic parameters over a wide range of pH. Activation energies for the lower bounds of the rate constants and enthalpy changes associated with ionization constants can be obtained. Such a study has been carried out for fumarase (*13*). However, even in this case an unequivocal interpretation of the activation parameters in mechanistic terms is not possible.

CONCLUSION

In summary, steady-state kinetics can give information about the overall reaction mechanism. The substrate specificity and stereochemical restrictions of the active site can be determined. Lower bounds to the rate constants can be calculated, although nothing can be said about the number and nature of the reaction intermediates. The effect of pH on the rate parameters allows qualitative statements to be made about the p*K* values of ionizable groups at the active site; however, such interpretations should be viewed with caution. Finally, information about the reaction pathway for multisubstrate reactions can be derived.

REFERENCES

1. G. G. Hammes, "Principles of Chemical Kinetics." Academic Press, New York, 1978.
2. I. H. Segel, "Enzyme Kinetics." Wiley (Interscience), New York, 1975.
3. A. Cornish-Bowden, "Fundamentals of Enzyme Kinetics." Butterworth, London, 1979.
4. W. W. Cleland, *in* "The Enzymes" (P. Boyer, ed.), 3rd ed., Vol. 2, p. 1. Academic Press, New York, 1970.
5. W. W. Cleland, *Adv. Enzymol.* **45**, 273 (1977).
6. L. Peller and R. A. Alberty, *J. Am. Chem. Soc.* **81**, 5907 (1959).
7. E. L. King and C. Altman, *J. Phys. Chem.* **60**, 1375 (1956).
8. V. Bloomfield, L. Peller, and R. A. Alberty, *J. Am. Chem. Soc.* **84**, 4367, 4375 (1962).
9. P. D. Boyer, *Arch. Biochem. Biophys.* **82**, 387 (1959).
10. R. A. Alberty, V. Bloomfield, L. Peller, and E. L. King, *J. Am. Chem. Soc.* **84**, 4381 (1962).
11. C. Frieden and R. A. Alberty, *J. Biol. Chem.* **212**, 859 (1955).
12. M. Dixon, *Biochem. J.* **55**, 161 (1953).
13. D. A. Brant, L. B. Barnett and R. A. Alberty, *J. Am. Chem. Soc.* **85**, 2204 (1963).

4

Transient Kinetic Methods for Studying Enzymes

INTRODUCTION

While steady-state kinetics provides a very useful approach for studying enzyme mechanisms, its inherent weakness is that little information about states occurring between the uptake of substrates and release of products is obtained. This is because reaction intermediates are not directly observable. The remedy for this difficulty is obvious, namely, higher concentrations of enzyme. However, enzymatic reactions then occur very rapidly, with typical half-times of milliseconds. In order to study such rapid reactions, special experimental methods are required. Also, at high enzyme concentrations the steady-state approximation cannot be utilized so that solution of the kinetic equations is difficult. A summary of the fast reaction techniques that have been used to study enzyme reactions and related processes and their time ranges of application is shown in Fig. 4-1. Detailed descriptions of these techniques are available (*1*), and only a brief discussion of some of the most useful methods for the study of enzymes is presented here. Magnetic resonance methods already have been discussed in Chapter 2. Approaches for solving the kinetic equations are presented following the discussion of techniques.

RAPID MIXING METHODS

The principle behind rapid mixing methods is simple: mix the reactants together as rapidly as possible and see what happens (cf. *2*). Liquids can be conveniently mixed in about 1 msec, although somewhat shorter mixing times can be obtained, and this is the time resolution of most rapid mixing

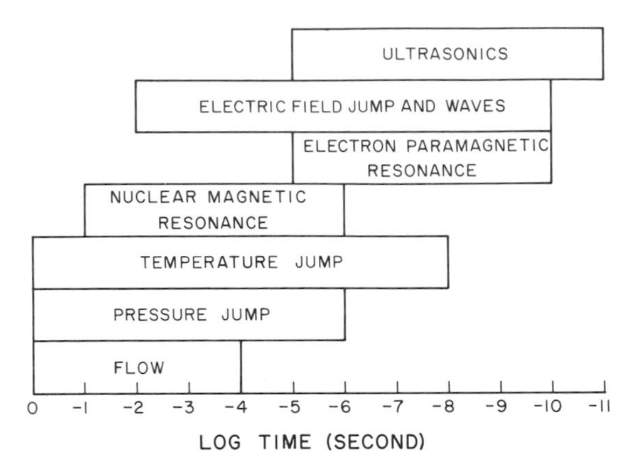

Fig. 4-1. Summary of the time ranges of fast reaction techniques which have been used to study enzyme reactions and related processes. The slow time limit has been arbitrarily terminated at 1 sec.

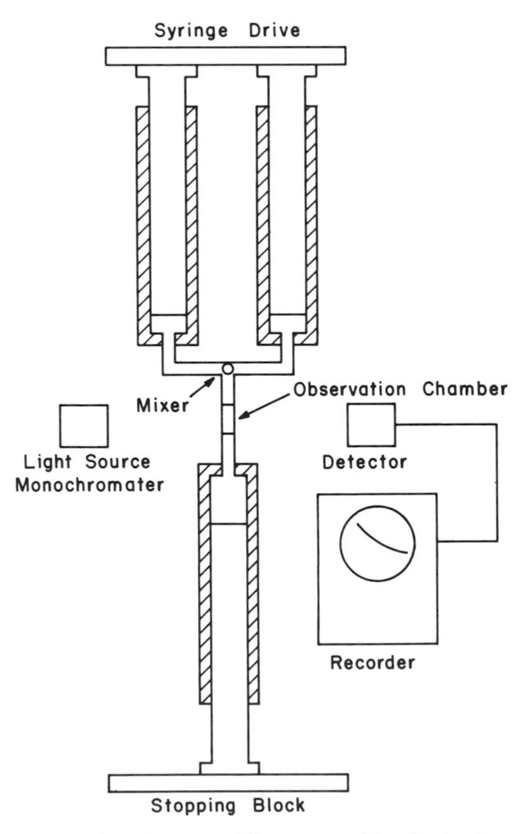

Fig. 4-2. Schematic drawing of a stopped-flow system. Usually the detector is a photomultiplier and the recorder an oscilloscope. [From G. G. Hammes, *in* "Methods for Determining Metal Ion Environments in Proteins" (D. W. Darnell and R. G. Wilkins, eds.), Elsevier/North Holland, New York, 1980.]

devices. Two types of rapid mixing apparatuses have proved most useful for studying enzyme reactions: the stopped flow and the rapid quench. With the stopped flow, the reactants are rapidly mixed, the flow is stopped within a few milliseconds, and the reaction progress is monitored, typically by absorption spectrophotometry, fluorescence, or light scattering. Such an apparatus is schematically depicted in Fig. 4-2. Rapid quench methods are useful if an optical method cannot be used conveniently to measure the reaction rate. With this method, the reactants are mixed rapidly and allowed to age for any desired time; the reaction mixture is then mixed rapidly with a third solution that quenches the reaction. The aging time is either fixed by the length of tubing between the mixed reactants and the addition point for quencher or by a timer (for longer aging times). The quencher used depends on the particular reaction; acid or base is typical. The time resolution of quench methods is about 10 msec, with 20 or 30 msec being more typical. A schematic diagram of a rapid quench apparatus is shown in Fig. 4-3 (cf. *3*).

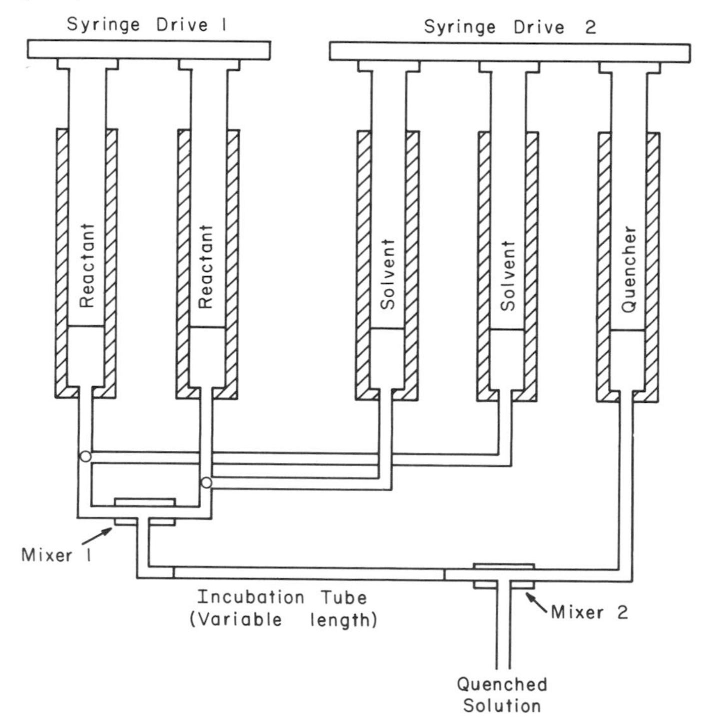

Fig. 4-3. Quenched flow apparatus. The reactants are mixed and driven into an incubation tube by syringe drive 1; after the desired time interval, syringe drive 2 drives the incubated mixture with solvent into mixer 2 where the reaction mixture is quenched. [From G. G. Hammes, *in* "Methods for Determining Metal Ion Environments in Proteins" (D. W. Darnell and R. G. Wilkins, eds.), Elsevier/North Holland, New York, 1980.]

An aspect of rapid mixing methods that merits special mention is the use of low temperatures, i.e., subzero, to slow the rate of chemical reactions. Obviously, chemical processes can be slowed by lowering the temperature; however the application of this principle to enzymes and proteins is not trivial. This is because biological processes occur in water, which, of course, freezes at 0°C. In order to study biological reactions at subzero temperatures, low temperature solvents must be found in which good biological activity occurs. Considerable success has been achieved toward this goal in recent years, and exciting results have been obtained (*4,5*). Construction of a rapid mixing device that works at low temperatures also is not trivial: special valves and syringes are necessary, and thermostating is difficult. However, if a reaction is sufficiently slow at subzero temperatures, conventional spectrophotometry may be sufficient. Low temperature rapid mixing experiments show considerable promise for exploring the molecular details of biological reactions.

To study reactions occurring in times shorter than about 1 msec, methods must be utilized that do not require mixing of the reactants. The primary techniques are chemical relaxation and magnetic resonance (which is discussed in Chapter 2).

CHEMICAL RELAXATION METHODS

Chemical relaxation techniques were developed in the 1950s by Eigen and co-workers (cf. *6*). Rather than trying to mix the reactants, a reaction mixture already at equilibrium is perturbed by varying an external parameter. This causes a shift to new equilibrium concentrations, and determination of the rate of this shift permits the kinetics of the reaction in question to be studied. The external parameters to be considered here are temperature, pressure, and electric field. They can be varied in several ways, e.g., periodically, by a rectangular pulse, or by a single step impulse. Because the reaction rates are finite, the actual shift of the concentrations does not follow the perturbation exactly; i.e., the equilibrium shift lags behind the perturbation. Examples of this phenomenon are illustrated in Figs. 4-4 and 4-5 for chemical reactions with a characteristic relaxation time τ (*7*).

By far the most useful perturbation is temperature because virtually all chemical reactions are temperature-dependent or can be coupled to temperature-dependent reactions. The temperature variation of the equilibrium constant, K_P, for a chemical reaction is given by the relationship

$$\frac{\partial \ln K_P}{\partial T} = \frac{\Delta H^\circ}{RT^2} \tag{4-1}$$

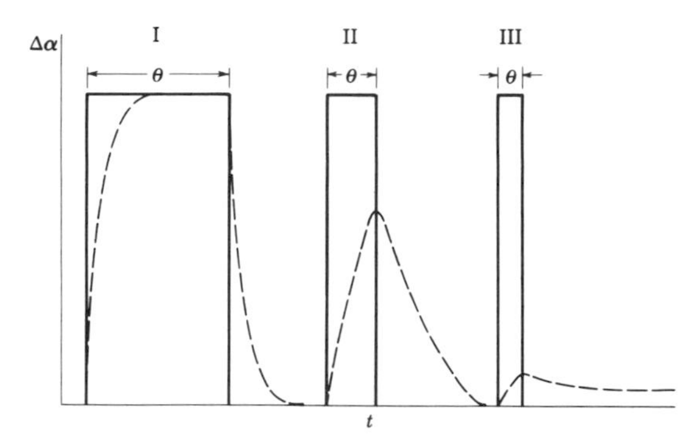

Fig. 4-4. Rectangular impulses; θ = impulse duration. I: $\tau/\theta = 0.1$; II: $\tau/\theta = 1$; III: $\tau/\theta = 10$. $\Delta\alpha$ is a measure of the equilibrium shift. —, external parameter; – – –, relaxation curves.

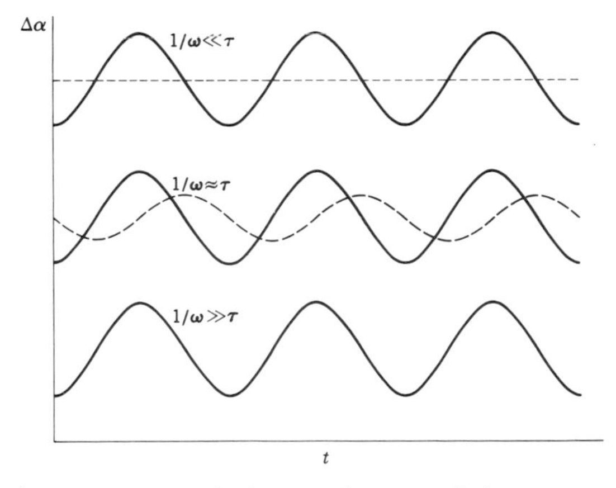

Fig. 4-5. Continuous wave perturbation. ω = frequency. $\Delta\alpha$ is a measure of the equilibrium shift. —, external parameter; – – –, $\Delta\alpha$. (For $1/\omega \gg \tau$, the two curves are identical.)

where ΔH° is the enthalpy of reaction. With the temperature jump method, the temperature typically is raised 5–10°C in 10^{-6} sec or less, and the rate at which the concentrations of reactants approach their new equilibrium values at the higher temperature is measured. The most common method for generating a temperature jump is by discharging a capacitor, initially charged to 10,000–100,000 volts, through the reaction mixture. Concentration changes are monitored with optical techniques: absorption spectroscopy, fluorimetry, polarimetry, or light scattering. After about 1 sec, convective mixing obscures the concentration changes, placing an effective limit on the slowness of reactions that can be studied. A schematic diagram of a temperature jump apparatus is shown in Fig. 4-6. Laser pulses and coaxial

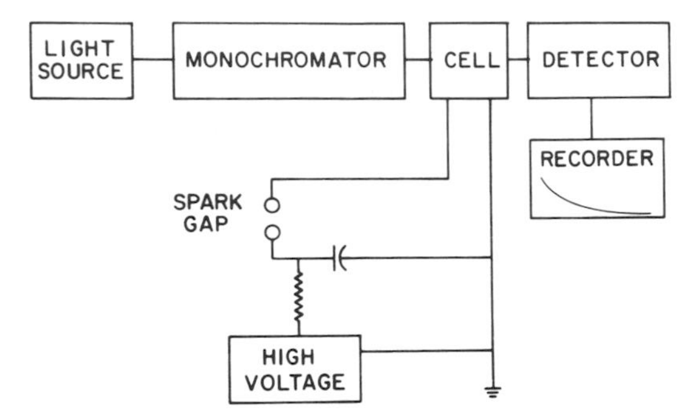

Fig. 4-6. Schematic drawing of a temperature-jump apparatus utilizing Joule heating by discharge of a high voltage through the solution. The detector is generally a photomultiplier and the recorder an oscilloscope. [From G. G. Hammes, *in* "Investigations of Rates and Mechanisms of Reactions," Part II. Wiley (Interscience), New York, 1974.]

electrical discharges can be used to extend the time resolution into the nanosecond region.

Clearly pressure changes also can be used to perturb chemical equilibria since

$$\frac{\partial \ln K_P}{\partial P} = -\frac{\Delta V^\circ}{RT} \tag{4-2}$$

where ΔV° is the standard volume change for the reaction. However, equilibrium constants normally are not very pressure-dependent. The most common type of pressure jump apparatus transmits a high pressure to the solution (~ 100 atm) through a diaphragm, which is then ruptured. Recently a device has been developed that applies a square or sine wave pressure perturbation by compressing the solution through a voltage-dependent extension of a piezo-electric crystal stack (*8*). Optical and conductometric methods are used to monitor the concentration changes. The time resolution of these methods is about 10^{-5} sec.

Electric fields perturb chemical equilibria involving a change in the number of ions or in the dipole moments of the reactants and products. Electric field methods are not useful for the study of enzymes but have been used to study simpler reactions that are relevant to understanding enzyme mechanisms. A similar statement can be made about ultrasonic techniques. An ultrasonic wave is an adiabatic pressure wave so that both the pressure and temperature vary as it passes through a solution. In water, the pressure perturbation generally is dominant, whereas in nonaqueous solutions, the temperature perturbation usually is of more importance. The primary advantage of electric field and ultrasonic methods is that they permit the study of very fast reactions (Fig. 4-1).

GENERAL SOLUTIONS OF KINETIC EQUATIONS

The steady-state approximation gives rise to many convenient solutions to the kinetic equations for enzyme mechanisms. Matters become considerably more complex without this approximation. For example, consider the formation of a complex between an enzyme, E, and a substrate, S

$$E + S \underset{k_{-1}}{\overset{k_1}{\rightleftharpoons}} X_1 \tag{4-3}$$

The rate equation governing this reaction is

$$\frac{d[X_1]}{dt} = -\frac{d[E]}{dt} = -\frac{d[S]}{dt} = k_1[E][S] - k_{-1}[X_1] \tag{4-4}$$

and two conservation relationships exist

$$\begin{aligned} [E_0] &= [E] + [X_1] \\ [S_0] &= [S] + [X_1] \end{aligned} \tag{4-5}$$

The rate equation can be solved to give explicit equations for the time dependence of the concentrations, but the solution is too complex to be of much use. If the total substrate concentration, $[S_0]$, is much greater than the total enzyme concentration, $[E_0]$, it can be assumed to remain constant during the course of the reaction. The reaction now becomes *pseudo first order*, and the rate equation can be integrated to a convenient form. Equation (4-4) can be written as

$$\frac{d[X_1]}{dt} = k_1[S_0]([E_0] - [X_1]) - k_{-1}[X_1] \tag{4-6}$$

Integration with $[X_1] = 0$ at $t = 0$ gives

$$\ln \frac{k_1[S_0][E_0]}{k_1[S_0][E_0] - (k_1[S_0] + k_{-1})[X_1]} = (k_1[S_0] + k_{-1})t$$

or

$$\ln \frac{[E_0] - [\bar{E}]}{[E] - [\bar{E}]} = (k_1[S_0] + k_{-1})t \tag{4-7}$$

where the overbar designates the equilibrium concentration and $k_1/k_{-1} = [\bar{X}_1]/([S_0][\bar{E}])$. If the equilibrium constant is known, both rate constants can be easily extracted from the data. This method of converting a second order reaction into a pseudo first order reaction is often done to simplify the kinetic analysis.

The next degree of complexity is to have two coupled reactions such as

$$E + S \underset{k_{-1}}{\overset{k_1}{\rightleftharpoons}} X_1 \underset{k_{-2}}{\overset{k_2}{\rightleftharpoons}} X_2 \tag{4-8}$$

The mass conservation equations are

$$\begin{aligned} [E_0] &= [E] + [X_1] + [X_2] \\ [S_0] &= [S] + [X_1] + [X_2] \end{aligned} \tag{4-9}$$

and *two* independent rate equations are needed to describe this system. (The number of independent rate equations is equal to the total number of concentration variables minus the number of conservation relationships between the concentrations.) Two such equations are

$$-\frac{d[E]}{dt} = k_1[E][S] - k_{-1}[X_1] \tag{4-10}$$

$$-\frac{d[X_2]}{dt} = k_{-2}[X_2] - k_2[X_1] \tag{4-11}$$

These equations cannot be integrated analytically. If the rate equations are made pseudo first order by requiring that $[S_0] \gg [E_0]$ and $[X_1]$ is eliminated by use of the mass conservation relationships, Eqs. (4-10) and (4-11) can be written as

$$\begin{aligned} -\frac{d[E]}{dt} &= (k'_1 + k_{-1})[E] + k_{-1}[X_2] - k_{-1}[E_0] \\ -\frac{d[X_2]}{dt} &= k_2[E] + (k_{-2} + k_2)[X_2] - k_{-2}[E_0] \end{aligned} \tag{4-12}$$

with $k'_1 = k_1[S_0]$. The solution to a set of simultaneous linear first order differential equations is always a sum of exponential functions with the number of exponential terms being equal to the number of independent rate equations. Thus

$$\begin{aligned} [E] &= [\bar{E}] + A_1 e^{-\lambda_1 t} + A_2 e^{-\lambda_2 t} \\ [X_2] &= [\bar{X}_2] + A_3 e^{-\lambda_1 t} + A_4 e^{-\lambda_2 t} \end{aligned} \tag{4-13}$$

The A_is are constants dependent on the initial conditions and

$$\begin{aligned} \lambda_{1,2} = \tfrac{1}{2}[&(k'_1 + k_{-1} + k_2 + k_{-2}) \\ &\pm \sqrt{(k'_1 + k_{-1} + k_2 + k_{-2})^2 - 4(k'_1 k_2 + k'_1 k_{-2} + k_{-1}k_{-2})}] \end{aligned} \tag{4-14}$$

where the plus and minus roots correspond to the two values of λ. Methods for obtaining the λ_is are presented in the discussion of rate equations near equilibrium.

From the discussion thus far it should be clear that analytical solutions of coupled rate equations that are not first order cannot be obtained. For example, the rate equations for the simple Michaelis–Menten mechanism with one intermediate cannot be solved analytically. On the other hand, if the rate equations can be made pseudo first order, analytical solutions always can be obtained, although they may be complex. The task of resolving several exponential terms from experimental rate data is not trivial. Another approach to solving complex coupled rate equations is the use of analog and digital computers. For example, the time dependence of the concentration variables for a single substrate–single product Michaelis–Menten mechanism with four intermediates is shown in Fig. 4-7. Specific values of the rate constants and initial concentrations have been assumed. After a brief induction period, all of the concentrations of intermediates approach constant values so that the steady-state approximation $d[X_i]/dt = 0$ then is valid; this is because $[E_0] = 0.1[S_0]$. With modern computer technology, the solution of simultaneous differential equations by numerical methods may be more convenient than with analog devices.

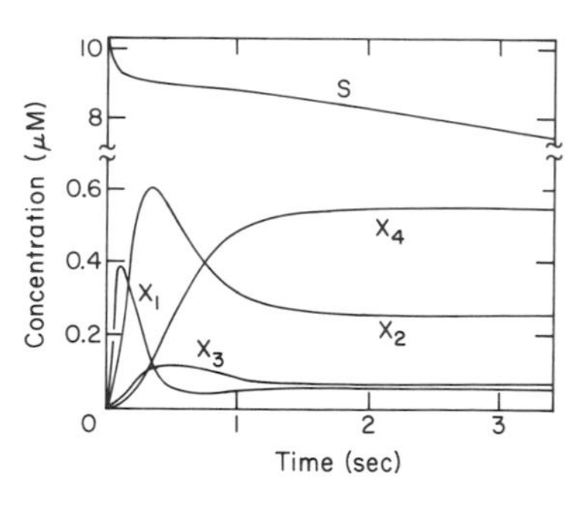

Fig. 4-7. Typical analog computer solutions for the four intermediate Michaelis-Menten mechanism. Here S is the substrate and the X_i are intermediates. Specific substrate and enzyme concentrations and rate constants have been assumed. [Courtesy of Dr. J. Higgins; cf. J. Higgins, *Tech. Org. Chem.* **8**, Part 1, 285 (1961).]

RATE EQUATIONS NEAR EQUILIBRIUM

With chemical relaxation methods, the system is displaced only a small amount from equilibrium. Analytical solutions to the rate equations near equilibrium always can be obtained regardless of the mechanistic complexity. To simplify matters, step function perturbations are considered first. In this case the conventional rate equations describe the system after the perturbation is complete. For example, let us reexamine the mechanism of Eq. (4-3) and the associated rate equation, Eq. (4-4). New concentration variables can be defined such that

$$[E] = [\bar{E}] + \Delta[E], [S] = [\bar{S}] + \Delta[S], [X_1] = [\bar{X}_1] + \Delta[X_1] \quad (4\text{-}15)$$

where the Δs designate deviations from the equilibrium concentrations.

Also from mass conservation

$$\Delta C = \Delta[\mathrm{X}_1] = -\Delta[\mathrm{E}] = -\Delta[\mathrm{S}] \tag{4-16}$$

Inserting Eqs. (4-15) and (4-16) into Eq. (4-4) gives

$$\begin{aligned} -\frac{d\Delta C}{dt} &= \{k_1([\bar{\mathrm{E}}] + [\bar{\mathrm{S}}]) + k_{-1}\}\Delta C \\ &\quad - k_1[\bar{\mathrm{E}}][\bar{\mathrm{S}}] + k_{-1}[\bar{\mathrm{X}}_1] - k_1(\Delta C)^2 \end{aligned} \tag{4-17}$$

However

$$\frac{k_1}{k_{-1}} = \frac{[\bar{\mathrm{X}}_1]}{[\bar{\mathrm{E}}][\bar{\mathrm{S}}]}$$

Furthermore, in the neighborhood of equilibrium $(\Delta C)^2$ is so small that the last term in Eq. (4-17) can be neglected. This equation, therefore, can be written as

$$-\frac{d\Delta C}{dt} = \{k_1([\bar{\mathrm{E}}] + [\bar{\mathrm{S}}]) + k_{-1}\}\Delta C = \frac{\Delta C}{\tau} \tag{4-18}$$

which defines the relaxation time τ. From this example, it is clear that all rate equations near equilibrium are linear first order differential equations. This follows by neglecting all terms higher than first order concentration deviations, that is $(\Delta C)^2$, etc. Integration of Eq. (4-18) gives

$$\Delta C = \Delta C_0 e^{-t/\tau} \tag{4-19}$$

where ΔC_0 is the concentration deviation at $t = 0$. The relaxation time can easily be extracted from the data by use of a plot of $\ln \Delta C$ versus t, which is a straight line with a slope of $-1/\tau$. If the equilibrium concentrations are varied or if the equilibrium constant is known, both rate constants can be determined.

Let us now examine the two-step mechanism of Eq. (4-8). The rate equations obtained by linearization of Eqs. (4-10) and (4-11), as before, and use of the mass conservation relationships $\Delta[\mathrm{E}] + \Delta[\mathrm{X}_1] + \Delta[\mathrm{X}_2] = 0$ and $\Delta[\mathrm{S}] + \Delta[\mathrm{X}_1] + \Delta[\mathrm{X}_2] = 0$ are

$$\begin{aligned} -\frac{d\Delta[\mathrm{E}]}{\mathrm{dt}} &= \{k_1([\bar{\mathrm{E}}] + [\bar{\mathrm{S}}]) + k_{-1}\}\Delta[\mathrm{E}] + k_{-1}\Delta[\mathrm{X}_2] \\ &= a_{11}\Delta[\mathrm{E}] + a_{12}\Delta[\mathrm{X}_2] \end{aligned} \tag{4-20}$$

$$\begin{aligned} -\frac{d\Delta[\mathrm{X}_2]}{dt} &= k_2\Delta[\mathrm{E}] + (k_{-2} + k_2)\Delta[\mathrm{X}_2] \\ &= a_{21}\Delta[\mathrm{E}] + a_{22}\Delta[\mathrm{X}_2] \end{aligned} \tag{4-21}$$

where the a_{ij}s are defined by these equations. The solution to this pair of first order linear homogeneous differential equations is a sum of exponentials. This can be shown by assuming that

$$\begin{aligned} \Delta[\mathrm{E}] &= A_1 e^{-t/\tau_1} + A_2 e^{-t/\tau_2} \\ \Delta[\mathrm{X}_2] &= A_3 e^{-t/\tau_1} + A_4 e^{-t/\tau_2} \end{aligned} \tag{4-22}$$

where the A_is are constants. Inserting these solutions into the differential equations gives

$$[(a_{11} - 1/\tau_1)A_1 + a_{12}A_3]e^{-t/\tau_1} + [(a_{11} - 1/\tau_2)A_2 + a_{12}A_4]e^{-t/\tau_2} = 0$$
$$[(a_{22} - 1/\tau_1)A_3 + a_{21}A_1]e^{-t/\tau_1} + [(a_{22} - 1/\tau_2)A_4 + a_{21}A_2]e^{-t/\tau_2} = 0$$

For these equations to be true, the coefficients of each of the exponential terms must equal zero. This gives two sets of simultaneous equations, one set for A_1 and A_3 and another for A_2 and A_4. A solution for these simultaneous equations exists only if

$$\begin{vmatrix} a_{11} - 1/\tau & a_{12} \\ a_{21} & a_{22} - 1/\tau \end{vmatrix} = 0 \tag{4-23}$$

Solving this determinant for $1/\tau_{1,2}$ gives

$$1/\tau_{1,2} = \frac{a_{11} + a_{22}}{2}\left\{1 \pm \left[1 - \frac{4(a_{11}a_{22} - a_{12}a_{21})}{(a_{11} + a_{22})^2}\right]^{1/2}\right\} \tag{4-24}$$

where τ_1 corresponds to the positive sign and τ_2 to the negative sign. [The reciprocal relaxation times, $1/\tau_{1,2}$, correspond exactly to $\lambda_{1,2}$ of Eq. (4-14).] This is a rather complex function to extract rate constants from, but convenient combinations of the relaxation times are

$$\frac{1}{\tau_1} + \frac{1}{\tau_2} = a_{11} + a_{22} = k_1([\bar{\mathrm{E}}] + [\bar{\mathrm{S}}]) + k_{-1} + k_2 + k_{-2} \tag{4-25}$$

$$\frac{1}{\tau_1}\frac{1}{\tau_2} = a_{11}a_{22} - a_{12}a_{21} = k_1(k_2 + k_{-2})([\bar{\mathrm{E}}] + [\bar{\mathrm{S}}]) + k_{-1}k_{-2} \tag{4-26}$$

Both the sum and product of the reciprocal relaxation times are linear functions of $([\bar{\mathrm{E}}] + [\bar{\mathrm{S}}])$, and all four rate constants can be obtained from the two linear plots.

Quite often the bimolecular step in Eq. (4-8) is much more rapid than the second step in the mechanism. In this case, $k_1([\bar{\mathrm{E}}] + [\bar{\mathrm{S}}]) + k_{-1} \gg k_2 + k_{-2}$ or $a_{11} \gg a_{22}$. The bracketed term in Eq. (4-24) may be expanded by use of the relationship $(1 - X)^{1/2} \approx 1 - X/2$, when $X \ll 1$. The result is

$$1/\tau_1 = k_1([\bar{\mathrm{E}}] + [\bar{\mathrm{S}}]) + k_{-1} \tag{4-27}$$

$$1/\tau_2 = \frac{k_2}{1 + k_{-1}/\{k_1([\bar{\mathrm{E}}] + [\bar{\mathrm{S}}])\}} + k_{-2} \tag{4-28}$$

The expression for $1/\tau_1$ is identical to that derived for the one-step mechanism [Eq. (4-3)]. While $1/\tau_1$ is a linear function of $([\bar{\mathrm{E}}] + [\bar{\mathrm{S}}])$, $1/\tau_2$ approaches a limiting value of $k_2 + k_{-2}$ at high concentrations of $([\bar{\mathrm{E}}] + [\bar{\mathrm{S}}])$. (This is the relaxation time for the isolated reaction $X_1 \rightleftharpoons X_2$.) At low values of $([\bar{\mathrm{E}}] + [\bar{\mathrm{S}}])$, $1/\tau_2$ also is a linear function of $([\bar{\mathrm{E}}] + [\bar{\mathrm{S}}])$ so that data must be obtained over a wide range of equilibrium concentrations to determine which relaxation time is being measured. The concentration dependencies of these two relaxation times are shown for a particular set of rate constants in Fig. (4-8).

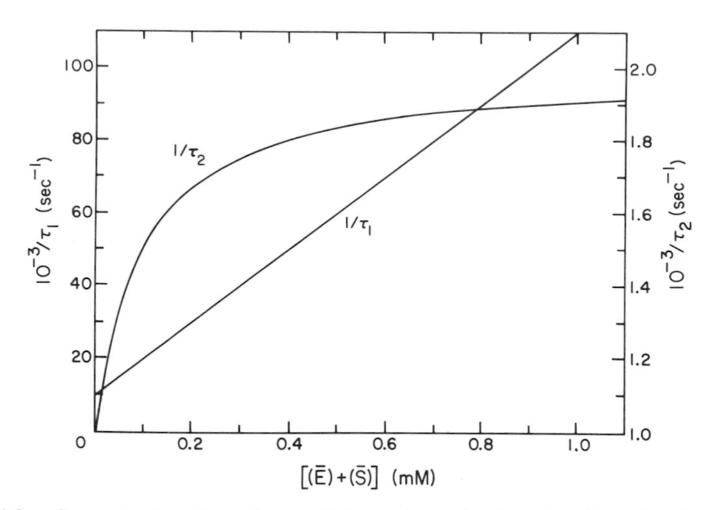

Fig. 4-8. Concentration dependence of the reciprocal relaxation times for the two-step mechanism of Eq. (4-8). Reciprocal relaxation times were calculated from Eqs. (4-27) and (4-28) assuming $k_1 = 10^8\ M^{-1}\ \mathrm{sec}^{-1}$, $k_{-1} = 10^4\ \mathrm{sec}^{-1}$, and $k_2 = k_{-2} = 10^3\ \mathrm{sec}^{-1}$.

The approximate expressions for the relaxation times given in Eqs. (4-27) and (4-28) can be derived without solving the secular determinant. The expression for τ_1 is readily obtained by assuming that the unimolecular step is sufficiently slow that the concentration of X_2 is fixed while the bimolecular step equilibrates. The second relaxation time may be derived by assuming that the fast bimolecular step is always at equilibrium while the second step equilibrates. The equilibrium constant relationship, $k_1/k_{-1} = [\bar{X}_1]/([\bar{E}][\bar{S}])$, then can be differentiated to give $\Delta[X_1] = \{k_1([\bar{E}] + [\bar{S}])/k_{-1}\}\,\Delta[E]$. Substituting this relationship and the mass conservation equation $\Delta[X_1] = -\Delta[E] - \Delta[X_2]$ into Eq. (4-21) gives

$$-\frac{d\Delta[X_2]}{dt} = \frac{\Delta[X_2]}{\tau_2} \tag{4-29}$$

For mechanisms described by n independent rate equations, the rate equations can be linearized to give a set of equations of the form

$$-\frac{d\Delta C_i}{dt} = \sum_{j=1}^{n} a_{ij}\,\Delta C_j \tag{4-30}$$

where the a_{ij}s are functions of the rate constants and equilibrium concentrations as before. This system of equations is analogous to that encountered in vibration spectroscopy where the equations of motion of the individual atoms in a molecule are described by coupled first order linear differential equations. This coupled set of equations of motion can be transformed to an uncoupled set of equations of motion in which the dependent variables are linear combinations of the coordinates characterizing the motions of the individual atoms, and the time dependence of the motion is described by normal mode frequencies which are functions of the individual bond frequencies. Similarly, the coupled rate equations [Eq. (4-30)], can be transformed to a set of independent linear first order rate equations which can be represented as

$$-\frac{d\Delta y_i}{dt} = \frac{\Delta y_i}{\tau_i} \tag{4-31}$$

The concentration variable Δy_i is a linear combination of the ΔC_i, and the τ_i are normal mode relaxation times that are functions of the a_{ij}. The time dependence of the individual concentrations is a sum of exponentials

$$\Delta C_i = \sum_{j=1}^{n} A_{ij} e^{-t/\tau_j} \tag{4-32}$$

and the n relaxation times are obtained by solving the determinant

$$\begin{vmatrix} a_{11} - 1/\tau & a_{12} \cdots & a_{1n} \\ a_{21} & a_{22} - 1/\tau \cdots & a_{2n} \\ \vdots & & \\ a_{n1} \cdots & & a_{nn} - 1/\tau \end{vmatrix} = 0$$

Thus far the amplitudes associated with chemical relaxation (A_{ij}) have been assumed to be constants. However, thermodynamic information about the reactions is contained in these constants. In principle, the equilibrium constants and a thermodynamic variable, such as enthalpy changes for the temperature jump method and volume changes for the pressure jump method, can be calculated from the amplitudes. For coupled reactions, the thermodynamic variables are normal mode quantities, that is linear combinations of the individual reaction thermodynamic parameters. The relaxation amplitudes also are functions of constants associated with the detection methods, for example, extinction coefficients and quantum yields. Thus far detailed analyses of relaxation amplitudes have only been carried out for relatively simple reactions (cf. *6*, *9*).

Step function perturbations have been assumed in solving the rate equations, but for some relaxation methods the equilibrium concentration itself varies in some known manner with time. For example, ultrasonic perturbations are usually applied as sine waves. The relaxation times are unaltered in such a case but the actual time dependence of the concentrations may not be simple exponentials. This situation can be accommodated by defining new concentration variables

$$\begin{aligned} \bar{C}_i &= C_i^0 + \Delta\bar{C}_i \\ C_i &= C_i^0 + \Delta C_i \end{aligned} \tag{4-33}$$

where now C_i^0 is a time-independent concentration. The deviation from equilibrium of the concentration now is

$$C_i - \bar{C}_i = \Delta C_i - \Delta\bar{C}_i$$

If this is now substituted in a rate equation such as Eq. (4-4) and the usual linearization procedure is used, the result is

$$\tau \frac{d\Delta C}{dt} + \Delta C = \Delta\bar{C} \tag{4-34}$$

If the perturbation is sinosoidal

$$\Delta\bar{C} = ae^{i\omega t}$$

where a is a constant, ω is the frequency of the perturbation, and $i = \sqrt{-1}$. The solution to Eq. (4-34) for this case is

$$\Delta C = \frac{ae^{i\omega t}}{1 + i\omega\tau} = \frac{\Delta\bar{C}e^{-i\Phi}}{[1 + (\omega\tau)^2]^{1/2}} \tag{4-35}$$

where $\tan\Phi = \omega\tau$. This result indicates the ΔC oscillates with the same frequency as $\Delta\bar{C}$ but lags behind by the phase angle Φ; moreover, the amplitude of ΔC is decreased by the factor $[1 + (\omega\tau)^2]^{-1/2}$. Equation (4-35) can be used to calculate the curves in Fig. 4-5. The primary purpose for discussing perturbations not applied stepwise is to indicate that the calculation of the relaxation times is unchanged and that the modification of the rate equations is straightforward.

The successful application of relaxation methods requires that the concentrations of the species in equilibrium be comparable. However, perturbations also can be applied to steady states. The mathematical analysis is similar to that for equilibrium perturbations. Essentially irreversible reactions can be studied by perturbing steady states of the reaction mixture as equilibrium is approached. The stopped flow and temperature jump methods have been coupled together and utilized in this manner for studying an enzymatic reaction (*10*).

More comprehensive mathematical accounts of chemical relaxation are available (cf. *6, 9*), but the treatment presented here should suffice for most practical applications.

REFERENCES

1. G. G. Hammes, ed., "Techniques of Chemistry," Vol. 6, Part 2. Wiley (Interscience), New York, 1974.
2. B. Chance, *in* "Techniques of Chemistry" (G. G. Hammes, ed.), Vol. 6, Part 2, pp. 5–62. Wiley (Interscience), New York, 1974.
3. A. R. Fersht, "Enzyme Structure and Mechanism" p. 106. Freeman, San Francisco, California, 1977.
4. P. Douzou, *Adv. Enzymol.* **51**, 1 (1980).
5. A. L. Fink, *Accts. Chem. Res.* **10**, 233 (1977).
6. M. Eigen and L. de Maeyer, *in* "Techniques of Chemistry" (G. G. Hammes, ed.), Vol. 6, Part 2, pp. 63–146. Wiley (Interscience), New York, 1974.
7. G. G. Hammes, "Principles of Chemical Kinetics," p. 189. Academic Press, New York, 1978.
8. R. M. Clegg and B. W. Maxfield, *Rev. Sci. Inst.* **47**, 1383 (1976).
9. C. F. Bernasconi, "Relaxation Kinetics." Academic Press, New York, 1976.
10. J. E. Erman and G. G. Hammes, *Rev. Sci. Inst.* **37**, 746 (1966).

5

Some Chemical Aspects of Enzyme Catalysis

CATALYSIS

Every chemical reaction proceeds by the reactants going through a series of different states until the products are reached. In molecular terms the intermediate states have changes in bond lengths, angles, etc. Very often this gradual transition from reactants to products is simplified by assuming the reactants are in equilibrium with a single *transition state*, and that products are produced by the breakdown of the transition state. The overall rate of the reaction then can be expressed in terms of the equilibrium constant characterizing the equilibrium between the reactants and the transition state and the rate of breakdown of the transition state into products. In the very simplest theory, the rate constant, k, for reactions in liquid solutions is expressed as

$$k = \frac{RT}{N_0 h} e^{-\Delta G^{\circ\ddagger}/RT} \tag{5-1}$$

$$= \frac{RT}{N_0 h} e^{\Delta S^{\circ\ddagger}/R} e^{-\Delta H^{\circ\ddagger}/RT} \tag{5-2}$$

where $\Delta G^{\circ\ddagger}$ $(= \Delta H^{\circ\ddagger} - T\Delta S^{\circ\ddagger})$ is the standard free energy change for the conversion of reactants to transition state (usually called the free energy of activation), $\Delta H^{\circ\ddagger}$ is the standard enthalpy change, $\Delta S^{\circ\ddagger}$ is the standard entropy change, R is the gas constant, T is the absolute temperature, N_0 is Avogadro's number, and h is Planck's constant (cf. *1* for a detailed description of the transition state theory).

Catalysis can be regarded as a stabilization of the intermediate states, or in terms of transition state theory as a stabilization of the transition state with respect to the reactants. In other words, $\Delta G^{\circ\ddagger}$ is decreased. (However, a statement such as this is only meaningful if the uncatalyzed and catalyzed reactions proceed through the same reaction pathway.) This is a formal way of stating that catalysts increase the rate of a chemical reaction but are not consumed during the reaction. Of course, a catalyst accelerates the forward and reverse reactions an equal amount since the equilibrium constant cannot be altered by a catalyst.

An important principle to remember is that of detailed balance (sometimes called microscopic reversibility). This principle states that if a reaction proceeds through a given pathway (mechanism) in going from reactants to products, then at equilibrium the reaction pathway in going from products to reactants must be exactly opposite that of the forward reaction. This means, of course, that the forward and reverse reactions have identical transition states. This principle also is valid for some nonequilibrium situations, for example, in many steady states. An enzyme is an incredibly efficient catalyst, usually many orders of magnitude more efficient than man-made catalysts for comparable reactions. However, the chemical basis of enzyme catalysis is no different from that found with simple catalysts. Therefore, a brief discussion of the three types of catalysis commonly encountered in chemical reactions is profitable, namely acid–base catalysis, nucleophilic catalysis, and electrophilic catalysis.

ACID–BASE CATALYSIS

This type of catalysis is involved in virtually every enzymatic reaction. As an example, consider the hydrolysis of esters, a reaction catalyzed by a number of enzymes and by acids and bases. The mechanism proceeds through a transition state of water and ester that has a partial negative charge on the carbonyl oxygen and partial positive charge on the oxygen of the attacking water

$O^{\delta-}$
R—C—OR′
$O^{\delta+}$
H H

The reaction rate can be catalyzed by a base through stabilization of the positive charge by transfer or partial transfer of a water proton to a base.

Similarly, acid catalysis can be accomplished by transfer or partial transfer of a proton to the carbonyl oxygen. While this obviously is a very simplistic view of acid–base catalysis, it illustrates the fundamental feature of all such reactions: stabilization of intermediate states by proton transfer reactions.

The rate laws and terminology associated with acid–base catalysis are quite simple. The rate law is written as rate $= k_{obs}[S]$ where S is the substrate and k_{obs} is the observed rate constant. If $k_{obs} = k_{OH}[OH^-]$ over a range of buffer concentrations, where k_{OH} is a rate constant, the catalysis by hydroxide ion is *specific base catalysis*. *Specific acid catalysis* means that hydrogen ion is the catalyst and $k_{obs} = k_H[H^+]$. For all other cases, the catalysis is referred to as *general acid catalysis* and *general base catalysis*. The rate laws can be represented as $k_{obs} = k_{BH}[BH]$ and $k_{obs} = k_B[B^-]$, respectively, where BH and B^- represent the general acid and general base. The various rate laws can be readily distinguished by determining the observed rate constant at constant pH and varying buffer concentration. If the rate constant is independent of buffer concentration, as illustrated in Fig. 5-1, either specific acid or specific base catalysis is occurring; variation of the pH readily permits a distinction to be made between these two possibilities. For specific acid catalysis, a plot of log k_{obs} versus pH is a straight line with a slope of -1; for specific base catalysis the slope is $+1$. Mention should be made of the fact that the rate constant also can have a noncatalytic component that is independent of pH and buffer concentration. If the observed rate constant is

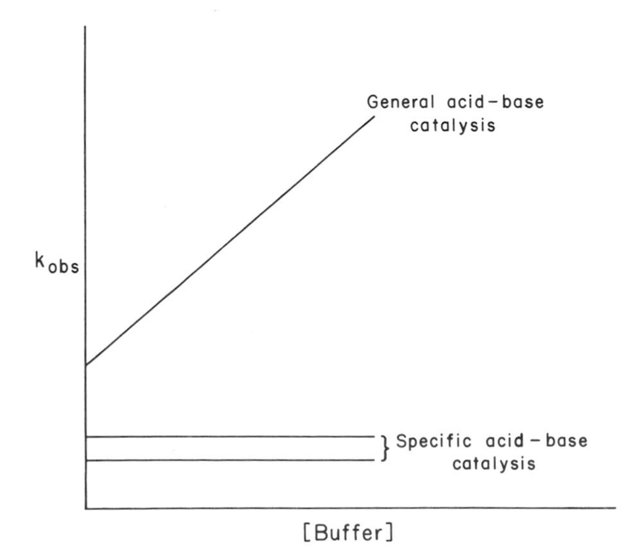

Fig. 5-1. Schematic plot of the catalytic rate constant, k_{obs}, versus the buffer concentration showing specific and general acid–base catalysis.

dependent on buffer concentration, a plot of k_{obs} versus buffer concentration often is a straight line (Fig. 5-1). The intercept of this line, k_0, is the sum of the rate constants associated with the uncatalyzed reaction and specific acid and base catalysis, whereas the slope contains the rate constants for general acid and general base catalysis. If a series of such linear plots is constructed at different pH values, all of the rate constants can be sorted out. The intercepts can be analyzed to determine the rate constants for the uncatalyzed, and specific acid and base catalyzed reactions. We then can write

$$k_{obs} - k_0 = k_{BH}[BH] + k_B[B^-] \tag{5-3}$$

If Eq. (5-3) is divided by the total buffer concentration $[B_T]$

$$\frac{k_{obs} - k_0}{[B_T]} = k_{BH}\frac{[BH]}{[B_T]} + k_B\frac{[B^-]}{[B_T]}$$

$$= k_{BH}\frac{[H^+]}{K_A + [H^+]} + k_B\frac{K_A}{K_A + [H^+]} \tag{5-4}$$

where K_A is the ionization constant of the buffer. If only general acid or only general base catalysis is occurring, plots of $(k_{obs} - k_0)/[B_T]$ versus pH, shown schematically in Fig. 5.2a,b, are typical titration curves, and both K_A and k_{BH} or k_B can be determined. If both general acid and general base catalysis occur, a plot of $(k_{obs} - k_0)/[B_T]$ versus $[B^-]/[B_T]$ is a straight line, which has an intercept of k_{BH} on the left ordinate, an intercept of k_B on the right ordinate, and a slope of $k_B - k_{BH}$ (remember that $[BH]/[B_T] = 1 - [B^-]/[B_T]$), as illustrated in Fig. 5-2c. An additional contribution to k_{obs} sometimes occurs due to *simultaneous* catalysis by BH and B; this term can be represented as $k_{BHB}[BH][B^-]$. In water, such terms are difficult to detect

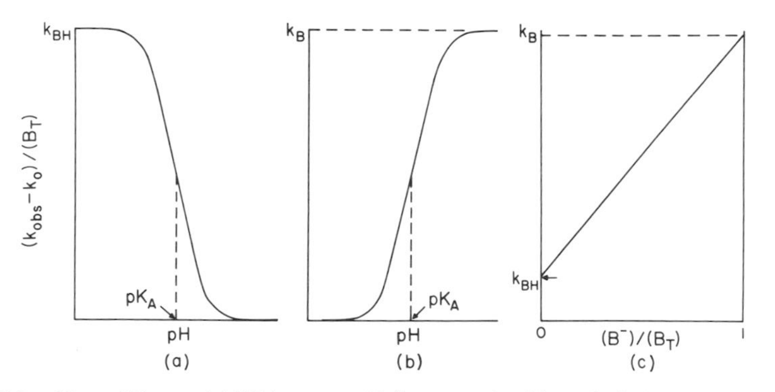

Fig. 5-2. Plot of $(k_{obs} - k_0)/(B_T)$ versus pH for general acid catalysis (a) and general base catalysis (b). The plot of $(k_{obs} - k_0)/(B_T)$ versus $(B^-)/(B_T)$ in c shows the occurrence of both general acid and general base catalysis.

because water itself can serve as either an acid or a base which participates in catalysis with a buffer component, and its concentration is 55 M. A contribution to k_{obs} such as $k_{BHB}[B^-][H_2O]$ cannot be distinguished from $k_B[B^-]$.

An inherent ambiguity exists in deriving mechanisms for acid–base catalysis from kinetic rate laws. This is because the rate law gives information only about the composition of the transition state; it gives no information about *how* the transition state is formed. Thus, for example, if the conversion of a substrate, S, to a product is general acid catalyzed, rate $= k_{BH}[BH][S]$. However, two kinetically equivalent expressions are, rate $= k_B[B^-][H^+][S]$ and rate $= k'_B[B^-][SH^+]$. In all three cases, the rate law has one B, one H, and one S, which is the minimal composition of the transition state. (An unknown number of solvent molecules also can be associated with the transition state.) To show that all of these rate laws are equivalent, the ionization equilibria for catalyst and substrate can be used

$$K_B = \frac{[B^-][H^+]}{[BH]}$$

$$K_S = \frac{[S][H^+]}{[SH^+]}$$

Substitution of these relationships into the rate laws shows that

$$\text{Rate} = k_B[B^-][H^+][S] = k_B K_B[BH][S]$$

and

$$\text{Rate} = k'_B[B^-][SH^+] = k'_B(K_B/K_S)[BH][S]$$

or

$$k_{BH} = k_B K_B = k'_B K_B/K_S$$

Therefore, the rate law is equally consistent with general acid catalysis, general base plus specific acid catalysis, and general base catalysis. These mechanisms cannot be differentiated through kinetic experiments unless one of the rate constants is larger than theoretically possible. This type of mechanistic ambiguity is quite general for acid–base catalysis, and nonkinetic arguments often must be used in deducing mechanisms. A specific example of this ambiguity is the hydration of an imine. A general acid catalyzed mechanism is

$$\mathrm{H_2O} \;\; \mathrm{C{=}N} \;\; \mathrm{HB} \rightleftharpoons \mathrm{H_2\overset{+}{O}{-}C{-}N{-}HB^-} \xrightleftharpoons{\text{fast}} \mathrm{H^+ + O(H){-}C{-}NH + B^-}$$

A general base catalyzed mechanism is

$$B^- + H_2O + \rangle C{=}N\langle + H^+ \overset{\text{fast}}{\rightleftharpoons} B^- \rightarrow H{-}\underset{H}{O} \curvearrowright \rangle C{=}\overset{H}{N^+}\langle \rightleftharpoons$$

$$B{-}H \cdots O{-}C{-}NH \rightleftharpoons BH^+ + \underset{H}{O}{-}C{-}NH$$

The rate law for the former mechanism is rate $= k_{BH}[H_2O][S][BH]$, whereas in the latter case it is rate $= k_B[H_2O][SH^+][B^-] = k_B(K_B/K_S) \cdot [H_2O][S][BH]$. Thus both mechanisms are consistent with the same rate law.

Proton transfer reactions are an essential feature of acid–base catalysis, and extensive studies of such reactions have been made using fast reaction techniques. For normal acids and bases, protonation and deprotonation are diffusion-controlled processes, with rate constants predicted by simple diffusion theory [Eq. (1-1)]. These reactions can be thought of as occurring in two steps: diffusion of the solvated proton (or hydroxyl ion) toward the base (or acid) followed by a rapid transfer of the proton through the hydrogen-bonded complex. This mechanism can be represented as

$$B^- + H^+ \underset{k_{-1}}{\overset{k_1}{\rightleftharpoons}} B^- \cdots H^+ \overset{k_2}{\rightleftharpoons} BH \tag{5-5}$$

where k_1 and k_{-1} are the rate constants for diffusion-controlled association and dissociation, and k_2 is the rate constant for the proton transfer. If the intermediate species is assumed to be in a steady state, the observed rate constant for the reaction in the forward direction is

$$k_f = \frac{k_1}{1 + k_{-1}/k_2} \tag{5-6}$$

The reaction is diffusion-controlled if the proton transfer is rapid relative to diffusion apart of the reactants, i.e., $k_2 \gg k_{-1}$. Since $k_{-1} \sim 10^{10}\ \text{sec}^{-1}$, k_2 must be about $10^{12}\ \text{sec}^{-1}$ if the reaction is diffusion-controlled. (This analysis is similar to that presented for hydrogen bonding in Chapter 1.) Rate constants for some typical protonation and deprotonation reactions in aqueous solution are summarized in Table 5-1. Most of the second order rate constants are in the range 10^{10}–$10^{11}\ M^{-1}\ \text{sec}^{-1}$. On the other hand, dissociation rate constants, k_r, vary considerably, reflecting large differences in the strength of the chemical bond broken.

Several factors can contribute to deviations from "normal" diffusion-controlled rates. If the reactants are highly charged or steric restrictions limit

Table 5-1

Protonation–Deprotonation Rates

	Reaction[a]	k_f (M^{-1} sec^{-1})	k_r (sec^{-1})	Reference
Normal	$H^+ + OH^-$	1.3×10^{11}	2.6×10^{-5}	*3*
	$H^+ + CH_3COO^-$	4.5×10^{10}	8×10^5	*4*
	$H^+ + NH_3$	4.3×10^{10}	24	*5*
	H^+ + imidazole	1.5×10^{10}	1.7×10^3	*6*
	$H^+ + CuOH^+$	$\sim 1 \times 10^{10}$	$\sim 1 \times 10^2$	*7*
	$OH^- + HCO_3^-$	6×10^9	1.4×10^2	*7*
	$OH^- + NH_4^+$	3.4×10^{10}	5×10^5	*4,8*
	OH^- + imidazole$^+$	2.3×10^{10}	2.5×10^3	*6*
	$OH^- + HPO_4^{2-}$	$\sim 2 \times 10^9$	$\sim 3 \times 10^2$	*2*
Hindered due to H bond perturbation	$OH^- + EDTA^{3-}$	3.8×10^7	6.9×10^3	*2*
	$OH^- + NTA^{2-}$	1.3×10^7	7.0×10^2	*2*
Hindered by internal chelation	OH^- + DMA	1.2×10^7	32	*9*
	OH^- + PAS	3×10^7	3×10^4	*2*

[a] EDTA, ethylenediaminotetraacetic acid; NTA, nitrilotriacetic acid; DMA, dimethylanthranilic acid; PAS, *p*-aminosalicylic acid.

the accessibility of the reacting site, the diffusion-controlled rates are altered, but usually by less than an order of magnitude. However, if the hydrogen-bonded solvent structure is disrupted, the rate of proton transfer can be sufficiently decreased so that the rate is no longer diffusion-controlled. Examples of this phenomenon in Table 5-1 are reactions of aminopolycarboxylic acids such as $EDTA^{3-}$ with OH^-, where the negatively charged carboxylic groups disrupt the hydrogen-bonded structure between the amine proton and water, and the reaction of OH^- with molecules having intramolecular hydrogen bonds.

In the case of pseudo acids such as carbon acids, changes in electronic structure usually are rate determining rather than diffusion. For example, acetylacetone can exist in keto (85%) and enol (15%) forms; the conjugate base can have two resonance structures

$$CH_3-\overset{O}{\overset{\|}{C}}-CH_2-\overset{O}{\overset{\|}{C}}-CH_3 \rightleftharpoons \left[\begin{array}{c} CH_3-\overset{O}{\overset{\|}{C}}-CH=\overset{O^-}{\overset{|}{C}}-CH_3 \\ CH_3-\overset{O}{\overset{\|}{C}}-\underset{-}{C}H-\overset{O}{\overset{\|}{C}}-CH_3 \end{array}\right] \rightleftharpoons \overset{O\cdots\cdots HO}{CH_3\overset{\|}{C}-CH=\overset{|}{C}-CH_3}$$

Conjugate base

Table 5-2

Protolysis and Hydrolysis of Acetylacetone[a]

Reaction	k_f (M^{-1} sec^{-1})	k_r (sec^{-1})
H^+ + enolate $\rightleftharpoons$ enol	3×10^{10}	1.7×10^2
H^+ + enolate $\rightleftharpoons$ keto	1.2×10^7	1.4×10^{-2}
OH^- + enol $\rightleftharpoons$ enolate + H_2O	1.6×10^7	28
OH^- + keto $\rightleftharpoons$ enolate + H_2O	4×10^4	3.5×10^{-1}

[a] 298°K, data from Ref. *2*.

All of the possible protolytic rate constants for this system are summarized in Table 5-2. Protonation at the C^- atom is slower than protonation at the O^- atom by a factor of 10^3. A similar effect is exhibited by the deprotonation rates. In this case, the enol form is stabilized by an internal hydrogen bond further reducing the rate. Pseudoacids generally are characterized by slow rates of protonation and deprotonation.

We now consider proton transfer between a proton donor D and a proton acceptor A. The mechanism can be written as

$$DH^+ + A \underset{k_{-1}}{\overset{k_1}{\rightleftharpoons}} DH^+ \cdots A \underset{k_{-2}}{\overset{k_2}{\rightleftharpoons}} D \cdots {}^+HA \underset{k_{-3}}{\overset{k_3}{\rightleftharpoons}} D + HA^+ \quad (5\text{-}7)$$

where the two intermediate species form and dissociate at diffusion-controlled rates. If these species are assumed to be in a steady state

$$k_f = \frac{k_1}{1 + (k_{-1}/k_2)(1 + k_{-2}/k_3)} \quad (5\text{-}8)$$

$$k_r = \frac{k_{-3}}{1 + (k_3/k_{-2})(1 + k_2/k_{-1})} \quad (5\text{-}9)$$

and

$$\frac{k_f}{k_r} = \frac{k_1 k_2 k_3}{k_{-1} k_{-2} k_{-3}} = \frac{[D][HA^+]}{[DH^+][A]} = \frac{K_D}{K_A} \quad (5\text{-}10)$$

where K_D and K_A are the ionization constants of the donor and acceptor, respectively. If $k_2 \gg k_{-1}$, $k_f = k_1$ or the reaction is diffusion-controlled in the forward direction. For this situation $k_f \gg k_r$ or $pK_A \gg pK_D$. In other words, the reaction is diffusion-controlled if the acceptor binds the proton much more tightly than the donor. Similarly, if the reaction is diffusion-controlled in the reverse direction, $k_{-2} \gg k_3$ and $pK_D \gg pK_A$. When $pK_A = pK_D$, the reaction is not diffusion-controlled in either direction. A plot of $\log k_f$ versus $\Delta(pK) = (pK_A - pK_D)$ is shown in Fig. 5-3 with imidazole and

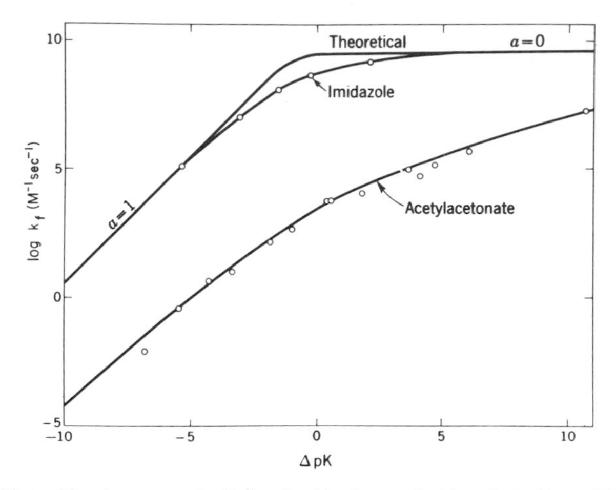

Fig. 5-3. Plot of $\log k_f$ versus ΔpK for the ideal case, imidazole ($pK_A = 6.95$), and acetylacetonate ($pK_A = 9.5$). The data are from Refs. (*2*), (*11*), and (*12*).

acetylacetonate as proton acceptors. By definition, $\log k_f - \log k_r \equiv \Delta pK$. When $\Delta pK > 0$, k_f is diffusion-controlled and, therefore, constant; when $\Delta pK < 0$, k_r is diffusion-controlled and $\log k_f$ decreases linearly with decreasing ΔpK with a slope of 1. This can be summarized as

$$\frac{\partial \log k_f}{\partial \Delta(pK)} = \alpha \tag{5-11}$$

where α is a constant equal to zero if $\Delta(pK)$ is positive and equal to 1 if $\Delta(pK)$ is negative. The broadness in the transition region around $\Delta pK = 0$ is expected for imidazole on the basis of the mechanism of Eq. (5-7). The broadness in the transition for acetylacetonate is because of the very slow proton transfer reactions. Presumably the limiting values of α would be reached at extreme values of $\Delta(pK)$. Finally, referring to Eqs. (5-7) and (5-10), estimates of the rate constants for intramolecular proton transfer for "normal" acids and bases can be made. Since $k_1 \approx k_{-3}$ and $k_{-1} \approx k_3$, $k_2 \sim 10^{12}\ \sec^{-1}$ and $k_{-2} \sim 10^{12}\ K_A/K_D$ when $K_A < K_D$. The rate constant k_2 will be smaller when $K_A \approx K_D$ and for pseudo acids and bases.

The catalytic behavior of a large number of structurally similar acids and bases is correlated remarkably well by relationships known as the Brönsted equations which state that

$$\log k_B = G_B + \beta(pK_A) \tag{5-12}$$

$$\log k_{BH} = G_{BH} - \alpha(pK_A) \tag{5-13}$$

where pK_A refers to the acid ionization constant of BH, G_B, G_{BH}, α and β are empirical constants and $\alpha + \beta = 1$. Both α and β are positive numbers, and the fact that $\alpha + \beta = 1$ is a direct consequence of the kinetic equivalence of general base and general acid catalysis previously discussed. These equations can be cast into an identical form to Eq. (5-11) since the pK of the proton acceptor, the substrate, is constant for acid catalysis, and the pK of the proton donor, the substrate, is constant for base catalysis. This is because proton transfer reactions often are rate determining in acid–base catalysis. If α or $\beta = 1$, this corresponds to almost complete proton transfer in the transition state, whereas if they are close to zero, very little proton transfer occurs. In most cases, Brönsted plots of $\log k_B$ or $\log k_{BH}$ versus pK_A are straight lines approximating intermediate linear portions of the plots in Fig. 5-3. This is usually because the range of pK_A values used is too small for observation of the limiting slopes. Moreover, since substrates are often extreme pseudo acids, the transition from a limiting slope of 1 to a limiting slope of 0 will occur so gradually that curvature is not detected. However, curvature of Brönsted plots is often observed.

An interesting reaction in which limiting slopes are observed in the Brönsted plot is the transfer of an acyl group from sulfur to nitrogen in *S*-acetylmercaptoethanolamine (*10*). The postulated mechanism is

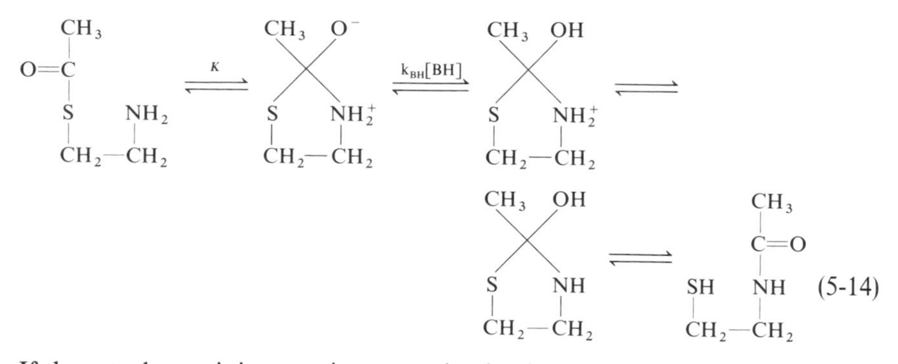

If the rate determining step is assumed to be the proton transfer in the second step, the rate of the overall reaction is

$$\text{Rate} = Kk_{BH}[BH][S]$$

A plot of $\log k_{BH}$ versus pK_A is shown in Fig. 5-4. (The value of K was determined by assuming k_{BH} is diffusion-controlled when k_{BH} is H_3O^+.) This plot very closely parallels that in Fig. 5-3. These results can be explained if the rate controlling step is the protonation of the second species in Eq. (5-14) by a diffusion-controlled reaction with BH when pK_A is less than the pK of the second species ($\alpha = 0$). A change in slope of the plot occurs approximately at the pK_A of the intermediate ($\sim$7.4) and at higher values of pK_A, the rate becomes diffusion-controlled in the reverse direction ($\alpha = 1$). The idea that a diffusion-controlled proton transfer is rate limiting may seem

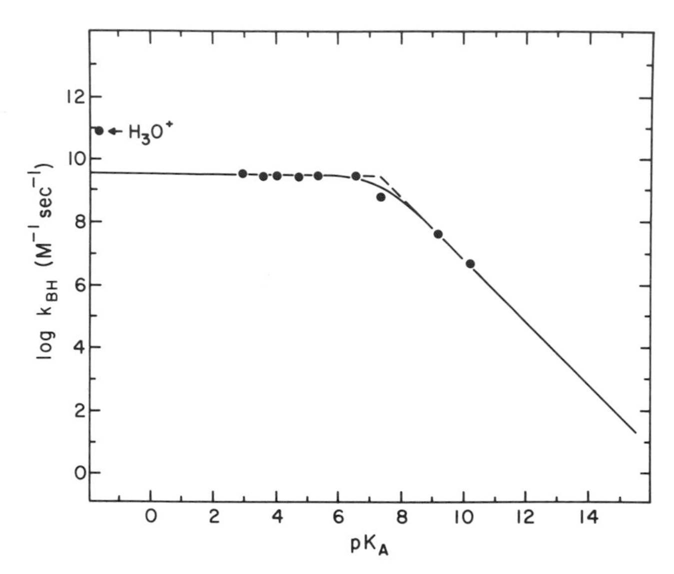

Fig. 5-4. Bronsted plot for general acid catalysis of the acetyl transfer of *S*-acetyl-mercapto-ethylamine at 50° and ionic strength 1.0 *M* based on a value of $k_{H_3O^+} = 6.5 \times 10^{10}\ M^{-1}\ sec^{-1}$. [Adapted with permission from R. E. Barnett and W. P. Jencks, *J. Am. Chem. Soc.* **91**, 2358 (1969). Copyright (1969) American Chemical Society.]

unreasonable, but the rate is a product of concentrations and a rate constant. In this instance, the concentration of the second species in Eq. (5-14) is extremely low. (A kinetically indistinguishable mechanism is general base catalysis with a protonated substrate. In this case, rate = $k_B[B^-][SH^+]$. General acid catalysis appears to correlate all of the known facts better, but general base catalysis cannot be rigorously excluded.)

Concerted acid–base catalysis involving simultaneous proton transfer reactions between the reactant and an acid and a base has the mechanistic advantage that a low concentration intermediate is not formed. A concerted reaction is suggested by a term such as $k_{BHB}[BH][B^-][S]$ in the rate law. The possibility of consecutive proton transfer reactions rather than a concerted process can sometimes be excluded because the rate constants calculated on the basis of consecutive transfers exceed theoretically possible values. The enolization of acetone catalyzed by acetic acid and acetate is a relatively well-documented example of a concerted proton transfer mechanism. The mechanism can be depicted as

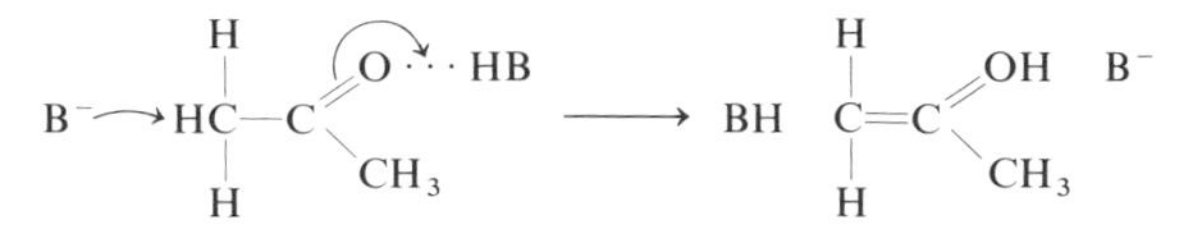

In general acid and general base catalysis, the solvent plays the role of the alternate proton donor or acceptor. Bifunctional catalysts can be very effective in concerted proton transfer mechanisms. For example, in the 2-pyridone catalyzed mutarotation of glucose, N can serve as a base and O as an acid. Concerted proton transfers are especially attractive possibilities for enzymatic mechanisms because of the many acidic and basic amino acid side chains available.

NUCLEOPHILIC CATALYSIS

Nucleophilic catalysis differs from general base catalysis in that an intermediate species is formed in which the catalyst and substrate are joined by a covalent bond. In general, a relatively unreactive species is converted to a more reactive species. For example, the hydrolysis of esters is catalyzed by nucleophiles, N, through the mechanism

$$\mathrm{N} + \mathrm{R{-}C}\begin{smallmatrix}\nearrow \mathrm{O} \\ \searrow \mathrm{OR'}\end{smallmatrix} \longrightarrow \mathrm{R{-}}\underset{\delta+\mathrm{N}}{\mathrm{C}}\begin{smallmatrix}\nearrow \mathrm{O}^{\delta-} \\ \searrow \mathrm{OR'}\end{smallmatrix} \xrightarrow{\mathrm{H_2O}} \mathrm{RC}(=\mathrm{O}){-}\mathrm{OH} + \mathrm{R'OH} + \mathrm{N}$$

The catalytic efficiency of a nucleophile depends on the nature of the reagent, not just on the pK_A as for general base catalysis. Nucleophilic and general base catalysis can be differentiated in several different ways: bulky substituents alter nucleophilic reactions more; nucleophilic reactions may give isolatable intermediates; and nucleophilic reactions have a small isotope rate effect.

Nucleophilic reactions can be categorized into two classes, namely, attacks on hard electrophilic centers and on soft electrophilic centers. Hard centers include carbonyl, phosphoryl, and sulfuryl groups. The base strength of the nucleophile is the dominant factor in determining reactivity, and the valency of carbon is not extended beyond its normal value of 4. Nucleophilic reactions with saturated carbon to form transition states with pentavalent carbon are typical of soft centers. Polarizability is the dominant factor in determining reactivity rather than pK. Thus, large polarizable atoms such as S and I are more reactive than small atoms such as O and N which are not very polarizable. Although the concept of hard and soft centers is rather qualitative in nature, it has proven useful when categorizing nucleophilic reactions.

If the logarithm of the rate constant is plotted versus the pK_As of the nucleophile, a straight line is obtained, providing the nucleophiles are chemically similar and the variation in pK_A is not too large. This is an example of a linear free energy relationship. In this case it indicates the standard free energy of activation is proportional to the thermodynamic standard free energy change for the transfer of a proton to the nucleophile. The slope of the plot, which is called the Brönsted coefficient, β, can be greater than unity. The simple interpretation often attributed to β is that if β is large the transition state closely resembles products, whereas if β is small the transition state closely resembles reactants. A Brönsted plot also is sometimes used in which the pK_A of the leaving group is used instead of that of the nucleophile. In the case of ester hydrolysis, a very negative value (< -1) of β when the leaving group pK_A is plotted corresponds to a very positive value of β (> 1.5) when the nucleophile pK_A is plotted. Chemically this means that the rate of nucleophilic attack increases with increasing nucleophile base strength and increasing electron withdrawal on the leaving group. The Hammett equation provides an alternative method for plotting the data. With this approach, the electron-donating or withdrawing power of a substituent in a benzene ring is derived empirically by fitting the pK_As of substituted benzoic acids to the equation

$$(pK_A)_X = (pK_A)_\circ - \sigma_X$$

where X denotes the substituent, $\circ$ benzoic acid, and σ_X is the substituent constant. The rate constants are then plotted according to the equation

$$\log k_X = \log k_\circ + \rho\sigma_X \tag{5-15}$$

where ρ is a constant for a given series of reactions. The Hammett and Brönsted equations are both statements of a linear free energy correlation. The use of these equations has proved very useful in systematizing and

Table 5-3

Some Nucleophiles and Intermediates in Enzymes[a]

Nucleophile	Enzyme	Intermediate
Serine (—OH)	Serine proteases, alkaline, and acid phosphatases	Acyl enzyme
Cysteine (—SH)	Thiol proteases, glyceraldehyde-3-phosphate dehydrogenase	Acyl enzyme
Aspartate (—COO^-)	Pepsin	Acyl enzyme
	ATPase (Na^+/K^+, Ca^{2+})	Phosphoryl enzyme
Lysine (—NH_2)	Acetoacetate decarboxylase, aldolase, pyridoxal enzymes	Schiff base
Histidine (imidazole)	Phosphoglycerate mutase, succinyl-CoA synthetase	Phosphoryl enzyme
Tyrosine (—OH)	Glutamine synthetase	Adenyl enzyme

[a] Adapted from Ref. *13*.

understanding organic reaction mechanisms. However, such equations are not often useful for enzyme reactions because steric restrictions at the catalytic site usually dominate in determining reactivity rather than small changes in the electronic structure of the substrate.

Some examples of nucleophiles in enzymatic reactions are given in Table 5-3.

ELECTROPHILIC CATALYSIS

In electrophilic catalysis, the rate is enhanced by withdrawal of electrons from the reaction center. This can occur by proton transfer from a general acid, by the presence of metal ions, or by the use of organic molecules which serve as electron sinks. Metal ions are often required in enzymatic reactions, and they appear to function in a variety of different ways. However, one obvious function is simply to stabilize negative charges that are formed during the course of the reaction. Probably the most important single mechanism of covalent electrophilic catalysis is one in which nitrogen participates either as a cationic electron sink or as an electron pair donor. A typical example is Schiff base formation between an amine and a carbonyl compound

$$\begin{matrix} R \\ R' \end{matrix}\!\!>C=O \quad H_2\ddot{N}-E \rightleftharpoons \begin{matrix} R \\ R' \end{matrix}\!\!>C=N-E + H_2O$$

If the Schiff base is protonated, an enamine can be formed, thus activating the methylene carbon as a nucleophile.

Pyridoxal phosphate (I) is an important coenzyme involved in many catalytic processes because of its ability to form Schiff bases with amino acids (II) and to utilize the pyridine ring as an electron sink. A key intermediate is formed by removal of the α hydrogen (III)

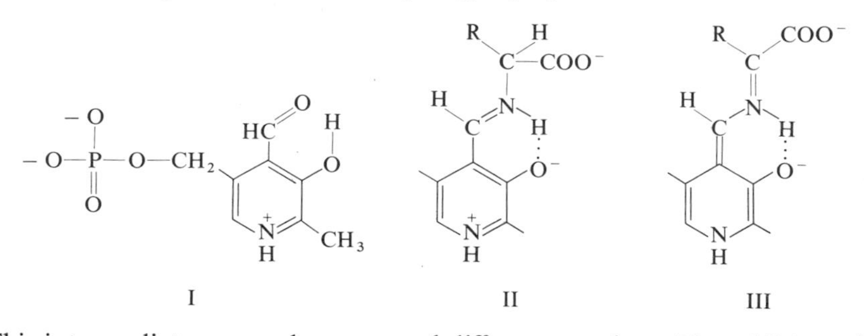

I II III

This intermediate can undergo several different reactions. The addition of a proton back to the same position of the amino acid from the opposite side yields racemization. Addition of a proton to the carbonyl carbon atom gives the Schiff base of pyridoxamine and an α keto acid, which can then undergo hydrolysis. This and the reverse process are the mechanistic bases of transamination reactions. If a good leaving group is present on the β carbon atom, it is readily eliminated to give an unsaturated product that can then undergo hydrolysis, rearrangement, or addition. This type of reaction occurs in the breakdown and interconversion of serine, threonine, cysteine, and tryptophan. Finally both α and β decarboxylation can proceed through the pyridoxal phosphate–Schiff base intermediate.

Thiamin pyrophosphate (IV) is another coenzyme that can stabilize a negative charge and covalently bind to a substrate.

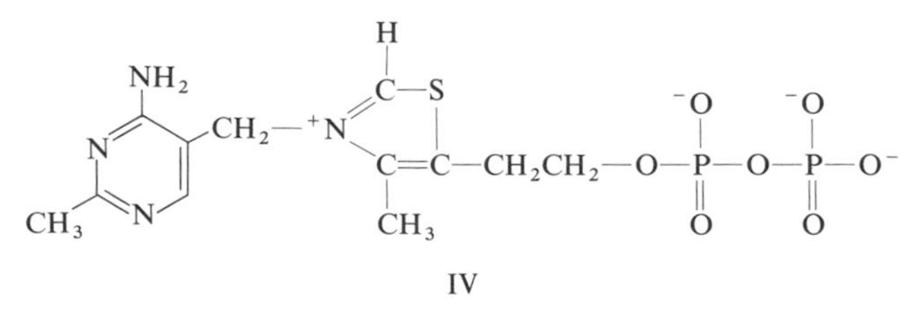

IV

In this case, the positive charge on the nitrogen assists in the removal of the proton from the C-2 carbon, and the resulting ionized carbon is a potent nucleophile.

The above discussion is intended only to give the basic principles of nucleophilic and electrophilic catalysis. For a more extensive discussion, Ref. *14* should be consulted.

STEREOCHEMISTRY

The realization that enzymes are stereospecific dates back to the classical work of Pasteur in the 1850s on the fermentation of tartaric acid. The stereochemistry of the carbon atom is well known: optical activity is caused by a carbon atom with four different groups around it. More generally a compound is optically active if it does not have a plane or center of symmetry, i.e., it is not superimposable on its mirror image. An asymmetric carbon is called a *chiral* center; a carbon atom of the type $CR_2R'R''$ is *prochiral*: although it is not chiral it can easily be converted to a chiral atom by replacement of one of the R groups.

The traditional notation for configuration uses the terms D and L, but this has been supplanted by the *RS* convention which assigns priorities to groups around a chiral atom by a series of rules (*15, 16*). Only a very brief introduction to these rules is presented here. The atom directly attached to the chiral atom is considered first: the higher the atomic number, the higher the priority. For isotopes, the higher atomic mass has priority. For groups that have the same atom attached to the chiral carbon the atomic and mass numbers of the next atom are used. Unsaturated carbons should be treated as though additional carbon atoms were attached to the carbon atoms in the unsaturated center, and —C=O carbon is considered as being bonded to two oxygens. For example, a priority assignment of some typical groups is as follows: $—SH > —OR > —NH_2 > —COOH > —CHO > —CH_2OH > —C_6H_5 > —CH_3 > —{}^3H > —{}^2H > —{}^1H$. To designate a chiral atom as *R* (*rectus* = right) or *S* (*sinster* = left), the center is written so that the lowest priority substituent occupies the bottom position in a Fisher projection. If the three remaining substituents describe a clockwise turn in order of decreasing priority, the center is *R*; if a counterclockwise turn is described, the center is *S*. For example, consider D-alanine

$$\begin{array}{ccc} & COOH & \\ & | & \\ H_2N— & C & —CH_3 \\ & | & \\ & H & \end{array}$$

Since $—NH_2 > —COOH > —CH_3$, this is *R*-alanine. A *prochirality* assignment can be made by arbitrarily giving one of the two identical groups priority over the other. For example, consider ethanol

$$\begin{array}{ccc} & CH_3 & \\ & | & \\ HO— & C & —H_a \\ & | & \\ & H_b & \end{array}$$

where H_a is given priority over H_b. In this case, H_a is said to be *pro-R* since $—OH > —CH_3 > —H_a$, and H_b is *pro-S*. (Note that if H_b is given priority over H_a, the same chirality assignments are obtained.)

For a trigonal carbon, the assignment of chirality is made by viewing the three groups attached to the carbon in the plane of the paper. If the priority sequence is clockwise, the top of the paper is the *re* face; if it is counterclockwise, it is the *si* face. For example, consider acetaldehyde

si face is on top

re face is on top

From the standpoint of enzymes, stereospecificity is only possible if at least three of the groups surrounding the chiral carbon are bound to the enzyme. This stereospecificity is a direct result of the protein structure that has a binding site which will only accommodate a particular stereoisomer. In fact, an enzyme can convert a symmetric carbon to an asymmetric carbon through binding. For example, if the R groups of $CR_2R'R''$ are bound in different environments, the carbon becomes optically active. Also an optically active substance can be formed from an optically inactive substance because the protein only allows access to one of the geometrically possible substitution positions on the carbon. Extensive discussions of enzyme stereochemistry are given in Ref. *17* and *18*.

The demonstration that the transfer of hydrogen between acetaldehyde and NADH catalyzed by alcohol dehydrogenase is stereospecific was a seminal experiment in the study of enzyme mechanisms (*19*). The reaction catalyzed can be written as

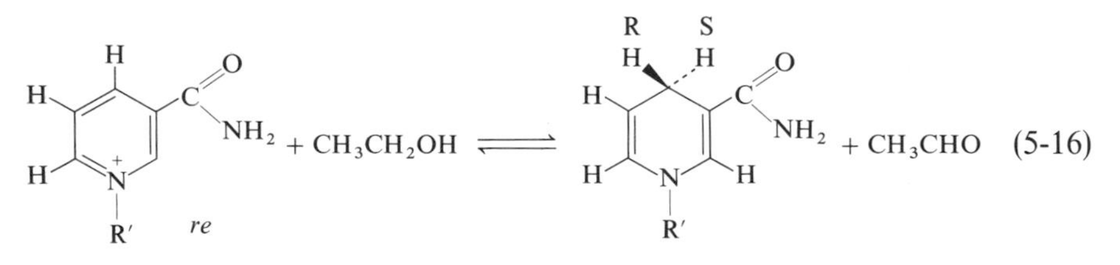

where R′ is

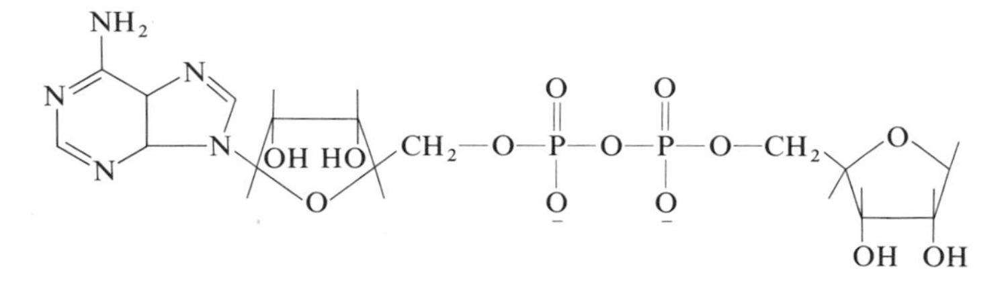

Position 4 in the dihydronicotinamide ring is prochiral as is the carbon in acetaldehyde previously discussed. The portion of the ring containing —$CONH_2$ is given group priority resulting in the chirality assignments shown in Eq. (5-16). If CH_3CD_2OH is reacted with NAD^+ in the presence of the enzyme, the resulting NAD^2H contains one atom of deuterium per mole. If the NAD^2H is incubated with enzyme and acetaldehyde, all of the deuterium is lost. This requires the deuterium to be transferred stereospecifically to one face of the NAD^+ and then transferred back to the *same* face. If this were not the case, only one-half of the deuterium would be transferred back to acetaldehyde. The stereospecificity of this reaction, of course, ultimately resides in the stereospecific binding of the substrates. Establishment of the absolute configuration required synthesis of *S* and *R*-1[^{2}H]ethanols and of 4*R*-[^{2}H]NADH. The stereochemical description of the reaction that has emerged is as follows: the *pro-R* hydrogen is transferred from C-4 of NADH to the *re* face of acetaldehyde, producing *pro-R*-hydrogen in the product; in the other direction, the *pro-R* hydrogen must be added to the

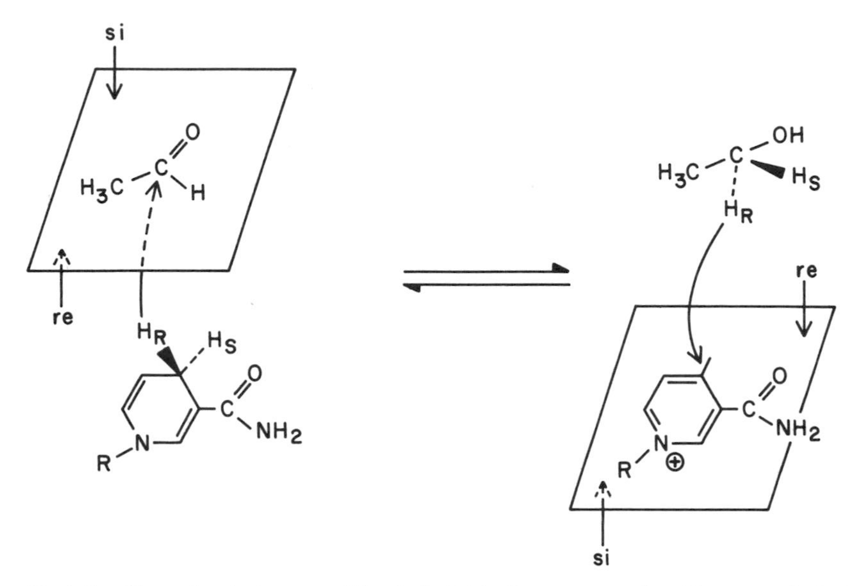

Fig. 5-5. Stereochemical course of the alcohol dehydrogenase reaction. The arrows indicate the direction of the hydrogen transfer. [Adapted from C. Walsh, "Enzymatic Reaction Mechanisms," p. 346. Freeman, San Francisco, California, 1979. Copyright © 1979.]

re face of NAD^+ (Fig. 5-5). As soon as the stereochemistry had been established for alcohol dehydrogenase, similar studies were carried out with many other dehydrogenases. Some enzymes transfer to and from the *re* face of NAD^+ and NADH, and some use the *si* face (cf. *18* for a summary of the stereochemistry of nicotinamide utilizing enzymes).

Sugars and sugar phosphates are important metabolic materials, and many enzymes utilize them as substrates. The stereochemistry of the sugars is of importance in understanding enzyme mechanisms. For example, glucose 6-phosphate can exist as the α and β anomers and in a free carbonyl form

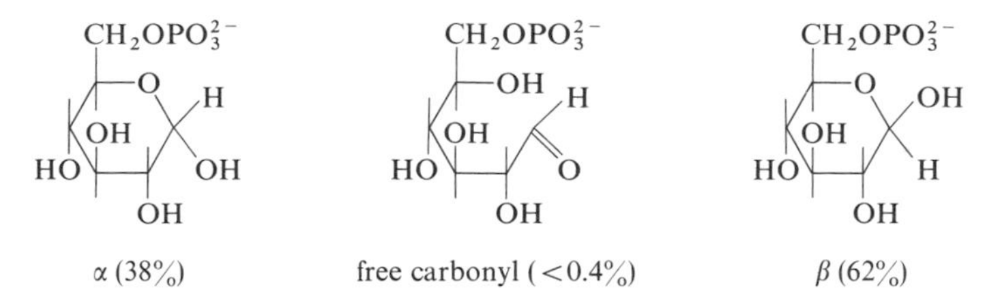

Enzymes have been shown to have different anomeric specificities, and in some cases the rate of anomerization may be of importance in the regulation of metabolism (*20*).

Stereochemical analysis of enzyme mechanisms is not confined to carbon. An atom of particular interest to enzymology is phosphorus because of the central importance of nucleotide phosphates in biology. Phosphorus is normally bonded to several equivalent oxygens and cannot be used for stereochemical mechanistic studies. However, substitution of sulfur for one of the oxygens can generate diastereomeric pairs. For example, the ATP-α-*S* diastereomers have been synthesized

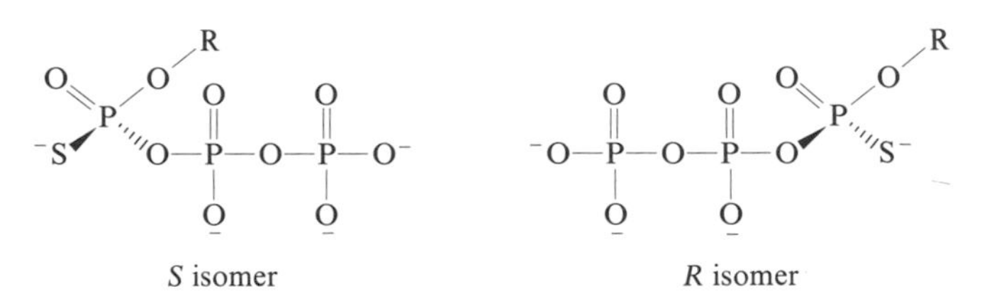

S isomer *R* isomer

Specific enzymes show kinetic preferences for one diastereomeric form, and once the absolute stereochemistry has been determined, the stereochemical course of the enzymatic reaction can be determined. A number of other interesting compounds with chiral phosphorus have been made including γ-[^{18}O]-γ-thio-ATP and monoesters with ^{16}O, ^{17}O, ^{18}O, and OR attached to phosphorus. The stereochemical course of many enzymatic phosphoryl transfers has now been studied. The cases of ribonuclease A is discussed in detail later, and a review of this field has recently appeared (*21*).

Analysis of the stereochemistry of reactants and products also permits inferences to be made about the nature of reaction intermediates. For example, the stereospecificity of enzymes suggests that inversion should occur with each displacement at a phosphorus atom. Thus, if net inversion occurs in the overall reaction, an odd number of reactions must have occurred at the phosphorus atom. Usually such a result is interpreted as a single direct displacement. If inversion does not occur, an even number of reactions must have occurred, and this is evidence for a reaction intermediate (cf. *21*). Stereochemical studies of enzyme mechanisms have been developed into a fine art form, and interesting results are being obtained at a prodigious rate.

KINETIC ISOTOPE EFFECTS

Kinetic studies of reactions in which isotopic substitutions have been made in the reacting molecules can provide insight into the molecular details of the reaction mechanism (cf. *22–24*). The most important property of a chemical bond that is being broken in a chemical reaction for kinetic isotope effects is the mass of the two atoms joined together covalently. Therefore, substitution of deuterium or tritium for hydrogen produces much larger rate changes than isotopic substitution of heavier elements, and hydrogen isotopes have been primarily used in kinetic studies. Isotope rate effects can be divided into three categories: primary, secondary, and solvent. All three of these are now considered for the isotopes of hydrogen.

Primary isotope rate effects are due to the fact that bonds involving deuterium have a smaller zero point energy than the analogous bond involving hydrogen. In many processes in which a covalent bond with hydrogen is broken, the zero point energy of the bond is frozen out in the transition state. This is shown schematically in Fig. 5-6. Since the ground state energy of the deutrium bond is lower than that of the hydrogen bond, the activation energy will be larger, and the breaking of the deuterium bond will be slower than breaking of the hydrogen bond. The magnitude of the isotope rate effect can be estimated from known bond energies. For example, the zero point energy for C—H is 4.15 kcal/mole and that for C—D is 3.00 kcal/mole. [This difference is very close to that predicted by assuming the bond is a simple harmonic oscillator with a characteristic frequency of $v = (\frac{1}{2})\sqrt{K/m}$ where K is a force constant and m is the reduced mass of the system which is approximately the mass of H or D.] If the breaking of the carbon–hydrogen bond is rate determining then

$$k_H/k_D \approx e^{(4150-3000)/RT} = 7$$

where the rate constants for breaking of the C—H and C—D bond are designated as k_H and k_D, respectively. This model is very approximate, and in fact, primary isotope rate effects for the breaking of carbon–hydrogen

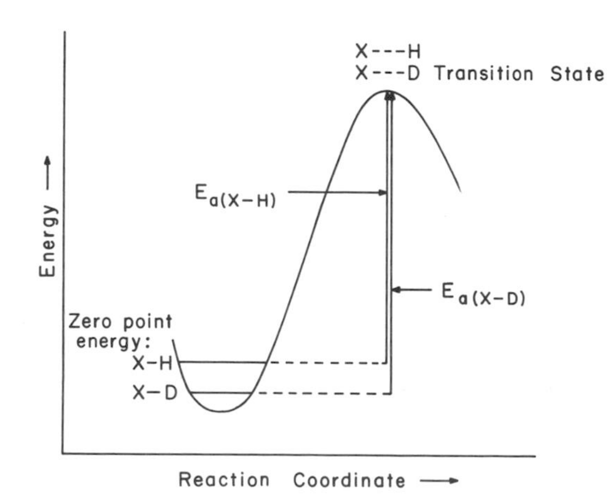

Fig. 5-6. Schematic representation of a primary isotope effect for X–H and X–D where E_a is the activation energy.

(deuterium) bonds range from about 2 to 15. Even larger effects may be observed if quantum mechanical tunneling occurs. Small isotope effects may reflect the fact that breaking of the carbon–hydrogen (deuterium) bond is only partially rate limiting or that isotopic substitution is altering equilibrium constants for reactions occurring prior to the rate determining step. A similar analysis can be made for other types of isotopic substitutions.

Secondary isotope rate effects are alterations in rate occurring when isotopic substitutions are made in a bond not directly involved in the reaction. The theoretical bases of secondary isotope effects are not simple; it is usually associated with a change in electronic structure such as going from sp^3 to sp^2 hybridization of carbon. Secondary isotope effects are quite small, typically ranging from 1.02 to 1.4 for hydrogen and deuterium. Tritium is especially useful in such studies because of the large mass difference between tritium and hydrogen and the ease in detecting tritium.

Solvent isotope effects are changes in rate caused by the substitution of deuterated solvents for protonated solvents. With enzyme reactions, the solvents are H_2O and D_2O. While such experiments are relatively easy to do compared to synthesizing substrates with specific isotopic substitutions, the interpretation of the results is difficult. Substitution of D_2O for H_2O causes changes in ionization constants; this obviously can have profound effects on both enzymes and substrates. In addition, changes in protein structure can occur due to alterations in hydrogen bonding, changes in bond lengths, etc. Because of these problems, solvent isotope rate effects must be interpreted with extreme care and should not be relied on strongly for mechanistic information. Of course, if the isotope of interest exchanges with the solvent, only solvent isotope effects can be studied. In such cases, isotopic partitioning in reaction intermediates can provide interesting mechanistic information (*25*, *26*).

For enzymatic reactions, the most common kinetic parameters determined are V_S and K_S. The observed primary kinetic isotope effects on V_S are not large, typically about 2 if present. This may be due to the fact that V_S is rarely determined by a single reaction involving a chemical change. Instead several interconversions of intermediates may be occurring at comparable rates, or conformational changes and/or product dissociation may be rate limiting. Isotope effects on K_S are even more difficult to interpret since K_S is a complex function of rate constants; if it represents a true equilibrium constant, little or no isotope effect would be expected. In fact, V_S/K_S is an easier parameter to interpret than K_S. The complexities of the situation can be illustrated by considering the mechanism (*27*)

$$E + S \underset{k_{-1}}{\overset{k_1}{\rightleftharpoons}} ES \underset{k_{-2}}{\overset{k_2}{\rightleftharpoons}} EP \xrightarrow{k_3} P + E \tag{5-17}$$

where k_2 is the rate constant for the bond breaking step which will have a primary isotope effect. For this mechanism

$$V_S/[E_0] = \frac{k_2 k_3}{k_2 + k_3}$$

$$K_S = \frac{k_3(k_{-1} + k_2)}{k_1(k_2 + k_3)}$$

$$V_S/([E_0]K_S) = \frac{k_1 k_2}{k_{-1} + k_2}$$

If $k_3 \gg k_2$, $V_S/[E_0] = k_2$ and a normal primary isotope effect is observed. Note that this inequality has no effect on V_S/K_S. If $k_{-1} \gg k_2$, a normal primary isotope effect is observed for V_S/K_S. This inequality has no effect on V_S. On the other hand, both inequalities are of importance for K_S, making this a more complex parameter to consider than V_S or V_S/K_S. Obviously if $k_3 \ll k_2$, a primary isotope effect is not observed for V_S, and if $k_{-1} \ll k_2$, a primary isotope effect is not observed for V_S/K_S. If $k_3 \approx k_2$ and/or $k_{-1} \approx k_2$, full primary isotope effects would not be observed. If more intermediates are present, the analysis obviously becomes much more complex. More detailed considerations indicate that when tritium is used, an isotope rate effect can be observed only in V_S/K_S (*27*). This arises because tritium is always used as a trace isotope, whereas deuterium is used as a stoichiometric replacement for hydrogen.

A classical example of a primary isotope effect has been observed with horse liver alcohol dehydrogenase [(Eq. (5-16)] whose stereochemistry has been discussed. If deuterium is put on NADH and CH_3CH_2OH, $V_{S[H]}/V_{S[D]} =$ 3.11 for reduction of acetaldehyde and 2.28 for oxidation of ethanol (*28*). For yeast alcohol dehydrogenase, this ratio was found to be 1.8 for both reactions (*29*). These results suggest hydride transfer is a significant factor in determining the catalytic rate. A small isotope effect also is found for the Michaelis constant ($K_H/K_D \sim 1.3$). The interpretation of this finding is not clear.

An example of a secondary isotope effect was found in the dehydration of L-malate catalyzed by fumarase. Isotopic substitution of ^2H for ^1H at the carbon containing the hydroxyl gave $V_{S[H]}/V_{S[D]} = 1.09$, and a similar substitution at the position not eliminated on the other carbon gave $V_{S[H]}/V_{S[D]} =$ 1.12 (*30*). These results suggest that as the hydroxyl group of L-malate leaves, considerable positive charge develops on the carbon to which it was attached. This is consistent with a carbonium ion type mechanism. No primary kinetic isotope effect was observed for this reaction (*31*).

REFERENCES

1. G. G. Hammes, "Principles of Chemical Kinetics." Academic Press, New York, 1978.
2. M. Eigen and W. Kruse, *Z. Naturforsch.* **186**, 857 (1963); and unpublished results.
3. M. Eigen and L. de Maeyer, *Z. Elektrochem.* **59**, 986 (1955).
4. M. Eigen and J. Schoen, *Z. Elektrochem.* **59**, 483 (1955).
5. M. T. Emerson, E. Grunwald, and R. A. Kromhout, *J. Chem. Phys.* **33**, 547 (1960).
6. M. Eigen, G. G. Hammes, and K. Kustin, *J. Am. Chem. Soc.* **82**, 3482 (1960).
7. M. Eigen and G. Schwarz, unpublished results.
8. M. Eigen, Z. *Phys. Chem.* (*Leipzig*) **NF1**, 176 (1954).
9. M. Eigen and E. M. Eyring, *J. Am. Chem. Soc.* **84**, 3254 (1962).
10. R. E. Barnett and W. P. Jencks, *J. Am. Chem. Soc.* **91**, 2358 (1969).
11. M. Eigen, *Angew. Chem.* **75**, 489 (1963); *Angew. Chem. Int. Ed. Engl.* **3**, 1 (1964).
12. G. Maass, Dissertation, Univ. of Göttingen, Göttingen, Germany, 1962.
13. A. Fersht, "Enzyme Structure and Mechanism," p. 59. Freeman, San Francisco, 1977.
14. W. P. Jencks, "Catalysis in Chemistry and Enzymology," McGraw-Hill, New York, 1969.
15. R. S. Cahn, C. K. Ingold, and V. Prelog, *Angew. Chem. Int. Ed.* **5**, 385 (1966).
16. W. L. Alworth, "Stereochemistry and Its Application in Biochemistry," Wiley, New York, 1972.
17. G. Popjak, *in* "The Enzymes" (P. Boyer, ed.), 3rd ed., Vol. 2, p. 115. Academic Press, New York 1970.
18. H. Simon and A. Kraus *in* "Isotopes in Organic Chemistry" (E. Buncel and C. Lee, eds.), Vol. 2, p. 153. Elsevier, Amsterdam, 1976.
19. F. Loewus, F. Westheimer, and B. Vennesland, *J. Am. Chem. Soc.* **75**, 5018 (1953).
20. K. J. Schrag and S. J. Benkovic, *Accts. Chem. Res.* **11**, 136 (1978).
21. J. R. Knowles, *Ann. Rev. Biochem.* **49**, 877 (1980).
22. Jencks, W. P. (1969). "Catalysis in Chemistry and Enzymology," Chapter 4. McGraw-Hill, New York.
23. W. W. Cleland, M. O'Leary, and D. Northrup, eds., *6th Steenbock Symp.*, *Isotope Effects in Enzyme Catalyzed Reactions*, (1976).
24. J. Klinman, *Adv. Enzym.* **46**, 415 (1978).
25. I. A. Rose, *in* "The Enzymes" (P. Boyer, ed.), 3rd ed. Vol. 2, p. 281. Academic Press, New York, 1970.
26. J. Albery and J. R. Knowles, *Biochemistry* **15**, 5627, 5631 (1976).
27. D. B. Northrup, *Biochemistry* **14**, 2644 (1975).
28. H. R. Mahler, R. H. Baker, Jr., and V. J. Shiner, Jr., *Biochemistry* **1**, 47 (1962).
29. H. R. Mahler and J. Douglas, *J. Am. Chem. Soc.* **79**, 1159 (1957).
30. D. E. Schmidt, Jr., W. G. Nigh, C. Tanzer, and J. H. Richards, *J. Am. Chem. Soc.* **91**, 5849 (1969).
31. H. Fisher, C. Frieden, J. S. M. McKee, and R. A. Alberty, *J. Am. Chem. Soc.* **77**, 4436 (1955).

6

Elementary Steps in Enzyme Catalysis

INTRODUCTION

In order to understand better the unusual catalytic efficiency of enzymes, we now consider some of the types of reactions that are part of most enzymatic reactions and how these reactions are coordinated to generate the catalytic process. Every enzymatic reaction begins with the binding of substrates, and this will be the first reaction type to be discussed. Conformational changes, acid–base catalysis, and the formation of covalent intermediates are common features of many enzymatic processes and receive attention next. Finally the advantages of enzymes over simpler catalysts are considered in terms of simple models. The many special reactions of cofactors are not considered. While these reactions are of great importance, they are in principle not different from the prototype reactions as far as understanding the basis of enzyme catalysis.

BINDING OF SUBSTRATES

The binding of substrates and substrate analogs to enzymes has been studied with a variety of techniques. Crystallographic data indicate the binding site on the enzyme has a structure that is exactly complementary to the substrate and that many weak noncovalent interactions are responsible for the binding. The tightness of binding varies with physiological function and probably has evolved over many years. Typical Michaelis constants are in the micromolar to millimolar range. Fast reaction techniques have been used to study the rates of many enzyme–substrate and enzyme–substrate analog reactions. Some of the rate constants obtained are presented in Table 6-1. In addition k_{cat}/K_S provides a lower bound for the second order

Table 6-1

Association and Dissociation Rate Constants of Enzyme–Substrate Reactions

Enzyme	Substrate	$10^{-8} k_1$ (M^{-1} sec^{-1})	k_{-1} (sec^{-1})	Reference
Aspartate aminotransferase	Glutamate, aspartate	>0.1–1	10^5–10^6	*1*
	Oxalacetate, ketoglutarate	>1	10^4	*1*
	α-Methyl aspartate	1.2×10^{-4}	130	*2*
	Erythro-β-Hydroxyaspartate	0.031	1.1×10^4	*3*
	NH_2OH	0.037	62	*4*
Chymotrypsin	Proflavin	1.1	2.15×10^3	*5*
	Furylacryloyl-L-tryptophanamide	6.2×10^{-2}	2.7×10^3	*6*
Catalase	H_2O_2	0.05	—	*7*
Catalase-H_2O_2	H_2O_2	0.15	—	*7*
Creatine kinase	ADP	0.22	1.8×10^4	*8*
	MgADP	0.053	5.1×10^3	*8*
	CaADP	0.017	1.2×10^3	*8*
	MnADP	0.074	4.1×10^3	*8*
Glyceraldehyde-3-phosphate dehydrogenase	NAD	0.19, 0.0137	1×10^3, 210	*9*
Lactate dehydrogenase (rabbit muscle)	NADH	~10	~10^4	*10*
Lactate dehydrogenase (pig heart)	NADH	0.546	39	*11*
	Oxamate	0.081	17	*11*
	3-Thio-NAD	0.058	410	*11*
Lysozyme	(N-Acetyl-D-glucosamine)$_2$	0.4	10^5	*12, 13*
Malate dehydrogenase	NADH	5	50	*14*
Old yellow enzyme	FMN	0.015	~10^{-4}	*15*
Peroxidase	H_2O_2	0.09	<1.4	*16*
	Methyl H_2O_2	0.015	<2.2	*16*
	Ethyl H_2O_2	0.036	—	*16*
Peroxidase-H_2O_2	Hydroquinone	0.023	—	*17*
	Cytochrome *c*	1.2	—	*18*
Pyruvate carboxylase-Mn^{2+}	Pyruvate	0.045	2.1×10^4	*19*
Pyruvate kinase-Mn^{2+}	Fluorophosphate	0.13	3.4×10^4	*20*
Ribonuclease	Cytidine 3′-phosphate	0.46	4.2×10^3	*21*
	Uridine 3′-phosphate	0.78	1.1×10^4	*21*

Table 6-1 (*continued*)

Enzyme	Substrate	$10^{-8}\,k_1$ ($M^{-1}\,\text{sec}^{-1}$)	k_{-1} (sec^{-1})	Reference
	Cytidine 2′,3′-cyclic phosphate	0.2-0.4	$1–2 \times 10^4$	*22*
	Uridine 2′,3′-cyclic phosphate	0.1	2×10^4	*23*
	Cytidylyl 3′,5-cytidine	0.14	7×10^3	*24*
Tyrosyl-tRNA synthetase	Tyrosine	0.024	24	*25*
	$tRNA^{Tyr}$	2.2	1.5	*26*
		$10^{-8}\,k_{cat}/K_S$ ($M^{-1}\,\text{sec}^{-1}$)		
Acetylcholin-esterase	Acetylcholine	1.6	—	*27*
Carbonic anhydrase	CO_2	0.83	—	*28*
	HCO_3^-	0.15	—	*29*
Fumarase	Fumarate	1.6	—	*30*
	Malate	0.36	—	*30*

rate constant of enzyme–substrate reactions; some of the larger values found are included in Table 6-1. The second order rate constants are typically in the range of 10^6–10^8 $M^{-1}\,\text{sec}^{-1}$ for substrates. These are large rate constants but probably usually somewhat below the values expected for diffusion-controlled reactions ($\sim 10^8$–10^9 $M^{-1}\,\text{sec}^{-1}$). Still considering the severe steric restrictions at the catalytic site, the rate constants are surprisingly large. The binding of poor substrates or analogs appears to be characterized by significantly smaller rate constants. The pH dependence of the second order rate constants can be used to obtain information about ionizable groups at the active sites. Since a relatively simple reaction is being studied, the interpretation of the pH dependence of the rate constants is less ambiguous than that of steady-state parameters. Finally, dissociation rate constants vary considerably and are some reflection of the strength of the enzyme–substrate interactions. Because of the large second order rate constants, the rate of association of enzyme and substrate is rarely rate limiting, i.e., only when the substrate concentration is very low. On the other hand, the dissociation of products appears to be sufficiently slow in some cases to be rate limiting. For the sake of comparison, the turnover numbers of some representative enzymes are presented in Table 6-2. They are a measure of the rate determining step in catalysis at high substrate concentrations ($\gg K_S$).

Table 6-2

Approximate Turnover Numbers of Representative Enzymes

Enzyme	k_{cat} (sec^{-1})	Reference
Chymotrypsin and some other serine proteases	10^2–10^3	*31*
Carboxypeptidase	10^2	*31*
Urease	10^4	*31*
Fumarase	10^3	*30*
Ribonuclease A	10^2–10^4	*32*
Transaminases	10^3	*33*
Carbonic anhydrase	10^6	*28, 29*
Acetylcholinesterase	10^4	*27*

CONFORMATIONAL CHANGES

Considerable evidence exists that conformational changes accompany the binding of substrates to enzymes. The occurrence of conformational changes has been deduced from changes in a variety of physical properties of the enzyme such as absorption spectra, fluorescence spectra, and nmr spectra. In a few cases, changes in structure have been observed directly by x-ray crystallographic methods. Kinetic studies also suggest conformational changes occur when the substrate binds to enzyme: observed rates of binding reach a limiting value at high substrate concentrations rather than increasing as anticipated for a simple bimolecular process (cf. Chapter 4). A conformational change is best defined in an operational sense, namely, by observation of a change in a physical property or by the form of a rate law. To actually define a conformational change in terms of the total protein structure is desirable, but almost never possible. The few x-ray crystallographic studies suggest that the conformational changes of importance in enzyme catalysis involve structural changes of the order of magnitude of 1 Å. Observation of a conformational change when a substrate binds to an enzyme is, of course, not proof that the conformational change is relevant to catalysis. A minimal requirement is that the rate constants associated with the conformational transitions be larger than or equal to the turnover number. The direct study of a number of reactions between enzyme and substrate suggests that the minimal mechanism is a rapid bimolecular reaction followed by a conformational change. This can be depicted as

$$E + S \rightleftharpoons X_1 \rightleftharpoons X_2 \tag{6-1}$$

Some of the observed time constants for such conformational changes are summarized in Table 6-3; the conformational changes occur sufficiently rapid to be of importance in catalysis. Nevertheless they are much smaller than the rate constants for the elementary processes of noncovalent bonding discussed in Chapter 1. Therefore, the conformational changes must be highly cooperative.

Table 6-3

Rate of Conformational Changes of Enzymes and Enzyme–Substrate Complexes

Enzyme	Substrate	Approximate time constants (sec)	Reference
Alkaline phosphatase	2-Hydroxy-5-nitrobenzyl phosphate	10^{-2}	*34*
Aspartate aminotransferase	α-Methyl aspartate	10^{-2}	*2*
	erythro-β-Hydroxyaspartate	10^{-3}–10^{-1}	*3*
Chymotrypsin	Proflavin, furylacryloyl-L-tryptophanamide	10^{-4}, 10^{-2}	*5, 6*
Creatine kinase	ADP, MgADP, CaADP, MnADP, and ATP	10^{-4}	*8*
Lysozyme	(*N*-acetyl glucosamine)$_n$	10^{-4}	*12, 13*
Lactate dehydrogenase (rabbit muscle)	NADH	10^{-3}	*10*
Liver alcohol dehydrogenase	NADH-imidazole	10^{-3}	*35*
Peroxidase	H_2O_2, methyl and ethyl H_2O_2	10^{-1}	*16*
Pyruvate kinase	None, Mg^{2+}, Mn^{2+}	10^{-4}	*36*
Ribonuclease	None, cytidine 3′-phosphate, uridine 3′-phosphate, cytidine 2′3′-cyclic phosphate, uridine 2′,3′-cyclic phosphate, cytidylyl 3′,5′-cytidine	10^{-3}–10^{-4}	*21–24*

What is the catalytic function of a conformational change? Three factors appear to be of primary importance. First, enzyme functional groups are oriented optimally for the next step in the catalytic sequence. If an enzyme exists in two conformations E and E′, and the substrate binds predominantly to one conformation, say E′, which normally is present at very low concentrations, the binding energy is being used in a sense to convert the enzyme to a new form that binds substrate better. This mechanism can be written as

$$\begin{array}{ccc} S + E & \xrightleftharpoons{K_3} & E' + S \\ K_1 \Big\updownarrow & & \Big\updownarrow K_4 \\ ES & \xrightleftharpoons[K_2]{} & E'S \end{array} \qquad (6\text{-}2)$$

where the four equilibrium constants are linked by the relationship $K_1K_2 = K_3K_4$. This is sometimes called an *induced fit* (*37*). The second important function of a conformational change is to form a hydrophobic pocket in which catalysis can occur. Such a hydrophobic pocket strengthens enzyme–substrate interactions such as hydrogen bonding and electrostatic and prevents water from interfering with proton transfer reactions. In some cases, a limited amount of water may be retained to assist in proton transfer. Finally conformational changes provide a mechanism for utilizing the entire protein molecule in the catalytic process. Evidence suggesting that virtually the entire protein structure is of importance in catalysis comes from several types of experiments. Attempts to eliminate substantial portions of the enzyme polypeptide chain without inactivation have not been successful. Also attempts to produce enzyme activity by synthesizing small molecules with active-site-like environments have yet to produce rates comparable to the catalytic efficiency of enzymes. Finally, chemical modification of amino acid residues far from the active site can alter the enzymatic activity. For example, this is the case when tyrosines on ribonuclease A not near the active site are iodinated (*38*). The free energy of activation for *breaking* or *making* bonds in substrates can be lowered by simultaneously *making* or *breaking* the many noncovalent interactions involved in cooperative conformational transitions (*39*). A simple example of this effect occurs when hydrogen and iodine combine to give hydrogen iodide in the gas phase. The activation energy for this reaction is about 40 kcal/mole even though the activation energy for dissociation of the hydrogen molecule is about 100 kcal/mole. The reason for this lowered activation energy is that the breaking of the H-H bond is compensated for by formation of H-I bonds. A schematic representation of how such a compensation might assist in enzyme catalysis is shown in Fig. 6-1. Even though very little energy is associated with each noncovalent interaction, so many interactions exist that the total energy change can be large. Energy compensation is a logical framework for coordinating cooperative conformational transitions and catalysis.

The functional bases of conformational changes also can be stated in terms of transition state terminology, namely conformational changes destabilize the bound substrate relative to the transition state. This destabili-

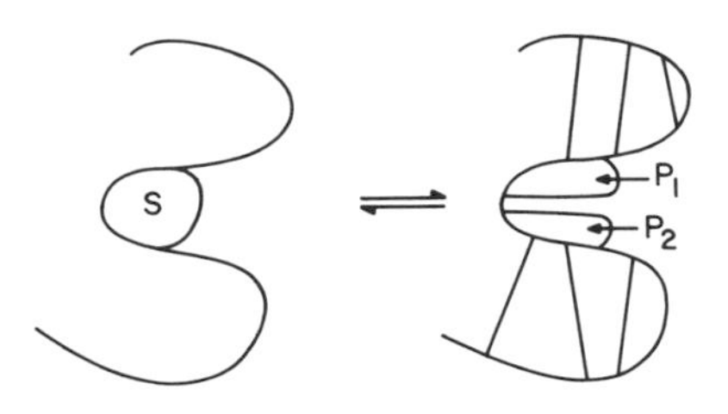

Fig. 6-1. Schematic representation of the lowering of the activation energy for the conversion of S to P_1 and P_2 by concommitant formation of noncovalent interactions between various regions of the enzyme.

zation can come about through deformation of the normal substrate structure, compression of the reactants to overcome van der Waals energy, and by desolvation and electrostatic stabilization. The idea that strain or distortion in the substrate is important was originally proposed as the "rack" hypothesis in which the enzyme was likened to the medieval rack used to pull humans apart (*40*). In terms of transition state theory, the enzyme active site can be viewed as complementary to the transition state rather than to substrates or products. In a sense, the binding energy that could be obtained with a deformed substrate is the driving force for catalysis. Since the transition state is bound more tightly than substrate, the reaction proceeds. This is a consequence of transition state theory and does not provide any particular insight into how enzymes work. However, transition state theory provides a convenient language for describing enzyme catalysis. Substances have been synthesized which are believed to resemble transition states; these *transition state analogs* bind very tightly to enzymes (*41*). Compression of reactants can be considered a special case of substrate deformation. Stabilization of the transition state through desolvation and electrostatic interactions is analogous to the hydrophobic pocket formation mentioned above. The transition state approach to describing enzyme catalysis has been discussed extensively (*42*).

ACID–BASE CATALYSIS

The rates and mechanisms of proton transfer reactions have been discussed in Chapter 5; the results obtained from model systems can be applied directly to enzyme catalysis. The mechanistic possibilities fall into three classes: solvent-mediated proton transfer, intramolecular proton transfer, and concerted proton transfer. Since true catalysis requires the acid or base to begin and end the catalytic cycle in the same protonation state, both protonation and deprotonation reactions must occur. For solvent-mediated proton transfer, the reactions occurring can be represented as

$$B^- + H^+ \underset{10^{10}\,K_A\,\text{sec}^{-1}}{\overset{10^{10}\,M^{-1}\,\text{sec}^{-1}}{\rightleftharpoons}} BH \tag{6-3}$$

$$BH + OH^- \underset{10^{10}\,K_W/K_A\,\text{sec}^{-1}}{\overset{10^{10}\,M^{-1}\,\text{sec}^{-1}}{\rightleftharpoons}} B^- + H_2O \tag{6-4}$$

The rate constants shown have been estimated by assuming the reactions of H^+ and OH^- with the base B^- and acid BH are diffusion-controlled; K_A is the ionization constant of the acid and K_W the ionization constant for water. Examination of the rate constants indicates that the maximum rate constant for the catalytic cycle, $10^3\ \text{sec}^{-1}$, occurs when $pK_A \approx 7$. For a larger pK_A, the hydrolysis rate constant is larger than 10^3 sec^{-1}, but the proton dissociation rate constant drops below 10^3 sec^{-1}. The converse occurs for a smaller pK_A. Thus the optimal amino acid side chain for solvent-mediated acid–base catalysis is one with a pK_A of about 7; likely possibilities in enzymes are imidazole and carboxyl groups. The turnover number for solvent-mediated acid–base catalysis cannot exceed about 10^3 sec^{-1}.

The rate constants for intramolecular proton transfer between a donor, D, and an acceptor, A, also can be estimated. The mechanism can be represented as

$$DH^+ + A \underset{10^{12}\, K_A/K_D \text{ sec}^{-1}}{\overset{10^{12}\text{ sec}^{-1}}{\rightleftharpoons}} D + AH^+ \quad (K_D > K_A) \tag{6-5}$$

For most enzymatic reactions, a large difference exists between the ionization constants of the amino acid side chains and the substrate that serves as a proton donor or acceptor. The result is that one of the intramolecular rate constants is considerably less than 10^{12} sec^{-1}, and an intermediate is formed in very low concentration. The low concentration intermediate must be utilized at a rapid rate to maintain a high turnover number. Consideration of some reasonable values for ionization and rate constants suggests that the turnover number is unlikely to exceed 10^5–10^6 sec^{-1}. Concerted acid–base catalysis eliminates the problem of a low concentration intermediate. The overall rate cannot exceed the slowest rate of intramolecular proton transfer so that if the ratio of ionization constants is about 10^{-6}, the maximum turnover number is 10^6 sec^{-1}.

The maximum rate constants calculated for acid–base catalysis can be compared with the turnover numbers of representative enzymes in Table 6-2. Turnover numbers as large as 10^7 sec^{-1} have been observed, but most are 10^3 sec^{-1} or less. This suggests the proton transfer reactions in enzyme catalysis are proceeding at close to the maximum possible rates. These maximum rates are based upon results obtained with "normal" acids and bases. Substrates generally are rather poor acids and bases so that slower catalytic rates might be anticipated. Apparently the binding of the substrate to the enzyme must cause major changes in the acid–base properties of the substrates. This could be readily accomplished through electrostatic and hydrophobic interactions.

COVALENT INTERMEDIATES

Many enzymatic reactions proceed through the formation of covalent intermediates. Some examples have been cited in the discussion of nucleophilic and electrophilic reactions in Chapter 5. In terms of catalytic efficiency, the function of covalent intermediates is to break down the catalytic cycle into a series of reactions with small activation energies, in contrast to a single reaction of high activation energy.

ADVANTAGES OF ENZYMES

The elementary steps involved in enzyme catalysis obviously are no different than found with nonenzymatic catalysis. Yet the catalytic efficiency of enzymes is not approached in nonenzymatic systems; apparently the individual steps in enzyme catalysis are proceeding at close to the maximum possible rates. What special features do enzymes possess to make this possible? The answer to this question seems to lie in the "entropic" advantage of enzymes and the flexible protein structure available. The entropy advantage of an enzyme stems from two factors. First, many functional groups are simultaneously present so intramolecular catalysis can occur. This can be viewed as an increase in the effective concentration of the catalyst. The maximum concentration of the catalyst would be expected to be similar to the concentration of water in pure water, i.e., 55 *M*. This is an appreciable contribution to the catalytic efficiency, but even more important is the restriction of substrate conformation and orientation that occurs when the substrate is bound to the enzyme (*43*). In other words, the substrate is restricted to a very small number of configurations that are optimal for catalysis. This can be viewed as a loss in the rotational entropy of the substrate. This was demonstrated in a study of model reactions by Bruice and co-workers (*43*). Some of the results are presented in Table 6-4. The rate of anhydride formation was studied for a series of compounds in which the rotational freedom was severely restricted. The intramolecular rate constants were compared with the corresponding bimolecular process by dividing the first order rate constant by the second order rate constant; the dimension of this ratio is concentration, and this can be viewed as the increase in the "effective" concentration. Effective concentrations as large as 10^7–10^8 *M* were found with the compounds having the most restricted configurations. A more formal treatment of the entropy restrictions in enzyme catalysis has been developed (*42–44*). However, in simple terms the catalytic groups and substrate are all part of the same molecule, and the energy associated with the binding of substrate to enzyme is used to restrict the translational and rotational freedom of the system to a configuration optimal for catalysis.

The macromolecular structure of the enzyme and its flexibility or ability to undergo conformational changes also are important. The observation of a large number of reaction intermediates appears to be a general feature of enzyme catalysis. Thus, the enzyme appears to achieve its catalytic efficiency by breaking down the reaction into several steps all of which proceed at

Table 6-4

Effects of Rotational Restriction on Reaction Rates (*43*)

Reactant	Relative rate of anhydride formation
$CH_3COO^- + CH_3COOC_6H_5Br(p)$	1.0
Glutarate mono-ester: $-COOC_6H_4Br(p)$, $-COO^-$	$\sim 10^3$ *M*
R, R-disubstituted glutarate: $-COOC_6H_4Br(p)$, $-COO^-$	3×10^3–10^6 *M*
Succinate: $-COOC_6H_4Br(p)$, $-COO^-$	2×10^5 *M*
Maleate: $-COOC_6H_4Br(p)$, $-COO^-$	10^7 *M*
Oxabicyclic: $COOC_6H_4Br(p)$, COO^-	5×10^7 *M*

similar rates close to their maximum possible values (*45*). This optimization can be achieved because of the "conformational adaptability" of the enzyme, which allows the structure to be modulated so that it can be slightly altered for each step in the mechanism. This structural optimization only can be realized with a flexible macromolecular structure. When the rates of the individual steps in enzyme catalysis become comparable, the enzyme in a sense has reached maximum efficiency (*46*).

To summarize, the chemical events in enzyme catalysis correspond to catalytic processes observed in model systems. However, the macromolecular nature of an enzyme allows many different chemical reactions to proceed at optimal rates, with the end result being very efficient catalysis.

REFERENCES

1. P. Fasella and G. G. Hammes, *Biochemistry* **6**, 1798 (1967).
2. G. G. Hammes and J. L. Haslam, *Biochemistry* **7**, 1519 (1968).

3. G. G. Hammes and J. L. Haslam, *Biochemistry* **8**, 1591 (1969).
4. G. G. Hammes and P. Fasella, *J. Am. Chem. Soc.* **85**, 3939 (1963).
5. B. H. Havsteen, *J. Biol. Chem.* **242**, 769 (1967).
6. G. P. Hess, J. McConn, E. Ku, and G. McConkey, *Phil. Trans. R. Soc. Ser. B* **256**, 27 (1969).
7. B. Chance, *in* "Currents in Biochemical Research" (D. E. Green, ed.), p. 308, Wiley (Interscience), New York, 1956.
8. G. G. Hammes and J. K. Hurst, *Biochemistry* **8**, 1083 (1969).
9. K. Kirschner, M. Eigen, R. Bittman, and B. Voight, *Proc. Natl. Acad. Sci. U.S.A.* **56**, 1661 (1966).
10. G. Czerlinski and G. Schreck, *J. Biol. Chem.* **239**, 913 (1964).
11. H. D'A. Heck, *J. Biol. Chem.* **244**, 4375 (1969).
12. E. Holler, J. A. Rupley, and G. P. Hess, *Biochem. Biophys. Res. Commun.* **37**, 423 (1969).
13. J. H. Boldo, S. E. Hulford, S. L. Patt, and B. D. Sykes, *Biochemistry* **14**, 1893 (1975).
14. G. Czerlinski and G. Schreck, *Biochemistry* **3**, 89 (1963).
15. H. Theorell and A. Nygaard, *Acta Chem. Scand.* **8**, 1649 (1954).
16. B. Chance, *Arch. Biochem.* **22**, 224 (1949).
17. B. Chance, *Arch. Blochem.* **24**, 410 (1949).
18. B. Chance, *in* "Enzymes and Enzyme Systems" (J. T. Edsall, ed.), p. 93. Harvard Univ. Press, Cambridge, Massachusetts, 1951.
19. A. S. Mildvan and M. C. Scrutton, *Biochemistry* **6**, 2978 (1967).
20. A. S. Mildvan, J. S. Leigh, and M. Cohn, *Biochemistry* **6**, 1805 (1967).
21. G. G. Hammes and F. G. Walz, Jr., *J. Am. Chem. Soc.* **91**, 7179 (1969).
22. J. E. Erman and G. G. Hammes, *J. Am. Chem. Soc.* **88**, 5607 (1966).
23. E. J. del Rosario and G. G. Hammes, *J. Am. Chem. Soc.* **92**, 1750 (1970).
24. J. E. Erman and G. G. Hammes, *J. Am. Chem. Soc.* **88**, 5614 (1966).
25. A. R. Fersht, R. S. Mulvey and G. L. E. Koch, *Biochemistry* **14**, 13 (1975).
26. A. Pingoud, D. Boehme, D. Riesner, R. Kownatski, and G. Maas, *Eur. J. Biochem.* **56**, 617 (1975).
27. T. I. Rosenberry, *Adv. Enzymol.* **43**, 103 (1975).
28. J. C. Kernohan, *Biochim. Biophys. Acta* **81**, 346 (1964).
29. J. C. Kernohan, *Biochim. Biophys. Acta* **96**, 304 (1965).
30. J. W. Terpel, G. M. Hass and R. L. Hill, *J. Biol. Chem.* **243**, 5684 (1968).
31. K. L. Laidler, *Discuss. Faraday Soc.* **20**, 83 (1955).
32. F. M. Richards and H. M. Wyckoff, *in* "The Enzymes" (P. Boyer, ed.), 3rd ed. Vol. 4, p. 647. Academic Press, New York, 1971.
33. S. F. Velick and J. Vavra, *J. Biol. Chem.* **237**, 2109 (1962).
34. S. E. Halford, N. G. Bennett, D. R. Trentham, and H. Gutfreund, *Biochem. J.* **114**, 243 (1969).
35. G. Czerlinski, *Biochim. Biophys. Acta* **64**, 199 (1962).
36. G. G. Hammes and J. Simplicio, *Biochim. Biophys. Acta* **212**, 428 (1970).
37. D. F. Koshland, Jr., *Adv. Enzymol.* **22**, 45 (1960).
38. G. G. Hammes and F. G. Walz, Jr., *Biochim. Biophys. Acta* **198**, 604 (1970).
39. G. G. Hammes, *Nature* (*London*) **204**, 342 (1964).
40. R. Lumry, *in* "The Enzymes" (P. Boyer, ed.), 2nd ed., Vol. 1, p. 157. Academic Press, New York, 1959.
41. R. Wolfenden, *Acc. Chem. Res.* **5**, 10 (1972).
42. W. P. Jencks, *Adv. Enzymol.* **43**, 219 (1975).
43. T. C. Bruice, *Annu. Rev. Biochem.* **45**, 331 (1976).
44. M. I. Page and W. P. Jencks, *Proc. Natl. Acad. Sci. U.S.A.* **68**, 1678 (1971).
45. G. G. Hammes, *Acc. Chem. Res.* **1**, 321 (1968).
46. J. R. Knowles, *Annu. Rev. Biochem.* **49**, 877 (1980).

7

Case Studies of Selected Enzyme Mechanisms

INTRODUCTION

Elegant mechanistic investigations of many enzymes have been carried out. To illustrate the principles discussed in the preceding chapters, a few selected enzymes are considered in detail now. The choice of enzymes is quite arbitrary and even with the enzymes selected, no attempt is made to give a complete review of the literature. The first enzymes studied with chemical methods were virtually all hydrolytic enzymes. This is no accident: these enzymes are extremely stable, are relatively easy to purify and crystallize, and are relatively small proteins.

PANCREATIC RIBONUCLEASE

Two main forms of ribonuclease are found in the pancreas, A and B. Both have the same polypeptide chain, but B is a glycoprotein with seven sugar residues covalently linked to the polypeptide. Only bovine pancreatic ribonuclease A is considered here. A comprehensive review of the enzyme has been given by Richards and Wyckoff (*1*). Ribonuclease A is a single polypeptide chain containing 124 amino acids and has a molecular weight of 13,683. Its complete amino acid sequence is known (*2*, Fig. 2-1) and, in fact, the enzyme has been synthesized from the amino acids (*3*, *4*). The crystal structure of ribonuclease A and ribonuclease S, which has a single peptide bond cleaved, are known with several substrate analogs bound to the catalytic site (*1*). As illustrated in Fig. 7-1, the enzyme catalyzes the breakdown of RNA in two steps. First the diester linkage is broken and a pyrimidine 2′,3′-

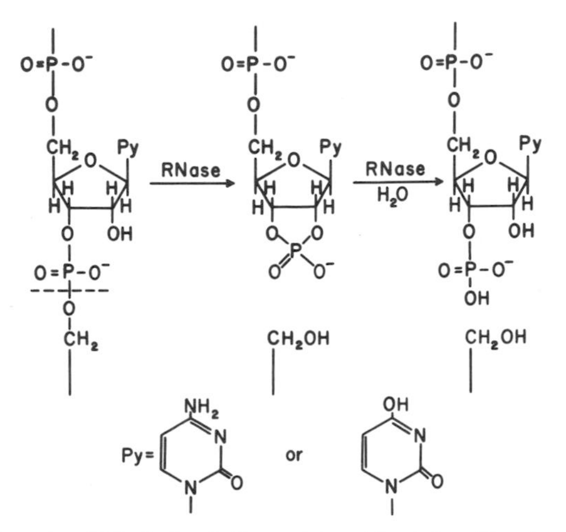

Fig. 7-1. The two-step hydrolysis of ribonucleic acid catalyzed by the enzyme pancreatic ribonuclease A.

cyclic phosphate is formed; then the cyclic phosphate is hydrolyzed to give a terminal pyrimidine 3′-phosphate.

Ribonuclease has been subjected to intensive structure–function studies using a variety of techniques (cf. *1*, *5*). The conclusions derived from these studies have proved to be remarkably consistent with the three-dimensional structure of the enzyme determined by x-ray crystallography. Ribonuclease has four intramolecular disulfide bonds. If these are reduced to sulfhydryl groups and the enzyme is denatured, the enzyme will refold to its native structure under the proper conditions. This demonstration that the three-dimensional structure of the enzyme is determined by its amino acid sequence is a classical experiment in the history of protein chemistry (*6*). If the enzyme is treated with the proteolytic enzyme subtilisin, a single peptide bond is cleaved, primarily between amino acid residues 20 and 21. This modified enzyme is fully active, but if the short (S) peptide (1–20) is removed from the enzyme, all catalytic activity is lost. A number of different portions of the S peptide have been synthesized and tested for binding to the S protein and for generation of catalytic activity (*1*, *7*, *8*). For example, the fragments 12–13, 8–13, 14–20, 1–7, 1–11, and 3–13 are inactive, whereas 1–13, 1–12-amide, and 2–13 are active. Thus amino acid residues 15–20 are unnecessary. Residues 2, 13, 14, and 8 are important for binding to the S protein; histidine 12 is essential for catalytic activity since if pyrazole is substituted for imidazole in S peptide 1–14, binding to S protein occurs, but the enzyme is inactive.

The four histidines in ribonuclease have been modified with several types of reagents (*1*). Some of the results obtained are described in Chapter 2.

Photooxidation in the presence of methylene blue oxidizes three histidines and inactivates the enzyme. Carboxyalkylation with iodoacetic acid and similar reagents yields an inactive enzyme with either histidine-12 or histidine-119 modified. From this work the correct conclusion was reached that these two residues are in close proximity and are essential for catalytic activity. If the inactive enzymes with histidine-12 or histidine-119 are mixed in equal proportion and lyophilized from 50% acetic acid, 50% of the enzyme activity is restored. Apparently a dimer is formed with the catalytic site containing unmodified histidine 12 and 119 from different enzyme molecules.

Modification of the tyrosines with various reagents, e.g., iodine, gives an active enzyme with slightly altered properties so tyrosine is not essential for activity. Three of the tyrosines titrate normally, whereas three cannot be titrated until the protein denatures indicating they are buried in the protein structure. All of the carboxyl groups can be methylated except for aspartate-14, -38, and -83, which are inferred to interact with buried tyrosine-92, -25, and -97, respectively. Modification of carboxyl groups with other reagents gives modified enzymes with varying degrees of activity depending on the reagent, the extent of modification and the exact assay. However, the general conclusion is that carboxyl groups are not necessary for activity.

If 9 of the 10 lysines are guanidated or 7 are polyalynated, the enzyme remains active so that the modified lysines on the exterior are not essential for catalysis. However, if lysine-41 is dinitrophenylated, an inactive protein is obtained. As discussed in Chapter 2, the enzyme remains active when lysine-31 and -37 or -37 and -7 are linked with bifunctional reagents. Chemical modification of 2–3 (of 4) arginine residues causes marked inactivation, and arginine-39 may be essential for catalysis. On the other hand, methionine residues can be modified without inactivating the enzyme.

Kinetic studies generally have not employed ribonucleic acid itself as a substrate because the system becomes inhomogeneous as ribonucleic acid is degraded, which makes the kinetic analysis difficult. Instead model substrates such as dinucleoside monophosphates and pyrimidine 2′,3′-cyclic phosphates have been used. The structural requirements for substrates have been determined (cf. *1*); the enzyme shows considerable specificity both with regard to the structure of the base and the sugar. The specificity for pyrimidine-like structures is not absolute as poly (A) and formycin derivatives show low activity. The turnover number, k_{cat}, changes markedly with modifications in the pyrimidine ring, whereas the Michaelis constant is relatively constant. The steady-state parameters for a series of dinucleosides and cytidine and uridine 2′,3′-cyclic phosphate are presented in Table 7-1. Cytidine derivatives always have higher turnover numbers than the corresponding uridine compounds, but the nature of the second base markedly influences the turnover number. In contrast, the Michaelis constants do not show great variations.

Table 7-1

Steady-State Parameters for Ribonuclease A at pH 7 (*9*)

Substrate	k_{cat} (sec^{-1})	K_S (mM)	Substrate	k_{cat} (sec^{-1})	K_S (mM)
CpA	3000	1.0	UpA	1200	1.9
CpG	500	3.0	UpG	—	—
CpC	240	4.0	UpC	40	3.0
CpU	27	3.7	UpU	11	3.7
C2′3′P	5.5	3.3	U2′3′P	2.2	5.0

The steady-state kinetics of cytidine and uridine cyclic phosphate hydrolysis have been studied in considerable detail (*10, 11*). The 3′-monophosphate is the exclusive product but the 2′-monophosphate is a potent inhibitor. Moreover, the 3′-monophosphate has a Michaelis constant 10 times lower than that of the substrate so that the substrates must be very pure and great care must be taken to obtain the initial velocity. The equilibrium constant of the hydrolysis reactions is about 500 $(1 + K_a/[H^+])$, where K_a is the second ionization constant of the 3′-monophosphate. The pH dependence of the kinetic parameters is well described by the simple mechanism

$$\begin{array}{ccccc} E & & X & & E \\ \upharpoonleft\downharpoonright & & \upharpoonleft\downharpoonright & & \upharpoonleft\downharpoonright \\ EH + S & \rightleftharpoons & XH & \longrightarrow & EH + P \\ \upharpoonleft\downharpoonright & & \upharpoonleft\downharpoonright & & \upharpoonleft\downharpoonright \\ EH_2 & & XH_2 & & EH_2 \end{array} \tag{7-1}$$

The pH dependence of k_{cat}/K_s is virtually identical for both cyclic phosphates and several dinucleotides, as shown in Fig. 7-2. The pK values of ionization constants derived from these data are 5.4 and 6.4, and these very probably are associated with the imidazole residues on histidine-12 and -119. This implies the same acid–base groups participate in both halves of the reaction for all substrates. The pK values are considerably raised when the substrate binds as judged from the pH dependence of the maximum velocities. Since the cytidine and uridine cyclic phosphates have the same pH profile for k_{cat}/K_s, protonation of the cytidine ring (p$K \sim 4.3$) apparently is not of importance. The question of which ionized form of the 3′-monophosphate binds cannot be resolved unambiguously. However, the pH dependence of

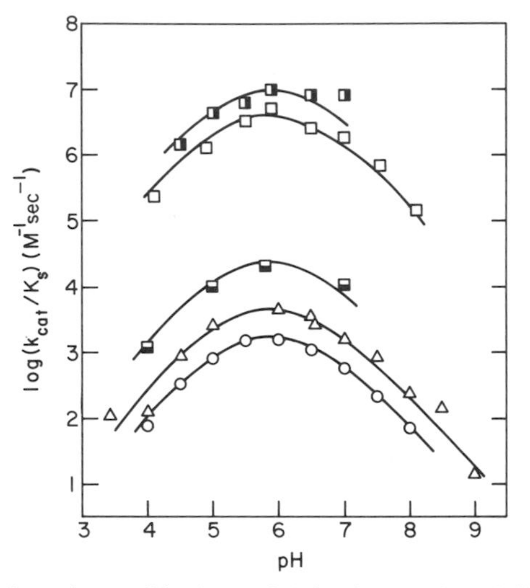

Fig. 7-2. The pH dependence of k_{cat}/K_S at 25° for the reaction of ribonuclease with 2′:3′ cyclic-CMP (Δ, *10*), 2′:3′ cyclic UMP (○, *11*), UpU (⬓, *9*), UpA (□, *9*) and CpA (◨, *9*). The curves were calculated assuming pK values of 5.4 and 6.4. [Adapted with permission from E. J. del Rosario and G. G. Hammes, *Biochemistry* **8**, 1884 (1969). Copyright (1969) American Chemical Society.]

the binding of uridine 3′-monophosphate (pK = 5.74), cytidine 3′-monophosphate (pK = 5.90), orthophosphate (pK = 6.67), and pyrophosphate (pK = 5.97) are identical if the monoanion is assumed to be the species binding to the enzyme (*12*).

Ribonuclease also has been studied with high resolution nuclear magnetic resonance (cf. *13, 14*). Some of the resonances of all of the imidazoles can be resolved, as well as those of several other specific groups. Titration of the histidines in D_2O gives pK values of 6.7, 6.2, 5.8, and 6.3 for histidine-105, -119, -12, and -48, respectively. The titration curves of histidine-12 and -119 are not quite normal and suggest the protonation of these groups is not independent. Some line broadening in histidine-48 was found as the pH was varied; this is paralleled by changes in the resonances of tyrosine-25. These two amino acids are postulated to be involved in a pH-dependent conformational change. Such a conformational change has been observed in kinetic studies to be discussed shortly. The resonances of histidine-119, -12, and -48 are shifted when inhibitors are bound to the enzyme; these changes can be used to calculate binding constants (*14*). The pK values of the imidazoles are markedly altered when inhibitors are bound. Both histidine-12 and -119 are postulated to be protonated in the complex containing cytidine 3′-monophosphate, but when a dinucleoside is present histidine-12 is not protonated. In the former case, one of the protons is probably more closely associated with the phosphate of the inhibitor rather than with an imidazole.

Transient kinetic studies have been made of the interaction of ribonuclease with cytidylyl-3′,5′-cytidine, cytidine and uridine 2′,3′-cyclic phosphate, and cytidine and uridine 3′-phosphates (cf. *15, 16*). The stopped flow–temperature jump was utilized for all but the 3′-monophosphates where the temperature jump method alone was sufficient. When a temperature jump is applied to a solution of enzyme and a pH indicator, a single relaxation process is observed whose relaxation time is independent of enzyme and indicator concentrations although it varies with pH. The pH dependence of the reciprocal relaxation time is illustrated in Fig. 7-3. This relaxation process is due to an isomerization (conformational change) of the enzyme since the dye does not bind to the enzyme under the experimental conditions. A simple mechanism consistent with the data is

$$E'H \underset{k_{-1}}{\overset{k_1}{\rightleftharpoons}} EH \overset{K_A}{\rightleftharpoons} E + H^+ \tag{7-2}$$

If the protolytic reaction is assumed to equilibrate rapidly relative to the isomerization

$$\frac{1}{\tau_1} = k_1 + \frac{k_{-1}}{1 + K_A/[H^+]} \tag{7-3}$$

The quantitative fit of the data to this equation is shown in Fig. 7-3, and at 25°C, $k_1 = 780\ \text{sec}^{-1}$, $k_{-1} = 2470\ \text{sec}^{-1}$, and $K_A = 10^{-6.1}\ M$. The rate constants are considerably smaller in D_2O than in H_2O suggesting hydrogen bonding may be of importance in the conformational change. A molecular

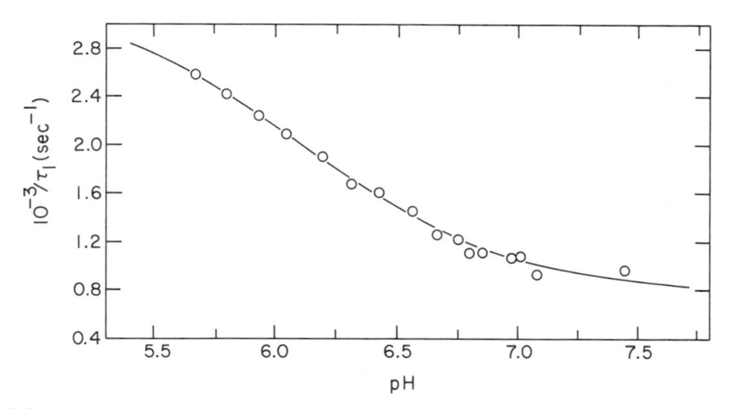

Fig. 7-3. The pH dependence of the reciprocal relaxation time characterizing the isomerization of ribonuclease at 25°C. The solid line is calculated according to Eq. (7-3) and the parameters given in the text. [Adapted with permission from T. C. French and G. G. Hammes, *J. Am. Chem. Soc.* **87**, 4669 (1965). Copyright (1965) American Chemical Society.]

interpretation of this conformational change is given below where the three-dimensional structure of ribonuclease is discussed.

The interaction of dinucleosides, pyrimidine 2′,3′-cyclic phosphates, or pyrimidine 3′-phosphates with the enzyme is characterized by two relaxation processes in addition to the process associated with the unliganded enzyme. These can be observed by monitoring absorption changes in the ultraviolet or by utilizing pH indicators. In all cases, the results obtained can be described by a two-step mechanism: a bimolecular combination of enzyme and substrate followed by a conformational change of the enzyme–substrate complex

$$\mathrm{E} + \mathrm{S} \underset{k_{-2}}{\overset{k_2}{\rightleftharpoons}} \mathrm{X}_1 \underset{k_{-3}}{\overset{k_3}{\rightleftharpoons}} \mathrm{X}_2 \tag{7-4}$$

If the conformational change of the unliganded enzyme is neglected, and if the first step is assumed to be rapid relative to the second, the reciprocal relaxation times for this mechanism are

$$\frac{1}{\tau_2} = k_2([\bar{\mathrm{E}}] + [\bar{\mathrm{S}}]) + k_{-2} \tag{7-5}$$

$$\frac{1}{\tau_3} = k_{-3} + \frac{k_3}{1 + k_{-1}/\{k_1([\bar{\mathrm{E}}] + [\bar{\mathrm{S}}])\}} \tag{7-6}$$

In all cases, a relaxation time has been observed that conforms to Eq. (7-5). The rate constants obtained are included in Table 6-1. A third relaxation time is seen in all cases, but often its amplitude is too small to detect until the enzyme is nearly saturated. When a concentration dependence can be observed, it conforms to Eq. (7-6). The rate constants for this conformational change range from approximately 10^3–10^4 $\sec^{-1}$. The role of the isomers of the unliganded enzyme in the mechanism is not certain. The assumption that the ligand binds equally well to both isomers fits the data, although a preference of the substrate for E′H cannot be excluded.

The pH dependence of k_2 for the reaction of enzyme with uridine 3′-monophosphate is identical to that of V_p/K_p. As expected, the same ionizable groups on the enzyme are implicated from both steady-state and transient kinetics, namely those with pK values of 5.4 and 6.4. The pH independent value of k_2 is 2×10^8 $M^{-1}\sec^{-1}$, which is very close to the value expected for a diffusion-controlled process. The data cannot distinguish whether the dianion or monoanion of uridine 3′-monophosphate reacts with the enzyme. In the former case, EH_2 would react with the substrate, whereas in the latter case EH would be the reactant. The pH dependence of $1/\tau_3$ at high substrate concentrations for uridine and cytidine 3′-monophosphate is very similar to that of $1/\tau_1$ at low pH values; in both cases an ionizable group on the enzyme with a pK of about 6.0 is implicated. A schematic representation of the minimal number of steps in the overall reaction, both transesterification and

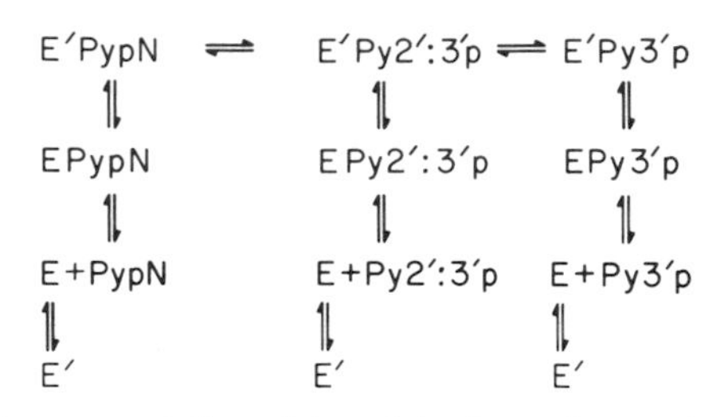

Fig. 7-4. Schematic representation of the minimal reaction mechanism for ribonuclease. The primed species are all different enzyme conformations; PypN is a pyrimidine dinucleoside, Py2′:3′p is a pyrimidine cyclic phosphate; and Py3′p is a pyrimidine 3′-phosphate.

hydrolysis, is shown in Fig. 7-4. The principal conclusions to be derived from the transient kinetic studies are (1) multiple enzyme conformations are important in the mechanism and (2) ionizable groups with pK values of 5.4 and 6.4 are of importance in catalysis and an ionizable group with a pK of 6.0 is of importance in the conformational changes observed.

The three-dimensional structure of ribonuclease S with and without an analog of uridylyl-3′,5′-adenosine bound to the catalytic site is shown in Fig. 1-6. The enzyme is kidney shaped, very compact, and has a long groove where the substrates bind. As predicted from the work described previously, histidine-12 and -119 are at the active site; lysine-41 is about 10 Å away. The x-ray structures show specific electrostatic, hydrogen bond, and hydrophobic interactions between enzyme and substrates. For a dinucleotide analog, specific sites can be seen for the two bases, the two riboses and the phosphate. All of the substrates are very close to histidine-12 and -119, but lysine-41 is not in direct contact with phosphorus. Histidine-48 lies at the top of the "hinge" of the groove, and its environment could be altered by an opening and closing of the groove associated with the active site. Such an opening and closing very likely is the conformational change detected in the nuclear magnetic resonance (nmr) and transient kinetic studies just described.

The overall mechanism of the reaction might be constructed as follows. The enzyme exists in dynamic equilibrium between two forms differing in the structure of the active site groove. The substrate binds to the groove at a rate that is nearly diffusion-controlled. Binding of the substrate causes a closing of the groove which brings lysine-41 close to the substrate and creates a hydrophobic pocket for the substrate. The chemical reaction then proceeds through rapid proton transfer reactions involving histidine-12 and -119. The conformational change is then reversed and the product dissociates. Both transesterification and hydrolysis can occur by this mechanism, which is illustrated schematically in Fig. 7-5.

The elementary steps associated with the proton transfer reactions cannot be observed directly. However, considerable insight into the mechanism of

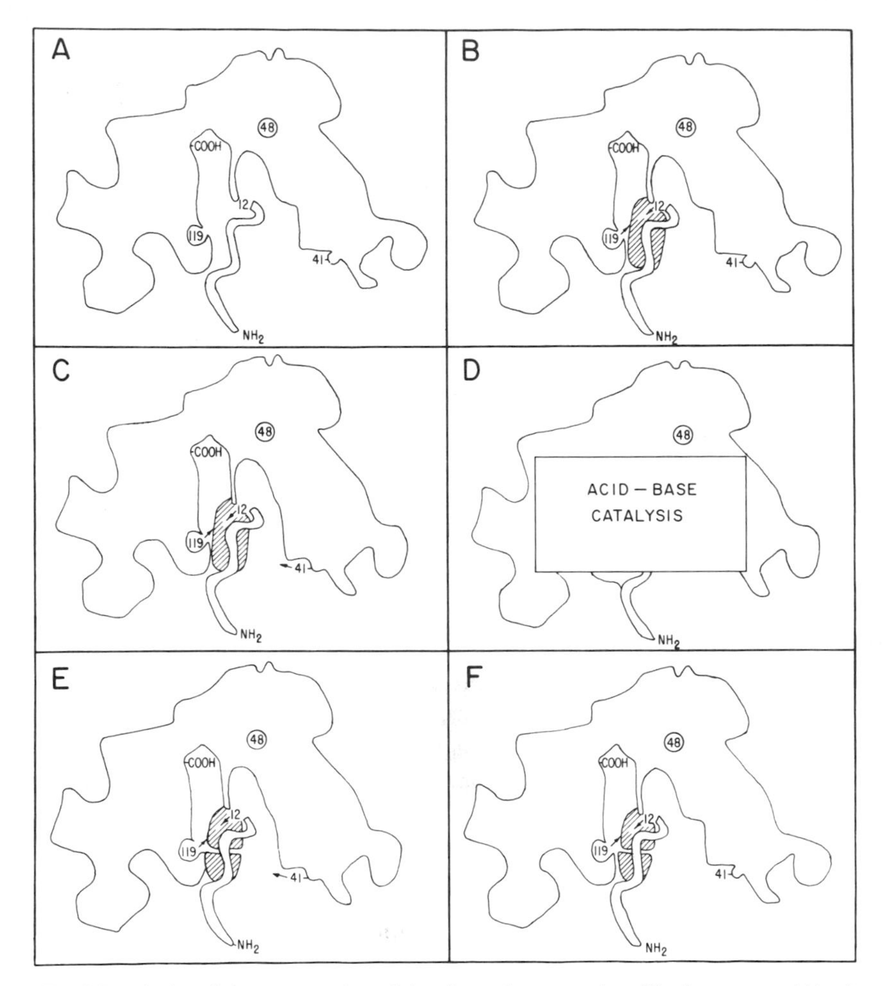

Fig. 7-5. A pictorial representation of the ribonuclease reaction. The free enzyme (A) exists in two conformational states differing by small movements of the hinge region joining the two halves of the molecule. The substrate is bound (B) and a conformational change occurs closing the hinge (C). Acid-base catalysis then occurs (D); products are formed (E); the conformational change is reversed (F); and products dissociate to give free enzyme.

the acid–base catalysis can be obtained through consideration of the detailed stereochemistry of the reaction. Through the study of the hydrolysis of many cyclic organophosphorus esters, a set of five mechanistic rules has been postulated (*17*).

1. Hydrolysis proceeds via a pentacoordinate species that has the geometry of a trigonal bipyramid.

P

2. A five-membered ring spans one basal and one apical position.

apical
basal

3. Groups enter and leave from apical positions only.

4. Position exchange can occur between apical and basal positions by means of a *pseudorotation.*

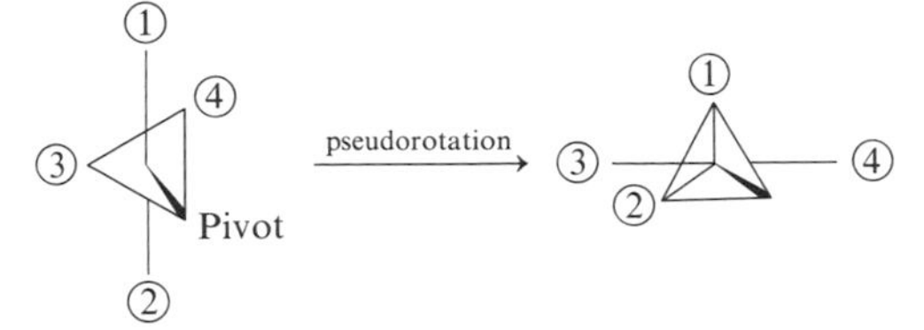

5. Electropositive groups tend to occupy basal positions (e.g., CH, O^-); electronegative groups tend to occupy apical positions (e.g., OH, F, OR).

The consequence of these rules for the ribonuclease reaction is that two possible types of mechanism can be envisaged, an in-line displacement where the group to be displaced is in an apical position and an adjacent mechanism whereby the group to be displaced is initially in a basal position so that a pseudorotation is required for the reaction to occur (Rule 3) (*18*). The structure of the intermediates for the first and second steps of the mechanism can be depicted as (Rule 2)

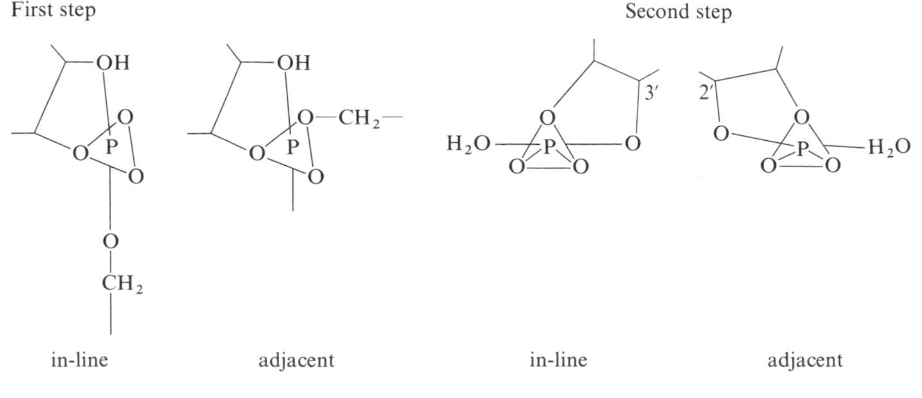

In looking at the ribonuclease structure, the most likely mechanisms are simultaneous acid and base catalysis by histidine-12 and -119 for the in-line mechanism and general acid and base catalysis by a single histidine residue for the adjacent mechanism. For example, for the first step in the mechanism, this can be represented as

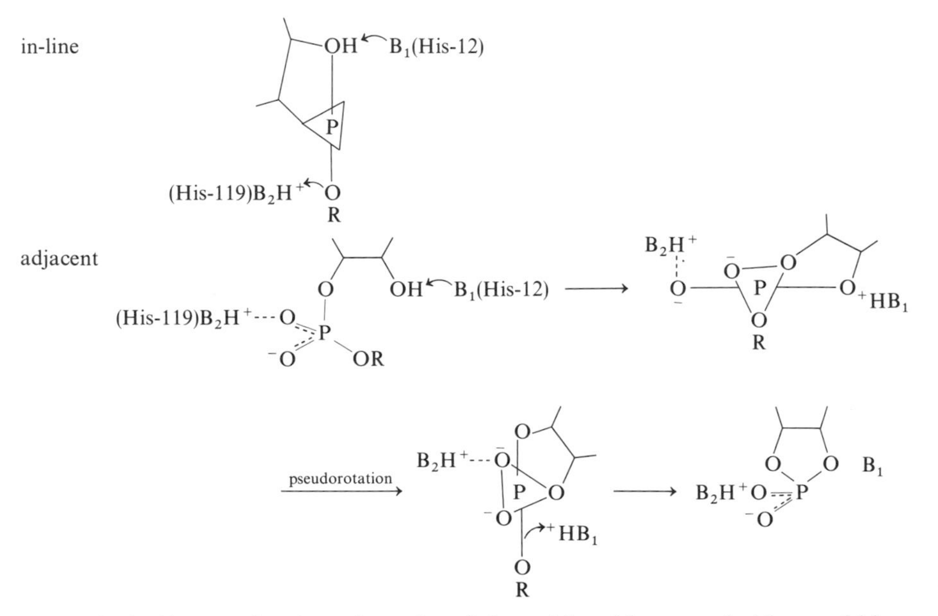

For the in-line mechanism, the roles of the acid and base probably would be reversed in the second step. The in-line mechanism would occur with inversion at the phosphorus, whereas the adjacent would occur with retention of configuration. The difference between these stereochemical possibilities cannot be detected with normal substrates. However, stereochemical isomers can be separated if a thiophosphoryl group is introduced. For the cyclic phosphate the two isomers are

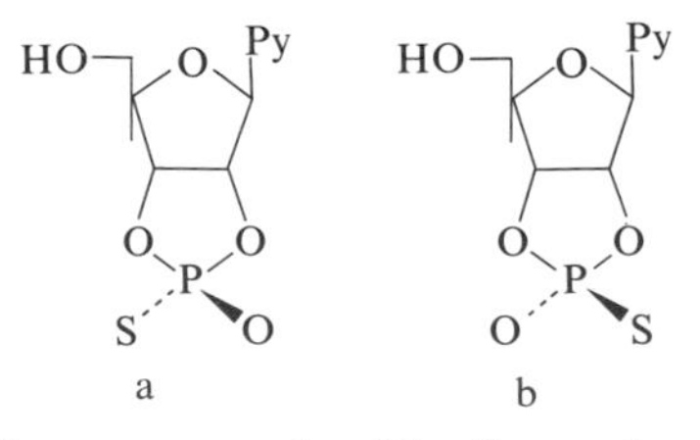

Isomer a (Py = uracil) was reacted with ribonuclease in $[^{18}O]H_2O$. The product was recyclized chemically with diethyl phosphochloridate, which gives equal populations of the two cyclic phosphate diastereomers. These isomers were then analyzed for ^{18}O content. The chemical closing is known

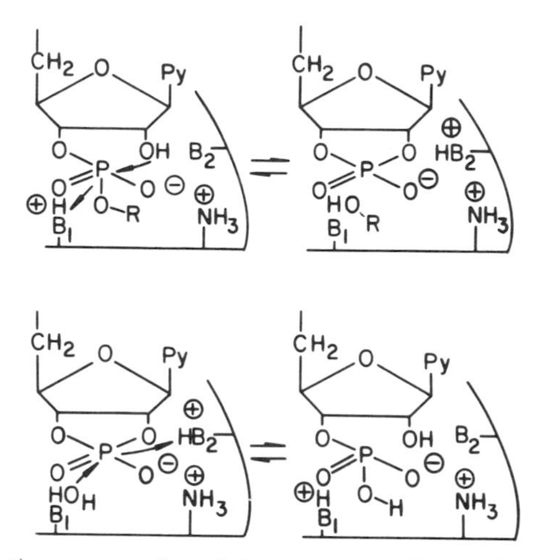

Fig. 7-6. Schematic representation of the proton transfer reactions involved in the two parts of the ribonuclease catalyzed hydrolysis of ribonucleic acid. The group B_1 is probably the imidazole residue of histidine-119 and B_2 is probably the imidazole residue of histidine-12. The amino group represents lysine-41. [From G. G. Hammes, *in* "Investigations of Rates and Mechanisms of Reactions," Part II. Wiley (Interscience), New York, 1974.]

to be in-line because the relatively electropositive $S^{(-)}$ and $O^{(-)}$ would preferentially be in basal positions. If the enzymatic opening is in-line, isomer a should have no ^{18}O, and isomer b should have one equivalent of ^{18}O. This was the result found, so a concerted proton transfer is most likely (*19*). The stereochemistry of the first step of the ribonuclease reaction was checked by use of the dinucleoside phosphorothioate Up(S)C in a similar experiment, and the mechanism in this case also appears to be in-line (*20*). A schematic representation of the probable mechanism for the acid–base catalysis of the ribonuclease reactions is shown in Fig. 7-6.

Thus the mechanism of the reactions catalyzed by ribonuclease is understood in molecular terms for the binding events and conformational changes involved in preparation for acid–base catalysis and for the stereochemistry of the acid–base catalysis.

CHYMOTRYPSIN

Chymotrypsin is a proteolytic enzyme that is secreted into the intestine as an inactive form (the zymogen) after synthesis in the pancreas. The zymogen, chymotrypsinogen, is a single polypeptide chain of 245 amino acids with a molecular weight of about 25,000. It is converted into an active protease by limited proteolysis with trypsin as depicted in Fig. 7-7. This

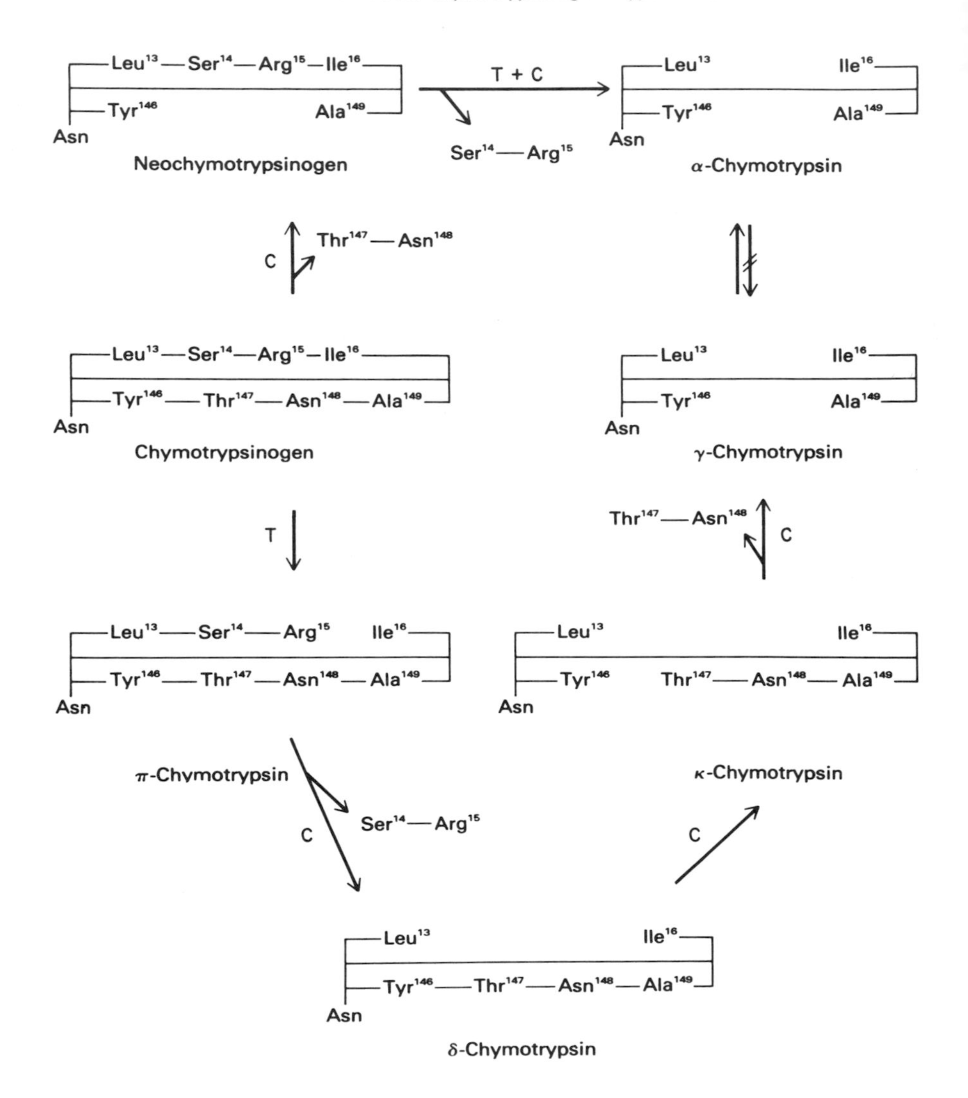

Fig. 7-7. The activation of chymotrypsinogen by chymotrypsin (C) and trypsin (T). Both chymotrypsinogen and neochymotrypsinogen are inactive. [Adapted from S. Blackburn, "Enzyme Structure and Function," p. 14, 1976. Courtesy of Marcel Dekker, New York.]

discussion focuses primarily on α-chymotrypsin which contains three polypeptide chains linked together by disulfide bonds. The amino acid sequences and crystal structures of both chymotrypsinogen and α-chymotrypsin are known (cf. *21, 22*). The inactivity of the zymogen appears to be due to the fact that part of the substrate binding site is missing in the zymogen. Chymotrypsin is one of a class of proteases, the serine proteases, that have very similar structures and mechanisms of action. Others in this class include trypsin, elastase, and thrombin. Chymotrypsin has been extensively studied and reviewed (cf. *21–24*).

Chymotrypsin is an endopeptidase that hydrolyzes bonds adjacent to the carboxyl group of aromatic amino acids (tryptophan, tyrosine, phenylalanine) very readily and less readily with other hydrophobic amino acids (e.g., leucine, methionine). The enzyme also is an esterase with similar specificities. As might be anticipated, well-defined synthetic substrates have been utilized to study the mechanism of action of the enzyme. These primarily are L aromatic amino acids with the α amino group blocked and with the C terminus modified to form an ester or amide, although many different groups have been put on the C terminus.

The first insight into the mechanism of ester and peptide hydrolysis by chymotrypsin came from a stopped flow study of the chymotrypsin catalyzed hydrolysis of *p*-nitrophenylacetate (*25–27*)

$$H_2O + CH_3-\overset{\displaystyle O}{\overset{\|}{C}}-O-C_6H_4-NO_2 \xrightarrow{\text{Chymotrypsin}} CH_3COO^- + {}^-O-C_6H_4-NO_2 \tag{7-7}$$

This reaction was selected for study because the phenolate ion is yellow and its appearance can be easily detected by optical techniques. What was observed was an initial rapid production of phenolate (a "burst"), the amount being approximately equal to the enzyme concentration, followed by a leveling off and steady-state production of phenolate. These biphasic time courses require the occurrence of at least two reaction intermediates so that a minimal mechanism is

$$E + S \underset{k_{-1}}{\overset{k_1}{\rightleftharpoons}} X_1 \xrightarrow{k_2} X_2 + P_1 \xrightarrow{k_3} E + P_1 + P_2 \tag{7-8}$$

In this mechanism P_1 was postulated to be phenolate and P_2 acetate. The rate equations describing this mechanism initially when $[S_0] \gg [E_0]$ are

$$\begin{aligned} -\frac{d[X_1]}{dt} &= (k_{-1} + k_2)[X_1] - k_1[E][S_0] \\ -\frac{d[X_2]}{dt} &= k_3[X_2] - k_2[X_1] \end{aligned} \tag{7-9}$$

(Note that $[E_0] = [E] + [X_1] + [X_2]$.) These coupled differential equations can be solved to give (cf. Chapter 4)

$$[X_1] = \frac{[E_0]k_1k_3[S_0]}{\lambda_2\lambda_3} + \frac{[E_0][S_0]k_1(\lambda_2 - k_3)}{\lambda_2(\lambda_3 - \lambda_2)} e^{-\lambda_2 t} + \frac{[E_0][S_0]k_1(k_3 - \lambda_3)}{\lambda_3(\lambda_3 - \lambda_2)} e^{-\lambda_3 t} \tag{7-10}$$

$$\lambda_2 = k_1[S_0] + k_{-1} + k_2 + k_3$$

$$\lambda_3 = \frac{k_3K'_S + (k_2 + k_3)[S_0]}{K'_S + [S_0]}$$

$$K'_S = (k_{-1} + k_2)/k_1$$

where the assumption has been made that $k_1[S_0] + k_{-1} \gg (k_2 + k_3)$ in obtaining the expressions for λ_2 and λ_3. Furthermore, $k_2 \gg k_3$ so $\lambda_3 \simeq k_2[S_0]/([S_0] + K'_S)$. Since $\lambda_2 \gg \lambda_3$, the term multiplied by $e^{-\lambda_2 t}$ rapidly becomes very small and can be neglected when deriving the rate of appearance of phenolate which is given by

$$\frac{d[P_1]}{dt} = k_2[X_1]$$

Integration of this equation gives

$$[P_1] = \frac{k_2k_3[E_0][S_0]t}{k_3K'_S + [S_0](k_2 + k_3)} + \frac{k_2^2[E_0][S_0]^2}{\{k_3K'_S + [S_0](k_2 + k_3)\}^2}(1 - e^{-\lambda_3 t}) \tag{7-11}$$

The second term corresponds to the burst and the first term is the steady-state velocity multiplied by time. This mechanism predicts that the initial burst is the formation of the intermediate X_2, which accumulates during the reaction since its breakdown is rate determining. Hartley and Kilby (*25*) correctly postulated that X_2 is a covalent acyl enzyme species

$$\text{Enzyme}\!-\!\overset{\overset{\displaystyle O}{\|}}{C}\!-\!CH_3$$

Numerous steady-state kinetic studies have been carried out to further delineate the reaction mechanism of Eq. (7-8) (cf. *23, 24*). The steady-state velocity, v, for this mechanism is

$$v = \frac{k_{cat}[E_0]}{1 + K_S/[S]} \tag{7-12}$$

$$k_{cat} = k_2k_3/(k_2 + k_3) \tag{7-13}$$

$$K_S = [(k_{-1} + k_2)/k_1][k_3/(k_2 + k_3)] = K'_Sk_3/(k_2 + k_3) \tag{7-14}$$

The hydrolysis of *p*-nitrophenyl acetate represents the limit that $k_2 \gg k_3$ so that $k_{cat} = k_3$ and $K_S = (k_3/k_2)K'_S$. The other limiting possibility is that $k_3 \gg k_2$ so that $k_{cat} = k_2$ and $K_S = K'_S$. Some representative steady-state data for chymotrypsin catalyzed hydrolysis of *N*-acetyl amides and esters are presented in Table 7-2 (*28*). Also included are the inhibition constants obtained for the corresponding D amino acids. In view of the results obtained with the D amino acids, the large difference in Michaelis constants for amide and ester hydrolysis must be due to the ratio $k_3/(k_2 + k_3)$. Also since $K_i^E/K_S^E \simeq 7$ and $K_i^A/K_S^A \simeq 0.4$, K_S^A must be close to K'_S. Direct measurements of K_S and K'_S indicate this is true (*23*). Therefore, acylation is rate controlling for amide hydrolysis and deacylation is rate controlling in ester hydrolysis for the series of compounds studied. This conclusion predicts that a series of ester substrates with the same acylamino acid moiety, but different leaving groups, should have the same values of $k_{cat} = k_3$. This is true in most cases. For example, *N*-acetyl tryptophan ethyl, methyl, and *p*-nitrophenyl esters hydrolyze with k_{cat} equal to 27, 28, and 31 sec^{-1}, respectively; k_{cat} for the corresponding amide is 0.026 sec^{-1} (*34*).

Table 7-2

Steady-State Parameters for Ester and Amide Hydrolysis by Chymotrypsin (*28*)[a]

Inhibitor	K_i^A (m*M*)	K_i^E (m*M*)	K_i^A/K_i^E	Reference
D-Tyrosine	12	5.	2.4	*29, 30*
D-Phenylalanine	12	2.4	5.0	*29, 30*
D-Tryptophan	2.3	0.8	2.9	*29, 30*
Substrate	K_S^A (m*M*)	K_S^E (m*M*)	K_S^A/K_S^E	**Reference**
L-Tyrosine	34	0.7	46	*31, 32*
L-Phenylalanine	31	1.2	26	*31, 33*
L-Tryptophan	5	0.09	55	*31, 32*

[a] K_i^A is the inhibition constant for amides; K_i^E is the inhibition constant for esters; K_S^A and K_S^E are the corresponding Michaelis constants.

Further evidence in support of the above conclusions comes from experiments in which a nucleophile (in addition to water) is used as an acyl group acceptor. If such a nucleophile is more efficient than water, the rate of catalysis (k_{cat}) should be increased for ester hydrolysis where deacylation is rate determining, whereas k_{cat} should be unchanged for amide hydrolysis where acylation is rate determining. However, evidence for acyl enzyme formation can be obtained even in the case of amide hydrolysis since the final products would be partitioned between water and the added nucleophile. Thus an analysis of the products would permit determination of the relative rates of reaction of the acyl enzyme and the nucleophiles. Furthermore, the product should be partitioned identically for the ester and amide of a given substrate since the same acyl enzyme is formed in both cases. The anticipated results have been obtained with several nucleophiles including hydroxylamine, methanol, and L-alanine amide (*24, 35*).

The pH dependence of the steady-state parameters is shown schematically in Fig. 7-8 (*23*, *24*). These results can be interpreted by the usual type of mechanism

$$
\begin{array}{ccccccc}
\mathrm{EH_2} & & \mathrm{X_1H_2} & & \mathrm{X_2H_2} & & \mathrm{EH_2} \\
K_1 \Updownarrow & & K_1' \Updownarrow & & K_1'' \Updownarrow & & \Updownarrow \\
\mathrm{EH + S} & \underset{}{\overset{K_{S0}'}{\rightleftharpoons}} & \mathrm{X_1H} & \xrightarrow{k_{20}} & \mathrm{X_2H + P_1} & \xrightarrow{k_{30}} & \mathrm{EH + P_2 + P_1} \\
K_2 \Updownarrow & & K_2' \Updownarrow & & K_2'' \Updownarrow & & \Updownarrow \\
\mathrm{E} & & \mathrm{X_1} & & \mathrm{X_2} & & \mathrm{E}
\end{array}
\tag{7-15}
$$

where the subscript 0 designates a pH-independent rate constant, K_{S0}' a pH-independent equilibrium constant, and the K_is are acid ionization

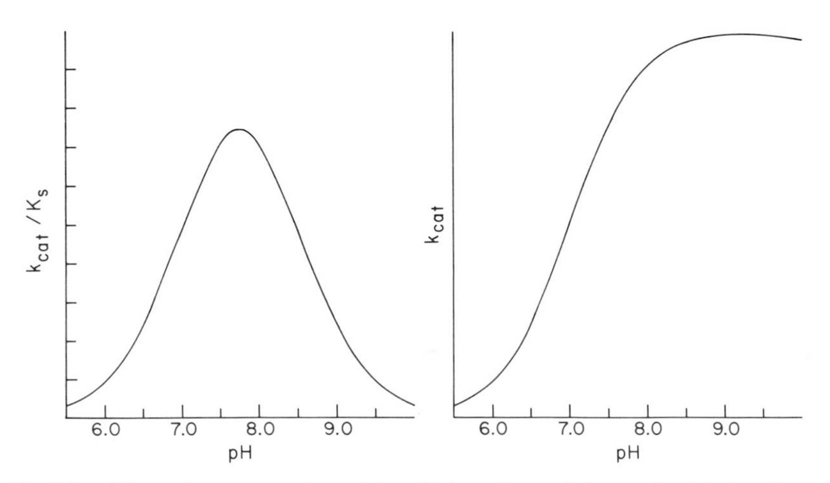

Fig. 7-8. Schematic representation of the pH dependence of the steady state kinetic parameters for the hydrolysis of esters and amides by α-chymotrypsin.

constants. For this mechanism

$$K'_S = K'_{S0} \frac{1 + [H^+]/K_1 + K_2/[H^+]}{1 + [H^+]/K'_1 + K'_2/[H^+]}$$

$$k_2 = k_{20}/(1 + [H^+]/K'_1 + K'_2/[H^+])$$

$$k_3 = k_{30}/(1 + [H^+]/K''_1 + K''_2/[H^+])$$

$$k_{cat} = k_2 k_3/(k_2 + k_3)$$

$$K_S = k_3 K'_S/(k_2 + k_3)$$

$$k_{cat}/K_S = (k_{20}/K'_{S0})/(1 + [H^+]/K_1 + K_2/[H^+])$$

As expected the pH dependence of k_{cat}/K_S is the same for esters and amides; the pH dependence of k_{cat} also is similar for esters and amides although different rate constants determine k_{cat}. The approximate pK values derived from the experimental data are p$K_1 \sim 7$, p$K_2 \sim 8.5$, p$K'_1 \sim 7$, p$K''_1 \sim 7$, and pK'_2, p$K''_2 \gg 8.5$. The two groups were postulated to be an imidazole residue (p$K \sim 7$) and an α amino group (p$K \sim 8.5$).

The necessity of an imidazole for catalysis and the identification of the specific residue as histidine-57 were derived from chemical modification studies. The use of the active site directed reagent, tosyl-L-phenylalanine chloromethyl ketone, to covalently label histidine-57 and abolish catalytic activity has been described in Chapter 2. The imidazole also can be specifically photooxidized with concurrent inactivation of enzyme. The identification of the specific α amino group was determined in an elegant experiment in which all of the free amino groups on chymotrypsinogen were acetylated (*23*). The zymogen was then activated to δ-chymotrypsin which is fully active and has one unmodified amino group, the N terminus isoleucine-16. If isoleucine-16 is acetylated, the enzyme is inactivated. This isoleucine was postulated to form a salt bridge with a carboxyl group, aspartate-194. Selective modification of aspartate-194 inactivates the enzyme, although all but two of the enzyme's carboxyl groups can be modified chemically with little loss in activity. The necessity of this salt linkage for the maintenance of structure can be seen in the x-ray structure. In addition, both inactive enzyme (high pH), acetylated enzyme, and chymotrypsinogen have the same circular dichroism and optical rotatory dispersion spectra, which are different than those of the protonated active form. The transformation between the two types of spectra is controlled by an ionizable group with a pK of 8.5. In the enzyme–substrate complexes, the salt linkage is protected so that the pK is raised substantially. Thus the α-amino group is essential for the correct structure, but is not involved in the catalysis. The imidazole, however, is probably involved in acid–base catalysis.

Chemical modification studies also have shown that serine-195 is essential for catalysis. (Hence the name serine proteases.) This residue is specifically and irreversibly labeled by diisopropylphosphofluoridate (DIPF) (*36*).

$$(CH_3)_2CH-O-\overset{\overset{\displaystyle O}{\|}}{\underset{\underset{\displaystyle F}{|}}{P}}-O-CH(CH_3)_2$$

The resultant phosphorylated enzyme is inactive. This serine has been labeled with a variety of other reagents leading to inactivation. Furthermore, to show inactivation is not blockage of the catalytic site by the groups attached to the enzyme, serine-157 was specifically dehydrated (*37*); the resultant enzyme is inactive. The kinetics of the inactivation of α-chymotrypsin by DIPF has been studied with the stopped flow method over a wide range of pH (*38*). The reaction was monitored through absorbance changes at 290 nm, hydrogen ion liberation, and enzyme inhibition. The minimal mechanism for the reaction is

$$E + DIPF \rightleftharpoons X_1 \rightleftharpoons X_2 \longrightarrow E\text{-}DIP + H^+ + F^- \qquad (7\text{-}16)$$

The first step is very rapid, the second step is a conformational change that causes the absorbance change at 290 nm, and the final step is formation of the covalent bond. Essential ionizable groups with pK values of ~6.7 and ~9 can be derived from the pH dependence of the kinetic parameters. These are identified with histidine-57 and isoleucine-16.

The elementary steps associated with the reaction of specific substrates and α-chymotrypsin have been studied using stopped flow and temperature jump methods (*23*). The binding of substrates to the enzyme is accompanied by absorbance changes at 290 nm. Finally, furoylacryloyl

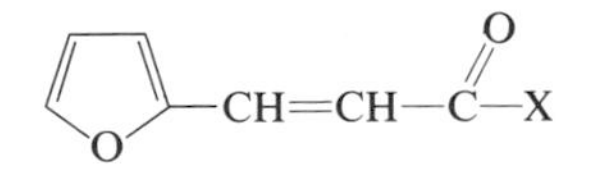

was linked to the amino group of substrates, and the absorption spectrum of this chromophore changes when the substrates interact with enzyme. For specific ester substrates, reaction with the enzyme was accompanied by an initial rapid increase in the absorbance at 290 nm. This change was complete in less than 3 msec and is due to the initial formation of the enzyme–substrate complex in Eq. (7-8); the kinetics of this process can be studied with the temperature jump method. A relatively slower increase in 290 nm absorbance occurs due to formation of EP_2, which is determined by the values of k_2 and k_3. (The observed rate constant is $k_2[S_0]/([S_0] + K_S') + k_3$.) The steady state then is established, which produces no absorbance change at 290 nm, and finally the absorbance decreases as the steady-state intermediate decays. The values of k_2, k_3, and K'_S obtained for several ester substrates are given in Table 7-3. Similar studies with amides demonstrated that $K_S \simeq K'_S$, and that an enzyme–substrate intermediate was formed prior to acyl enzyme formation. For example, the mechanism and associated rate constants for the hydrolysis of *N*-furylacryloyl-L-tryptophan amide at pH 6.7, 15°C determined by stopped flow and temperature jump methods is (*23*)

$$E + S \underset{10^4\ sec^{-1}}{\overset{6 \times 10^7\ M^{-1}\ sec^{-1}}{\rightleftharpoons}} X_1 \underset{30\ sec^{-1}}{\overset{1.5\ sec^{-1}}{\rightleftharpoons}} X_2 \xrightarrow[0.043\ sec^{-1}]{P_1} X_3 \xrightarrow{\sim 50\ sec^{-1}} E + P_2$$

The reaction of chymotrypsin with *N*-acetyl-L-phenylalanine-*p*-nitroanilide also has been studied in aqueous dimethylsulfoxide at −90°C (*39*). Four reactions were observed prior to rate determining acylation. These are attributed to formation of the initial enzyme–substrate complex, two pH-independent conformational changes, and a pH-dependent conformational change associated with the ionization state of histidine-57. Some of the observed changes may be due to solvation processes, but cryoenzymology may provide additional insight into the catalytic mechanism.

Table 7-3

Kinetic Parameters for Chymotrypsin Catalysis of Ester Hydrolysis (*23*)

Substrate	k_2 (sec^{-1})	k_3 (sec^{-1})	K'_S (mM)
N-acetyl-L-Trp ethyl ester	35	0.84	2.1
N-acetyl-L-Phe ethyl ester	13	2.2	7.3
N-acetyl-L-Tyr-ethyl ester	83	3.1	18.
N-acetyl-L-Leu-methyl ester	3.2	0.19	93.
N-furoylacryloyl-L-Tyr ethyl ester	53	1.5	0.7

The three-dimensional structure of chymotrypsin is shown in Fig. 7-9, and a localized view of a complex of chymotrypsin with formyl-L-tryptophan is illustrated in Fig. 7-10. A well-defined binding pocket for the aromatic side chains of the specific substrates of the enzyme is present; it is lined with nonpolar side chains of amino acids so that it is very hydrophobic. A short range of antiparallel β sheet is available for the interaction of up to three amino acids coupled to the N terminus of the aromatic amino acid. A hydrophobic site also is available for an amino acid on the C terminus of the aromatic amino acid; this probably prevents the C terminus of proteins

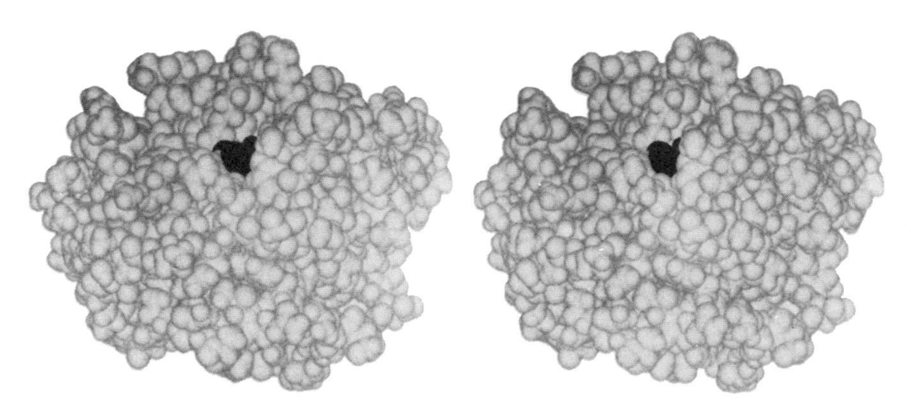

Fig. 7-9. The three-dimensional structure of chymotrypsin. The active site pocket is at the top of the structure; the black residues are His-57 and Ser-195. Stereo computer simulation courtesy of Dr. R. J. Feldmann, National Institutes of Health.

from binding to the enzyme, making chymotrypsin an endonuclease. Serine-195 and histidine-57 are in the positions anticipated for their critical roles in catalysis, and the salt linkage between isoleucine-16 and aspartate-194 in a hydrophobic region can be seen. A triad of amino acids, serine-195, histidine-57, and aspartate-102, is observed in the active site region, and similar triads are observed for other serine proteases (cf. *21*). This triad originally was postulated to exist as a "charge relay" system whereby hydrogen bonding between imidazole and aspartate reversed the p*K* values of the two ionizable groups, with the entire system becoming a more effective nucleophile because of this special structure

$$-C\langle^{O}_{O} - \cdots H-N \quad N \cdots H-O$$

Enhanced catalysis was not found in model systems with very similar structures (*40*). Recent ^{15}N-nmr studies show that the histidine, in fact, has a quite normal p*K* (~ 7) (*41*), and the above hydrogen bonding scheme is not borne out by refined x-ray maps (*42*). The lesson to be learned from this case is that the chemistry utilized by enzymes is not different from that found in ordinary chemical reactions; enzymes simply optimize the reaction conditions and thereby are spectacularly efficient.

A mechanism for the chymotrypsin reaction is presented in Fig. 7-11. Nucleophilic attack of the serine hydroxyl on the substrate is facilitated by the imidazole acting as a general base catalyst. Formation of a tetrahedral intermediate almost certainly occurs, although this species and its formation

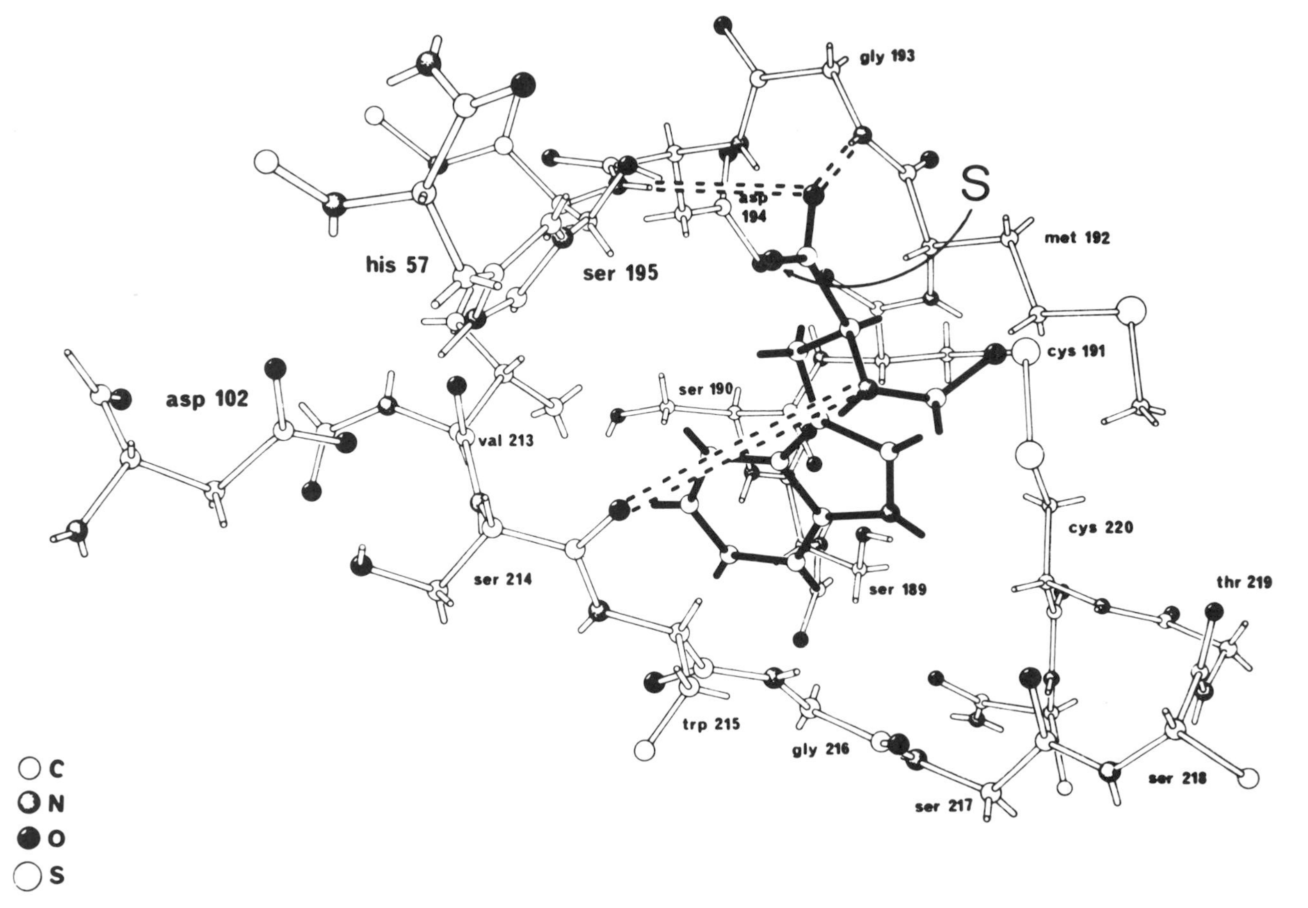

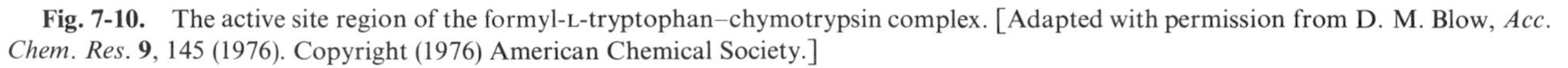

Fig. 7-10. The active site region of the formyl-L-tryptophan–chymotrypsin complex. [Adapted with permission from D. M. Blow, *Acc. Chem. Res.* **9**, 145 (1976). Copyright (1976) American Chemical Society.]

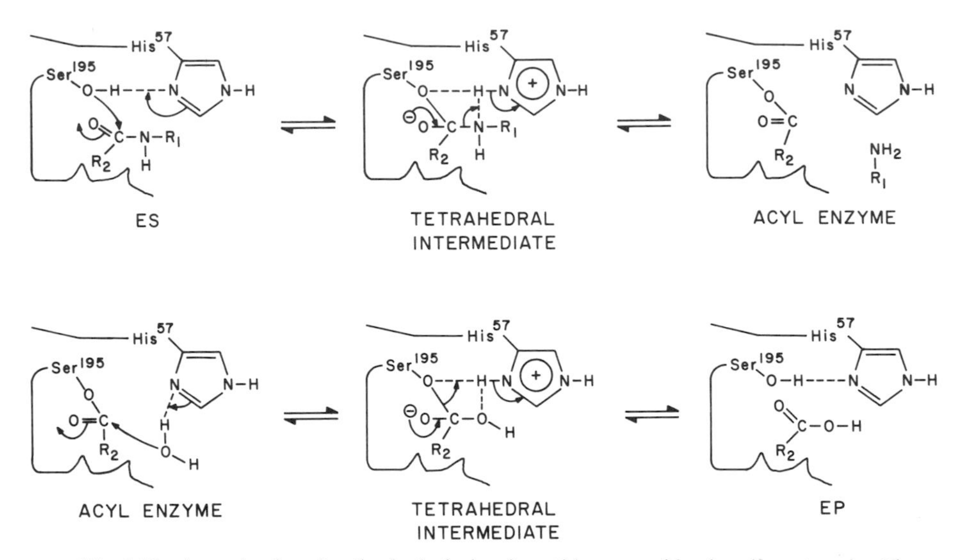

Fig. 7-11. A mechanism for the hydrolysis of peptides or amides by chymotrypsin. The imidazole acts as a general base to assist the nucleophilic attacks of serine and water on the substrate and acyl enzyme, respectively.

and breakdown have not yet been directly observed. One or more conformational changes (not shown) occur before formation of the acyl enzyme. Hydrolysis of the acyl enzyme occurs by a similar mechanism except that water now is the nucleophile. As with ribonuclease, the use of conformational changes to precisely orient the substrate and to create a hydrophobic environment appears to play an important role in catalysis.

LYSOZYME

Although lysozyme was discovered in 1922, a detailed study of its mechanism of action was not undertaken until the 1960s. The impetus for revival of interest in lysozyme was the crystallization of the hen egg white enzyme and determination of its three-dimensional structure (*43–45*), the first ever of an enzyme. For most enzymes, careful chemical and physical studies preceded knowledge of the three-dimensional structure, but in the case of lysozyme the situation was reversed. The reason for the early lack of interest in lysozyme was that it catalyzes a reaction with ill-defined substrates, the hydrolysis of bacterial cell wall polysaccharides. The basic repeating unit of bacterial cell walls is the disaccharide *N*-acetylglucosamine-*N*-

acetylmuramic acid in a β-1,4 linkage

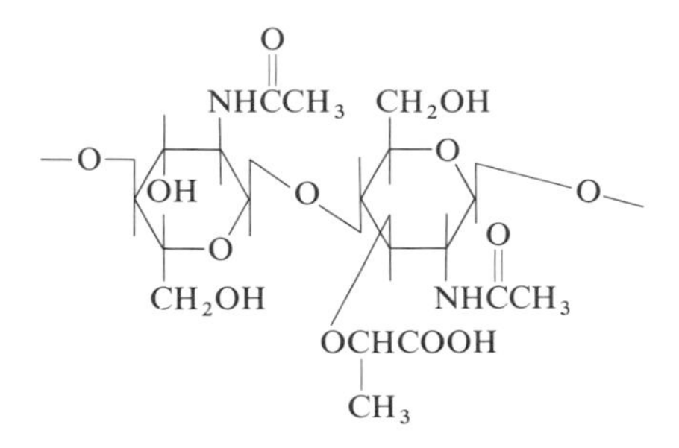

With this substrate, lysozyme acts as an *N*-acetylmuramidase transferring the *N*-acetylmuranyl group to water. However, the enzyme cleaves chitin (β-1,4-poly-*N*-acetylglucosamine) and model substrates of well-defined structure. In addition, transfers to acceptors other than water are catalyzed. Unlike enzymes such as ribonuclease and chymotrypsin, model substrates are difficult to obtain: for example, the smallest *N*-acetylglucosamine oligosaccharide appreciably hydrolyzed is the hexamer. This problem considerably delayed interest in lysozyme, but this delay has been more than compensated for (cf. *22, 44, 45*). A variety of model substrates has been developed; all require a β configuration, but the *N*-acetylglucosamine (GlcNAc) and *N*-acetylmuramic acid side chains are not essential.

Hen egg white lysozyme has 129 amino acid residues and contains four disulfide residues. The amino acid sequence is known (*45*), and a representation of the three-dimensional structure is shown in Fig. 7-12. The enzyme is very compact with a groove for the substrate. It contains three extensive helical regions (residues 5–15, 24–34, 88–96) and two pleated sheet regions (residues 41–45, 50–54). The three-dimensional structure of the enzyme with bound $(GlcNAc)_3$ also was determined. A very small conformational change accompanies binding (shifts of less than 1 Å in the enzyme structure), and the oligosaccharide binding is stabilized by formation of six hydrogen bonds with the enzyme plus hydrophobic interactions. The three sugar binding sites are called the ABC sites. The hexasaccharide $(GlcNAc)_6$ is cleaved between residues 4 and 5, and model building permitted postulation of three additional sugar sites on the enzyme (DEF). In the structure derived from model building, the sugar residue D is distorted from its normal chain configuration due to steric crowding. This steric strain was postulated to be an important factor in the catalytic mechanism (Chapter 6). The only ionizable groups (with pK values in a reasonable range) near the bond to be cleaved are aspartate -52 and glutamate -35, and a mechanism involving their participation was postulated. We now shall see how these early predictions were borne out by subsequent studies.

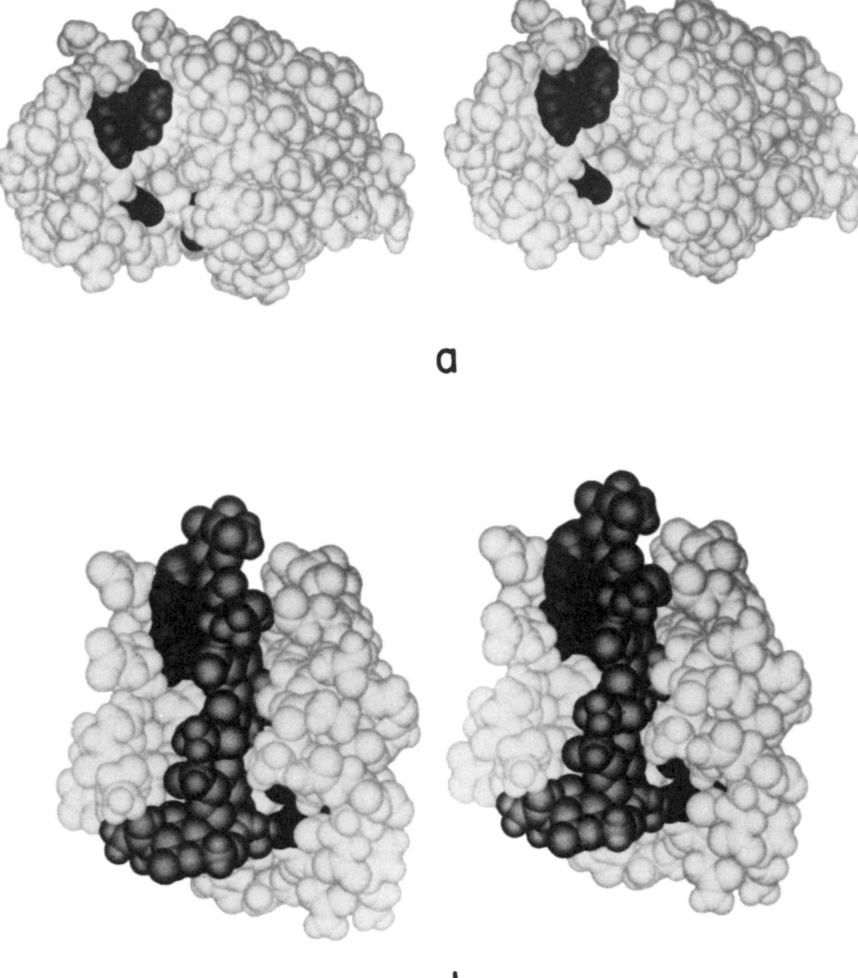

Fig. 7-12. The three-dimensional structure of lysozyme (a) and the active site region of the complex of lysozyme with $(GluNAc)_6$(b,c). The former is the structure determined by x-ray diffraction, while the latter has been calculated by free energy minimization procedures. In (a) the active site groove runs from top to bottom. The black residues at the top are Trp-62 and -63; the black residue on the bottom left is Asp-52 and on the bottom right is Glu-35. In (b), Trp-63 and Asp-52 are obscured by the substrate. The stereo computer simulations in a and b were provided by Dr. R. J. Feldmann, National Institutes of Health. The structures in b and c are adapted with permission from M. R. Pincus and H. A. Scheraga, *Macromolecules* **12**, 633 (1979). Copyright (1979) American Chemical Society. The enzyme subsites in c are designated by A–F.

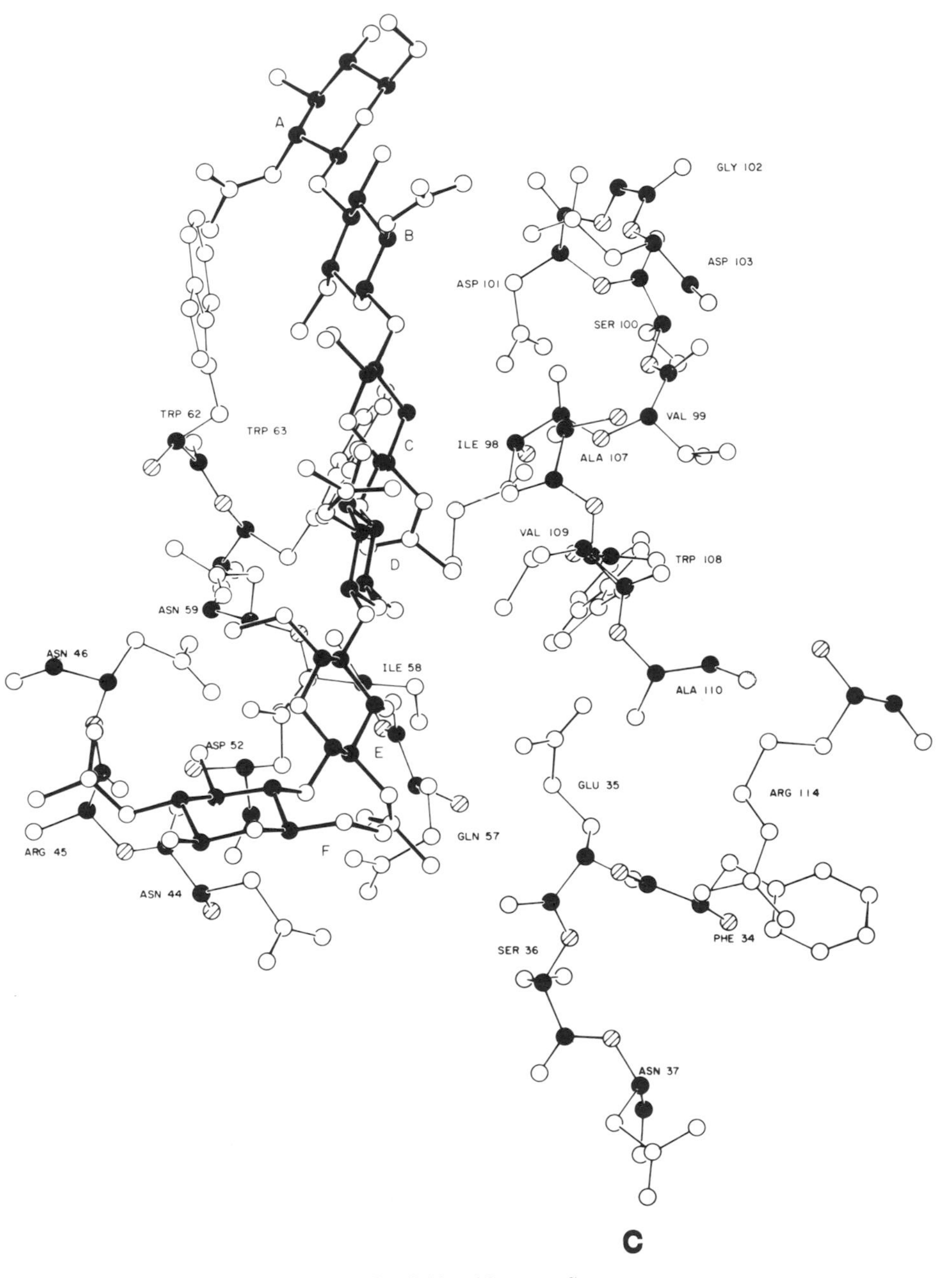

Fig. 7-12. *(Continued)*

Many chemical modification studies of lysozyme have been carried out (*45*). The six ε amino groups have been modified with a variety of reagents: for example, acetylation of all six results in an active enzyme with reduced substrate affinity and a shifted pH profile. This and other results clearly indicate lysine residues are not essential for catalytic activity. The single histidine also is not essential since alkylation does not inactivate the enzyme, and the three tyrosines and two methionines have been modified without substantially altering the catalytic activity. The enzyme has six tryptophan residues, and three are at the active site and participate in substrate binding. Oxidation of the tryptophans with *N*-bromosuccinimide inactivates the enzyme. Tryptophan-62 is converted to oxindolealanine, and this specifically modified enzyme has a greatly reduced activity (*45,46*). Apparently, the tryptophans are important for maintenance of the correct conformation, but are not essential for catalysis. If tryptophan-108 is specifically oxidized with I_2, the enzyme is inactivated, but this was shown to occur because an indoyl ester with glutamate-35 is formed. If all of the carboxyl groups are esterified, the enzyme is inactivated. If all carboxyl groups except aspartate-52 and glutamic acid-35 are modified, the enzyme is still active; this derivitization is possible because the binding of inhibitors and substrates to the enzyme prevents modification of aspartate-52 and glutamic acid-35. If aspartate-52 is esterified, the enzyme still binds inhibitors but is catalytically inactive; aspartate-52 also is inactivated by an epoxide aglycone affinity label (*47*)

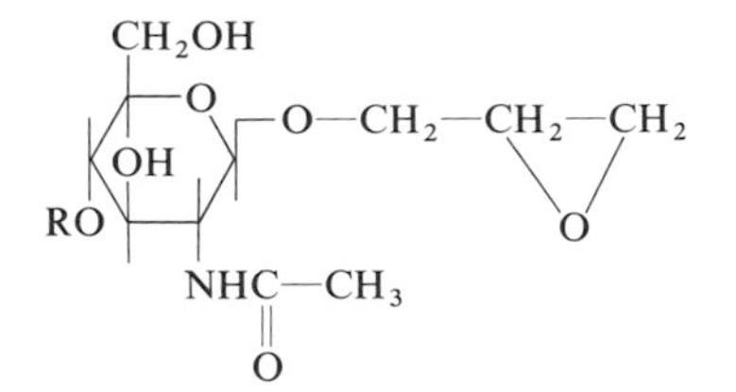

Thus, the two carboxyl groups are essential for activity. The results of chemical modification of the protein are gratifyingly consistent with the three-dimensional structure of the enzyme.

The first systematic study with model substrates was carried out by Rupley and Gates (*48*) who studied the cleavage pattern and rates of hydrolysis of a series of GlcNAc oligosaccharides. The results obtained are summarized in Table 7-4. The hexamer is cleaved only at one bond and is hydrolyzed much more rapidly than smaller oligosaccharides. The smaller oligosaccharides (trimer and pentamer) have more than one cleavage site suggesting more than one binding mode in the six subsites (ABCDEF) of the enzyme. Surprisingly the Michaelis constants are about the same for all of the oligosaccharides. The initial explanation for this was that a large amount of unproductive binding occurs in which subsites D and E are not occupied, thereby pre-

Table 7-4

Hydrolysis Rates and Cleavage Patterns of *N*-Acetylglucosamine Oligosaccharides Catalyzed by Lysozyme (*48*)

Oligosaccharide	Cleavage positions	Relative rate
Dimer		0.003
Trimer	$X_1 \overset{\downarrow}{—} X_2 \overset{\downarrow}{—} X_3$	1
Tetramer	$X_1 — X_2 — X_3 \overset{\downarrow}{—} X_4$	8
Pentamer	$X_1 — X_2 — X_3 \overset{\downarrow}{—} X_4 \overset{\downarrow}{—} X_5$	4,000
Hexamer[a]	$X_1 — X_2 — X_3 — X_4 \overset{\downarrow}{—} X_5 — X_6$	30,000

[a] $k_{cat} = 0.25\ sec^{-1}$.

venting hydrolysis; however, this was later shown not to be the case. The pH dependence of oligosaccharide binding implicates two ionizable groups on the enzyme with p*K* values of 6.3–6.7 and 3.5–4.2. These were tentatively ascribed to glutamic acid-35 and aspartic acid-52, respectively. An ionizable group with a very low p*K* (1.2–1.8) also was suggested to be of importance and may be aspartate-101 or -103. Similar p*K* values have been derived from other steady-state kinetic studies (*45*).

Further insight into the binding of oligosaccharides was sparked by the discovery that the dye biebrich scarlet

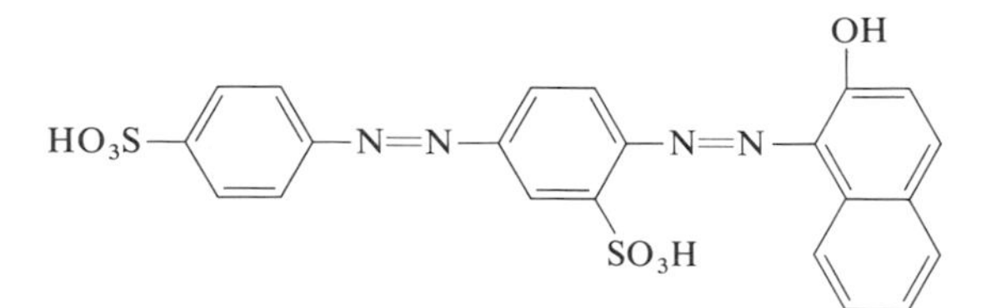

binds specifically to the F site (*49*). Productive binding of the hexose requires displacement of the dye so that various modes of unproductive and productive binding can be distinguished. Some of the results are summarized schematically in Fig. 7-13. About one-half of the binding is productive; in addition, the results obtained suggest the pyranose ring is not distorted or strained in the binding process. The kinetics of the binding of biebrich scarlet to the enzyme were studied with the temperature jump method. Biebrich scarlet binds in a two-step process typical of enzyme–substrate reactions (Eq. 7-4). Two relaxation times are observed, and the bimolecular rate constant obtained from the concentration dependence of the shorter relaxation time is $6 \times 10^6\ M^{-1}\ sec^{-1}$. The first order rate constants for the isomerization following binding are about $10^3\ sec^{-1}$. A combination of stopped flow and temperature jump methods was used to study the kinetics

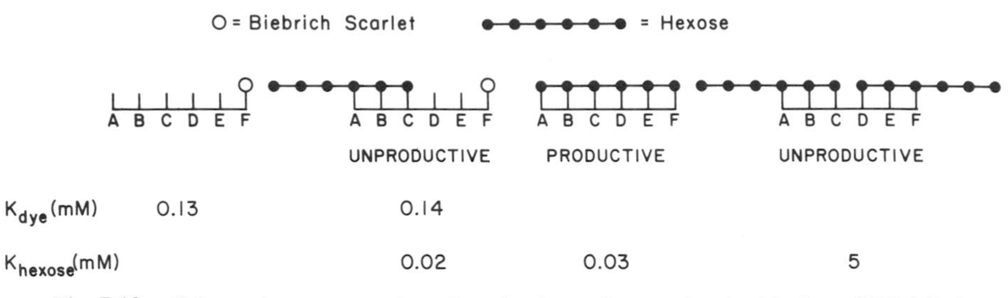

Fig. 7-13. Schematic representation of productive and unproductive binding of $(GlcNAc)_6$ and biebrich scarlet to lysozyme with associated dissociation constants. [Adapted with permission from E. Holler, J. A. Rupley, and G. P. Hess, *Biochemistry* **14**, 1088 (1975). Copyright (1975) American Chemical Society.]

of binding of tri-, tetra-, and hexa-GlcNAc to lysozyme (*50*). Two distinct binding processes were observed: one was found with all three oligosaccharides and was attributed to nonproductive binding; the other was found only with the hexasaccharide and was attributed to binding on the reaction pathway. As shown in Fig. 7-14, the single observed rate constant associated with each process reaches a limiting value at high saccharide concentrations, which indicates a conformational change following a rapid bimolecular reaction is occurring. Two slower processes are observed along the reaction pathway, following the conformational change. The suggested mechanism, and the rate constants for the hexasaccharide interaction with lysozyme at pH 6.3 and 25°C are shown in Fig. 7-15. This breakdown of the reaction mechanism into elementary steps again illustrates the multiple conformations of enzyme–substrate complexes.

The binding of various saccharides to lysozyme also has been studied with difference spectroscopy and nuclear magnetic resonance by Raftery and coworkers (cf. *45, 51, 52*). Chemical shifts accompany binding, and evidence was found for three sugar binding sites (ABC), with the reducing end of the pyranose always directed toward C. Very little difference was found between α and β anomers. The pK values of ionizable groups associated with the binding process were found to be 6.1, 4.2, and 4.7. The first is probably glutamate-35; one of the latter two is probably aspartate-52, and the other aspartate-101 or -103. Acid–base titrations of lysozyme with bound saccharides and/or aspartate-52 esterified suggest pK values of 4.5 and 5.9 for aspartate-52 and glutamate-35. All evidence points to glutamate-35 having an abnormally high pK value for a carboxyl group. Since the three-dimensional structure of the enzyme shows glutamate-35 to be in a very hydrophobic environment, this is not unreasonable.

We now return to a more detailed consideration of the mechanism of action of lysozyme. The finding that model building predicted a strained sugar structure in subsite D where hydrolysis occurs was acclaimed as proof

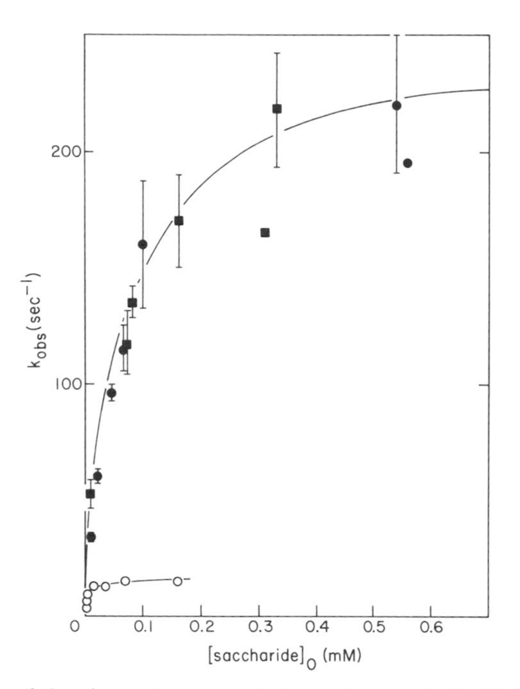

Fig. 7-14. Plots of the observed rate constant associated with the binding of $(GluNAc)_6$ to lysozyme versus the concentration of $(GluNAc)_6$. The upper curve is the rate constant for unproductive binding, while the lower curve is the rate constant for productive binding. [Adapted with permission from E. Holler, J. A. Rupley, and G. P. Hess, *Biochemistry* **14**, 2377 (1975). Copyright (1975) American Chemical Society.]

$$EU_2 \underset{20\,sec^{-1}}{\overset{240\,sec^{-1}}{\rightleftharpoons}} EU_1 \overset{8\times10^{-5}\,M}{\rightleftharpoons} E + S \overset{11\times10^{-5}\,M}{\rightleftharpoons} ES_1 \underset{4\,sec^{-1}}{\overset{11\,sec^{-1}}{\rightleftharpoons}}$$

$$ES_2 \underset{0.06^{-1}}{\overset{\leq 0.06\,sec^{-1}}{\rightleftharpoons}} ES_3 \xrightarrow{0.01\,sec^{-1}} E + P$$

Fig. 7-15. Elementary steps and rate constants associated with the binding of $(GlcNAc)_6$ to lysozyme (pH 6.3, 25°C). The EU_i are unproductive complexes, and the ES_i are complexes on the catalytic pathway. [Data are from E. Holler, J. A. Rupley, and G. P. Hess, *Biochemistry* **14**, 2377 (1975).]

of the importance of strain in enzyme catalysis. This was enhanced by the synthesis of the substrate analog

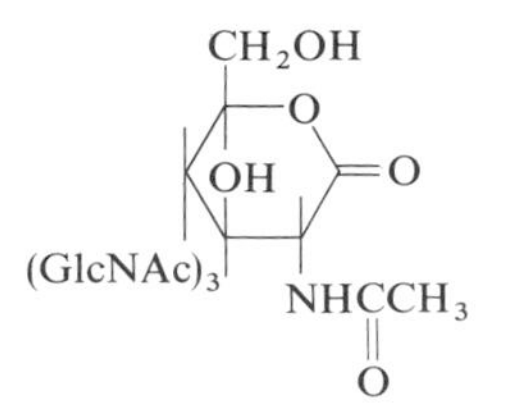

which binds 3000 times tighter to the enzyme than the substrate (*53*). However, the binding and kinetic studies with oligosaccharides provide no evidence for a strained structure. A further test of this idea was made by studying the nuclear magnetic resonance of

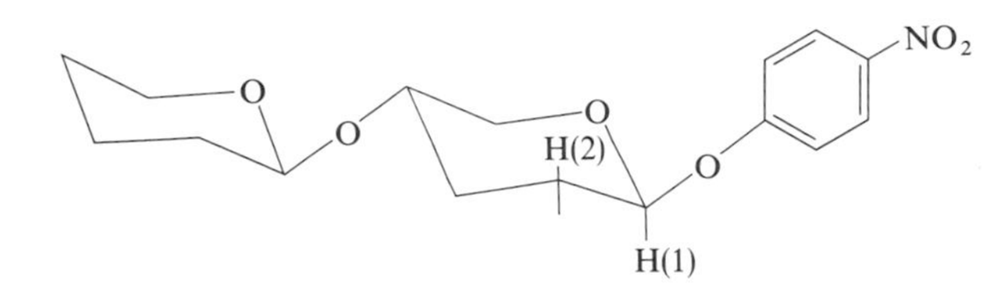

bound to the enzyme (*54*). Fast exchange was found to occur, and the spin–spin coupling constant between H(1) and H(2) was measured. This depends on the dihedral angle between them and should change if a flattening occurs. However, no change in the coupling constant accompanies binding so either flattening does not occur or no binding to the correct subsite occurs. Model systems were devised to test the importance of strain (*55*). However, for example, the unstrained planar system

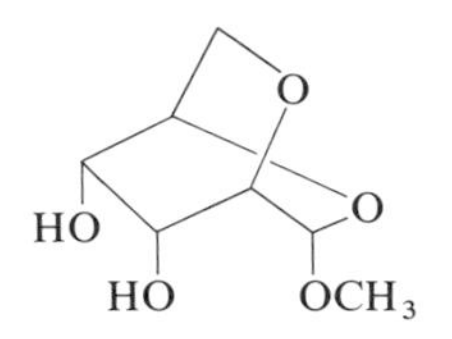

is not particularly unstable and general acid catalysis is not observed. The structure of the $(GlcNAc)_6$–lysozyme complex has been calculated by minimizing the free energy of the system (*56*). The binding to the ABC sites was consistent with that found by x-ray crystallography. However, the most favorable (lowest free energy) structure for binding to the DEF sites was slightly different than that found in the original model building. The final structure has no strained sugar residues and is shown in Fig. 7-12.

The absence of strain in saccharide binding also has been shown by binding studies with a variety of synthetic substrates with varying side chains attached to the sugars (*57*). These studies also indicate the flattened lactone mentioned above binds only 30 times more tightly to the enzyme if a comparison with the correct substrate analog is made. Thus, the current situation appears to be that no evidence exists for significant strain being introduced into the substrate by binding to lysozyme. This, of course, does not prove that strain does not occur during enzyme catalysis, but it remains an open question.

The chemical mechanism proposed for the lysozyme reaction is shown in Fig. 7-16. The carboxyl group of glutamate-35 serves as a general acid catalyst to promote the formation of a carbonium ion intermediate, which is

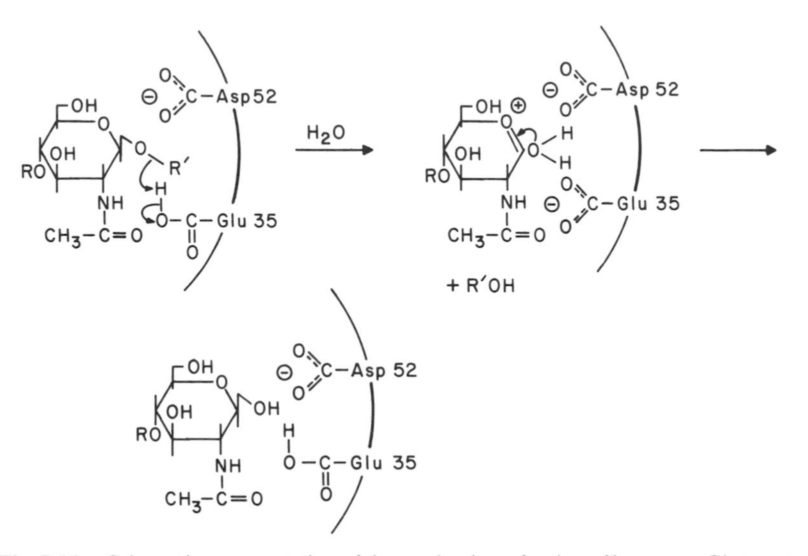

Fig. 7-16. Schematic representation of the mechanism of action of lysozyme. Glutamate-35 serves as a general acid catalyst for the formation of the intermediate carbonium ion, which then dissociates to product. The role of aspartate-52 is uncertain; it may assist in stabilizing the carbonium ion.

stabilized by an electrostatic interaction with the carbonyl group of aspartate-52. The carbonium ion then collapses to form the final product. This mechanism is consistent with the overall retention of substrate configuration and with the secondary isotope effect observed with disaccharides of the type (*58*)

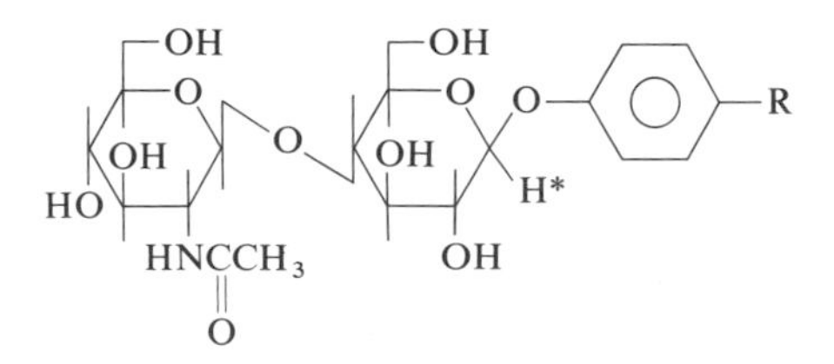

The atom marked by an asterisk was either deuterium or hydrogen, and the observed isotope effect was $k_H/k_D = 1.11$. In model studies, an isotope effect of 1.13 was observed for a reaction known to involve carbonium ion formation and an effect of 1.03 was observed for a reaction involving base catalysis. Thus, the conclusion was reached that general acid catalysis occurs.

Many model studies have been carried out to examine the catalytic mechanism of lysozyme (*55*). For example, the hydrolysis reaction

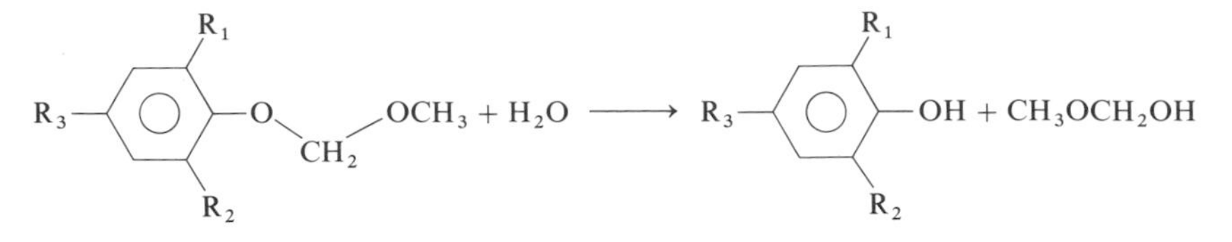

has been studied. If no ortho carboxyl is present, only specific hydrogen ion catalysis is found. However, with an ortho carboxyl, the observed rate constant is

$$k_{obs} = k_H a_H[a_H/(K + a_H)] + k' a_H[K/(K + a_H)]$$

where a_H is the hydrogen ion activity and K is the ionization constant of the carboxyl group. The rate constant k_H represents the specific acid catalysis of the undissociated species, whereas k' represents specific acid catalysis of the dissociated species. As previously discussed (Chapter 5), the second term in k_{obs} also can represent general acid catalysis (k_{ga}) with $k'K = k_{ga}$. Through a study of the substituent effects (R_3), the conclusion was reached that general acid catalysis occurs with electrostatic stabilization of the intermediate

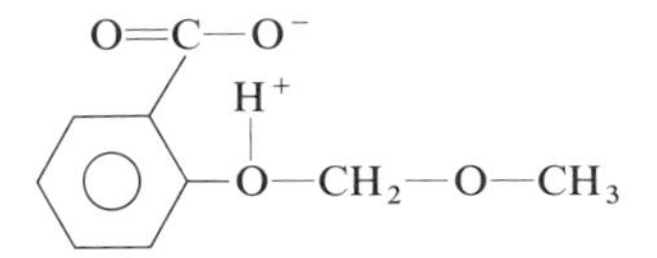

The rate enhancement of intramolecular to intermolecular catalysis was substantial, 350–10,000 *M*. If two carboxyl groups are ortho, an additional rate enhancement occurs, but the pH rate profile is insensitive to ionization of the second carboxyl group. Since acceleration also is observed with $R_1 = NO_2$ and $R_2 = COOH$, the observed enhancement can be attributed to steric compression. These results indicate that general acid catalysis can occur, but no evidence for stabilization of the intermediate by a second carboxyl group is observed.

The possible stabilization of a carbonium ion by a neighboring carboxyl group was investigated by studying vinyl ether hydrolysis of (*59*)

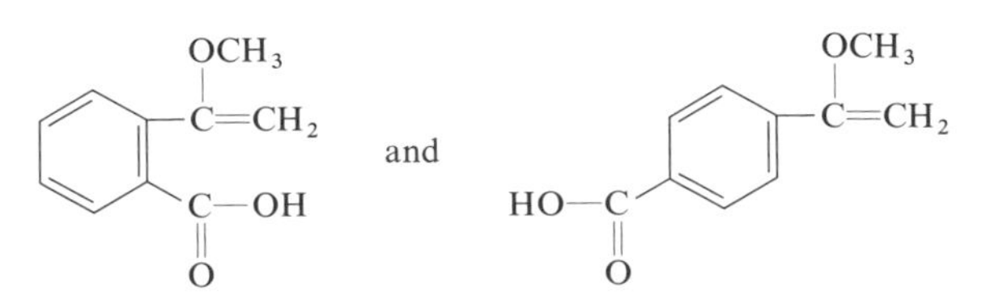

The enhancement of general acid catalysis in the ortho compound relative to the para is very small. If the carboxyl group is put in a hydrophobic environment with limited solvent accessibility such as with (*60*)

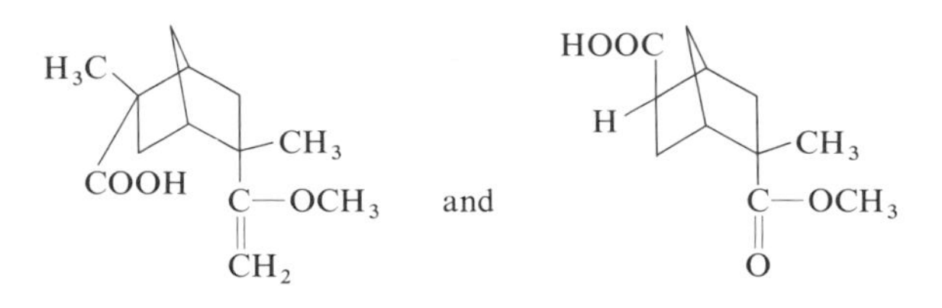

a large rate enhancement of the acid-catalyzed pathway is not found with the cis compound. Thus, the most likely role of aspartate-52 appears to be structural rather than direct involvement in catalysis. However, general acid catalysis by glutamate-35 and formation of a carbonium ion intermediate are consistent with both enzymatic and model studies. The specific interactions of oligosaccharides with the enzyme and the conformational changes associated with the reaction also are firmly established, providing a good picture of the overall catalytic process.

CREATINE KINASE

As a final example, the mechanism of action of creatine kinase is discussed (cf. *61*). This enzyme catalyzes the reaction

$$\text{MgATP} + \text{creatine} \rightleftharpoons \text{MgADP} + \text{phosphocreatine} + \text{H}^+ \qquad (7\text{-}17)$$

The equilibrium constant for this reaction at 30°C is

$$K_{\text{eq}} = \frac{[\text{MgADP}][\text{phosphocreatine}][\text{H}^+]}{[\text{MgATP}][\text{creatine}]} = 2.8 \times 10^{-10}\ M$$

Therefore, at neutral and slightly alkaline pH values, both the forward and reverse reactions can be easily studied. The enzyme can be obtained from many sources, the most common being rabbit skeletal muscle. The properties of the rabbit muscle enzyme are considered here. The enzyme has a molecular weight of 82,600 and consists of two identical dissociable polypeptide chains with no disulfide bridges. Enzymes from other sources, for example, ox brain, have polypeptide chains that can be distinguished electrophoretically from muscle enzymes. The structures are designated as BB and MM, respectively, and MB hybrids also are found. Hydrodynamic studies indicate the enzyme is highly asymmetric with an axial ratio of 4.4 and that the dimer can be viewed as two end-to-end cigars. Unlike the enzymes previously discussed, only a very limited amount of amino acid sequence data is available, and the crystal structure is not yet known. Therefore, the mechanism of action of this enzyme must be pieced together without knowledge of the enzyme structure.

The substrate specificity has been examined extensively. For example, if creatine is represented as

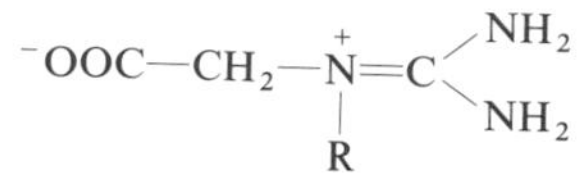

$$^{-}OOC{-}CH_2{-}\overset{+}{\underset{\displaystyle R}{\underset{|}{N}}}{=}C\begin{matrix} \diagup NH_2 \\ \diagdown NH_2 \end{matrix}$$

then increasing the chain length of R ($R = CH_3$ for creatine) decreases the maximum velocity and increases the Michaelis constant. Also the use of model substrates indicates the nitrogen *trans* to the methyl group is phosphorylated and that the planarity of the guanido group is very important. The data suggest that the methyl group may be important in orienting the guanido group properly. The creatine binding site can be viewed as a narrow slot with the primary amino group trans to the *N*-methyl group precisely aligned toward the incoming phosphoryl group of the nucleotide. The nucleotide specificity does not appear to be great as variations in the sugar structure have little effect on catalytic activity, and the base specificity is very broad. For example, ADP, IDP, CDP, UDP, and GDP are all good substrates. A metal ion activator is required for creatine kinase: Mg^{2+}, Mn^{2+}, Ca^{2+}, and Co^{2+} all are quite effective, whereas Ba^{2+}, Sr^{2+}, Be^{2+}, Ni^{2+}, Cr^{2+}, Cd^{2+}, and Zn^{2+} are inactive or inhibitory. (Ba^{2+} and Sr^{2+} may be very weak activators.)

The enzyme has been modified with many different chemical reagents (*61*). If the enzyme is alkylated with iodoacetamide, two groups react at the same rate and the enzyme is completely inactivated. Since the enzyme contains two identical subunits, this finding was interpreted as the modification of sulfhydryl groups at two catalytic sites that are essential for catalysis. However, a partially active enzyme is obtained if the sulfhydryl group is methylated, and nmr studies of the modified enzyme suggest that modification of the sulfhydryl group restricts the conformation of the enzyme (*62*). In the case of modification with iodoacetamide, the enzyme cannot attain the conformation necessary for catalysis. Peptide fragments containing the modified sulfhydryl group have been isolated and sequenced (*61*), but the peptides may not be part of the catalytic site. As expected, the presence of substrates greatly inhibits the reactivity of the thiols. A second pair of sulfhydryl groups that are not important for catalytic activity also have been modified (*63*).

If lysine is specifically acetylated or dansylated, the catalytic activity is inhibited; two lysines per molecule are labeled. By the use of radioactive labels and peptide analysis of the digested enzyme, the lysine was shown not to be adjacent to the critical thiol. However, the lysine and thiol can be cross-linked by dinitrodifluorobenzene. An equilibrium mixture of substrates protects the critical lysines, apparently through a conformational change that raises the pK of the amino group. If the enzyme is reacted with diethylpyrocarbonate, two histidines per molecule are labeled, and the enzyme is inactivated. However, the substrate binding properties and conformational changes of the enzyme are not altered by modification. The modification of arginine residues, one per polypeptide chain, by a variety of reagents leads to inactivation, and this residue has been implicated in nucleotide binding (*61,64*). Elegant confirmation of an arginine residue at the catalytic site comes from nmr experiments in which the resonances of protons on the nucleotide ring are specifically irradiated with a magnetic field; this induces resonances in nearby protein nuclei (Nuclear Overhauser Effect). The protein nuclei excited can be attributed to arginine and an aromatic residue (*65,66*). The quenching of tryptophan fluorescence by nucleotides suggests a tryptophan residue is nearby (*66*). Finally, an epoxycreatine has been utilized as an active site directed reagent; it forms a covalent adduct with a glutamate or aspartate carboxyl near the catalytic site (*67*).

To summarize, lysine, arginine, histidine, and aromatic residues appear to be near the catalytic site and of importance in catalysis and/or substrate binding. A sulfhydryl residue plays a crucial role in the protein conformation and may be near the catalytic site.

Steady-state kinetic studies indicate that the enzyme mechanism involves formation of a ternary complex between enzyme and substrate with the substrate equilibria being adjusted rapidly relative to the interconversion of intermediates (*61*). Metal nucleotides are the true substrates, and Mg^{2+} binds very weakly to the enzyme. The addition of substrates is random for MgADP and phosphocreatine, but a preference is shown for MgATP to bind before creatine (*68*). This mechanism is confirmed by isotope exchange experiments at equilibrium in which the rate of exchange between creatine and phosphocreatine is approximately equal to the rate of exchange between ADP and ATP. Also equilibrium binding experiments indicate all four substrates can bind to the enzyme independently (*61*). The approximate binding constants are 1 mM for MgATP, 0.2 mM for MgADP, 50 mM for creatine, and 45 mM for phosphocreatine. Finally, studies using quenched flow indicate the phosphoryl transfer step is rate limiting in both directions (*69*). Steady-state kinetic studies are complicated by the formation of the dead-end complex enzyme–MgADP–creatine. The inhibition by this complex is enhanced by certain anions, particularly nitrate. The dead-end complex with nitrate has been postulated to be a transition state analog with NO_3^- adopting the trigonal structure of PO_3^- assumed to be present in the catalytic process (*70*). The pH dependence of the steady-state kinetic parameters suggests that an unprotonated imidazole and carboxyl are necessary (*71*).

Evidence for conformational changes accompanying substrate binding comes from a variety of sources including alteration of the reactivity of groups on the protein already mentioned. Temperature jump studies of the binding of ATP, ADP, and their metal complexes indicate that a conformational change follows the initial binding process (*72*). The rate constants characterizing this conformational change are identical for all forms of the nucleotide so a metal is not crucial for the conformational change. A conformational change is not observed when creatine binds to the enzyme. Electron paramagnetic resonance studies of the binding of creatine by enzyme–MnADP indicate a conformational change occurs; nitrate ion causes an additional conformational change similar to that found with enzyme–creatine–MnATP. Measurements of the enhancement of the water proton relaxation rate (Chapter 2) indicate that Mn^{2+} does not bind directly to the enzyme; however, a strong enhancement is observed when ADP or ATP is added to Mn^{2+} and the enzyme, indicating the metal-nucleotide is the true substrate (*73*). The metal does not, however, act as a bridging ligand between enzyme and nucleotide. A stopped flow study of the reaction of creatine kinase–MgATP with creatine (Cr) down to $-15°C$ revealed a complex time course: a lag phase was followed by a "burst" and then steady-state catalysis (*74*, *75*). The results could be quantitatively interpreted in terms of the mechanism

$$\mathrm{E\cdot MgATP + Cr} \rightleftharpoons \mathrm{E\cdot MgATP\cdot Cr} \rightleftharpoons \mathrm{E^{*}\cdot MgATP\cdot Cr} \rightleftharpoons \mathrm{E\cdot MgADP\cdot PCr} \longrightarrow \mathrm{E + MgADP + PCr}$$

The conformational change following creatine binding is consistent with the nmr measurements and the usual finding of conformational changes accompanying substrate binding to enzymes. Since no other intermediates could be detected, a direct phosphoryl transfer was postulated.

The three-dimensional structure of enzyme–substrate complexes has been studied with magnetic resonance methods (*76*). Both Mn^{2+} and nitroxide acetamide attached to the critical sulfhydryl were used as paramagnetic probes, and their effect on phosphorus and proton magnetic resonances of substrates was determined. Some of the distances measured between the paramagnetic species and specific protons are summarized in Table 7-5. Both T_1 and T_2 measurements were utilized (Chapter 2). The formate anion was found to bind tightly to the enzyme–MnADP–creatine complex, and the species formed is assumed to be a transition state analog. In this work, the Mn^{2+} was postulated not to be coordinated directly to the γ-phosphate or to guanidine. However, an infrared investigation of the binding of anions (SCN^-, N_3^-, NO_3^-) to the enzyme–MgADP–creatine complex suggests the metal ion is interacting with the anion so the metal ion may be slightly

Table 7-5

Distances from Mn^{2+} or Nitroxide to Substrates on Creatine Kinase[a]

Substrate	Group	r (Å)
Phosphocreatine	CH_2	6.7
	CH_3	8.6
Creatine	CH_2	4.5
	CH_3	5.3
ADP	H(2)	4.2
	H(8)	6.0
	P_α	3.4
	P_β	2.9
ADP-creatine	CH_2	8.7
	CH_3	9.4
ATP-creatine	CH_2	8.7
	CH_3	9.4
ADP-creatine-formate	CH_2	≤ 8.8
	CH_3	≤ 8.2
ADP	H(2)	8.1
	H(8)	6.8
	H(1′)	8.1
MgADP	H(2)	8.1
	H(8)	7.2
	H(1′)	8.1
MgADP-creatine	CH_2	9.3
	CH_3	9.5
	H(2)	7.9
	H(8)	7.3
	H(1′)	7.9

[a] The spin label is a nitroxide on the critical sulfhydryl group of creatine kinase. The last three substrate entries are for the spin label and the first six substrate entries are for Mn^{2+} (*76*).

displaced toward the γ phosphorus in the transition state (*77*). Experiments utilizing Cr(III)ADP and Cr(III)ATP support the conclusion that Mg^{2+} transfers from the γ phosphate of ATP to the α and β phosphates of ADP before MgADP is released from the enzyme; the Cr^{3+} forms isomeric species with nucleotides which are not rapidly interconvertible as with Mg^{2+} (*78*). The addition of formate to the enzyme–MnADP–creatine complex alters the position of the guanido group, and although the guanido group is not liganded to Mn^{2+}, the electron paramagnetic spectrum of Mn^{2+} is strongly dependent on the guanidino substrate. The coordination of Mn^{2+}

in the enzyme–MnADP–formate complex has been examined by looking at the perturbations of the electron paramagnetic resonance spectra caused by ^{17}O in the ligands (*79*). The conclusion was reached that coordination occurs to oxygens on the α and β phosphates of ATP, to oxygen on formate, and to three water molecules. Therefore, binding to all three phosphates very likely occurs in the transition state and the coordination of the metal to an oxygen on the α phosphorus may be part of the rate limiting step. The water molecules liganded to Mn^{2+} are not in rapid equilibrium with the solvent, indicating a closing of the active site when the transition state is formed. The equilibrium constant between products and substrates on the enzyme has been measured with ^{31}P nmr and is about unity (pH 7.8, 20°C), as contrasted to about 0.1 in solution (*80*). The finding of multiple conformations, comparable amounts of reaction intermediates, and a closing of the enzyme site during catalysis is consistent with the general principles of enzyme catalysis discussed in Chapter 6.

The mechanism of the phosphoryl group transfer is, of course, a matter of speculation. A postulated structure of the transition state is shown in Fig. 7-17. In this structure, lysine and/or arginine also may stabilize the trigonal planar structure of the phosphoryl group, and a carboxyl group may interact with the creatine guanido group. Imidazole may serve as a general base catalyst. However, the specific role of protein side chains in the mechanism is not known. The stereochemistry of the phosphoryl transfer has been studied by utilizing ATP with three isotopes of oxygen and sulfer bonded to the γ phosphate (*81*). The reaction proceeds with inversion of the configuration. A similar result has beèn found with other kinases (*82*). Thus the mechanism appears to be an in-line displacement. The metal ion may

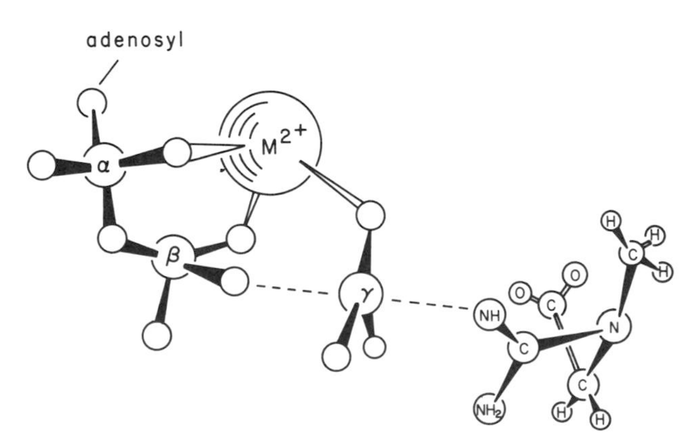

Fig. 7-17. Postulated structure of the transition state of the enzyme–substrate complex for creatine kinase. [From G. H. Reed, C. H. Barlow, and R. A. Burns, Jr., *J. Biol. Chem.* **253**, 4153 (1978).]

serve as a steric guide for the phosphate chain, may polarize the bond being broken, and/or may withdraw negative charge thereby assisting in the nucleophilic attack. Obviously, the relation between mechanism and protein structure is not as well established as in the cases where the three-dimensional structure of the enzyme is known. Nevertheless, considerable insight into both structure and mechanism has been obtained. The vision of future prospects for understanding the molecular details of the mechanism of action of creatine kinase provides an appropriate conclusion to our case studies.

REFERENCES

1. F. M. Richards and H. W. Wyckoff, *in* "The Enzymes" (P. Boyer, ed.), 3rd ed., Vol. 4. p. 647. Academic Press, New York, 1971.
2. D. J. Smyth, W. H. Stein, and S. Moore, *J. Biol. Chem.* **238**, 227 (1963).
3. B. Gutte and R. Merrifield, *J. Am. Chem. Soc.* **91**, 501 (1969).
4. R. R. Hirschmann, R. Nutt, D. Veber, A. Vitali, S. Varga, T. Jacob, F. Holly, and R. Denkewalter, *J. Am. Chem. Soc.* **91**, 507 (1969).
5. H. A. Scheraga and J. A. Rupley, *Adv. Enzymol.* **24**, 161 (1962).
6. C. B. Anfinson, E. Haber, M. Sela, and F. H. White, Jr., *Proc. Natl. Acad. Sci. U.S.A.* **47**, 1309 (1961).
7. K. Hoffman, J. P. Visser, and F. M. Finn, *J. Am. Chem. Soc.* **92**, 2900 (1970).
8. R. Rocchi, F. Marchiori, L. Moroder, G. Borin, and E. Scoffone, *J. Am. Chem. Soc.* **91**, 3927 (1969).
9. H. Witzel, *Prog. Nucleic Acid Res. Mol. Biol.* **2**, 221 (1963).
10. D. G. Herries, A. P. Mathias, and B. R. Rabin, *Biochem. J.* **85**, 127 (1962).
11. E. J. del Rosario and G. G. Hammes, *Biochemistry* **8**, 1884 (1969).
12. D. G. Anderson, G. G. Hammes, and F. G. Walz, Jr., *Biochemistry* **7**, 1637 (1968).
13. J. L. Markley, *Biochemistry* **14**, 3546, 3554 (1965); J. L. Markley and W. R. Finkenstadt, *Biochemistry* **14**, 3562 (1965).
14. D. H. Meadows and O. Jardetzky, *Proc. Natl. Acad. Sci. U.S.A.* **61**, 406 (1968).
15. G. G. Hammes, *in* "Investigation of Rates and Mechanisms of Reactions," (G. G. Hammes, ed.), 3rd ed. Part II, p. 147. Wiley (Interscience), New York, (1974).
16. G. G. Hammes and F. G. Walz, Jr., *J. Am. Chem. Soc.* **91**, 7179 (1969).
17. F. H. Westheimer, *Acc. Chem. Res.* **1**, 70 (1968).
18. D. A. Usher, *Proc. Natl. Acad. Sci. U.S.A.* **62**, 661 (1969).
19. D. A. Usher, D. I. Richardson, Jr., and F. Eckstein, *Nature (London)* **228**, 663 (1970).
20. D. A. Usher, E. S. Ehrenrich and F. Eckstein, *Proc. Natl. Acad. Sci. U.S.A.* **69**, 115 (1972).
21. D. M. Blow, *Acc. Chem. Res.* **9**, 145 (1976).
22. S. Blackburn, "Enzyme Structure and Function." Dekker, New York, 1976.
23. G. P. Hess, *in* "The Enzymes" (P. Boyer, ed.), 3rd ed. Vol. 3, p. 213. Academic Press, New York, 1970.
24. M. L. Bender and F. Kezdy, *Annu. Rev. Biochem.* **34**, 49 (1965).
25. B. S. Hartley and B. A. Kilby, *Biochem. J.* **56**, 288 (1954).
26. H. Gutfreund and J. M. Sturtevant, *Biochem. J.* **63**, 656 (1956).
27. L. D. Faller and J. M. Sturtevant, *J. Biol. Chem.* **241**, 4825 (1966).
28. B. Zerner and M. L. Bender, *J. Am. Chem. Soc.* **86**, 3669 (1964).

29. R. J. Foster, H. J. Shine, and C. Niemann, *J. Am. Chem. Soc.* **77**, 2378 (1955).
30. R. J. Foster and C. Niemann, *J. Am. Chem. Soc.* **77**, 3370 (1955).
31. R. J. Foster and C. Niemann, *J. Am. Chem. Soc.* **77**, 1886 (1955).
32. L. W. Cunningham and C. S. Brown, *J. Biol. Chem.* **221**, 287 (1956).
33. B. R. Hammond and H. Gutfreund, *Biochem. J.* **61**, 187 (1956).
34. B. Zerner, R. P. M. Bond, and M. L. Bender, *J. Am. Chem. Soc.* **86**, 3704 (1964).
35. A. R. Fersht, D. M. Blow and J. Fastrez, *Biochemistry* **12**, 2035 (1973).
36. E. F. Jansen, M. F. Nutting, and A. K. Balls, *J. Biol. Chem.* **179**, 201 (1949).
37. D. H. Strumeyer, W. N. White, and D. E. Koshland, Jr., *Proc. Natl. Acad. Sci. U.S.A.* **50**, 931 (1963).
38. A. Y. Moon, J. M. Sturtevant, and G. P. Hess, *J. Biol. Chem.* **240**, 4204 (1965).
39. A. Fink, *Biochemistry* **15**, 1580 (1976).
40. G. A. Rogers and T. C. Bruice, *J. Am. Chem. Soc.* **96**, 2473 (1974).
41. W. W. Bachovchin and J. D. Roberts, *J. Am. Chem. Soc.* **100**, 8041 (1978).
42. J. Kraut, *Annu. Rev. Biochem.* **46**, 331 (1977).
43. C. C. F. Blake, D. F. Koenig, G. A. Mair, A. C. T. North, D. C. Phillips and V. R. Sarma, *Nature (London)* **206**, 4986 (1965).
44. D. C. Phillips, *Sci. Am.* **215**, 78 (1966).
45. T. Imoto, L. N. Johnson, A. C. T. North, D. C. Phillips and J. A. Rupley, *in* "The Enzymes" (P. Boyer, ed.), 3rd ed. Vol. 7, p. 665. Academic Press, New York, 1972.
46. A. Schrake and J. A. Rupley, *Biochemistry* **19**, 4044 (1980).
47. E. W. Thomas, J. F. McKelvy and N. Sharon, *Nature (London)* **222**, 485 (1969).
48. J. A. Rupley and V. Gates, *Proc. Natl. Acad. Sci. U.S.A.* **57**, 496 (1967).
49. E. Holler, J. A. Rupley and G. P. Hess, *Biochemistry* **14**, 1088 (1975).
50. E. Holler, J. A. Rupley and G. P. Hess, *Biochemistry* **14**, 2377 (1975).
51. F. Millett and M. A. Raftery, *Biochemistry* **11**, 1639 (1972).
52. S. M. Parsons and M. A. Raftery, *Biochemistry* **11**, 1623 (1972).
53. I. I. Secemski, S. S. Lehrer and G. E. Lienhard, *J. Biol. Chem.* **247**, 4740 (1972).
54. B. D. Sykes and D. Dolphin, *Nature (London)* **233**, 421 (1971).
55. B. M. Dunn and T. C. Bruice, *Adv. Enzymol.* **37**, 1 (1973).
56. M. R. Pincus and H. A. Scheraga, *Macromolecules* **12**, 633 (1979).
57. M. Schindler, Y. Assaf, N. Sharon and D. M. Chipman, *Biochemistry* **16**, 423 (1977).
58. F. Dahlquist, T. Rand-Meir, and M. Raftery, *Biochemistry* **8**, 4214 (1969).
59. G. M. Loudon, C. K. Smith and S. E. Zimmerman, *J. Am. Chem. Soc.* **96**, 465 (1974).
60. G. M. Loudon and D. E. Ryono, *J. Am. Chem. Soc.* **98**, 1900 (1976).
61. D. C. Watts, *in* "The Enzymes" (P. Boyer, ed.), 3rd ed., Vol. 8, p. 383. Academic Press, New York, 1973.
62. G. D. Markham, G. H. Reed, E. T. Maggro and G. L. Kenyon, *J. Biol. Chem.* **252**, 1197 (1977).
63. M. C. Lane and F. A. Quiocho, *Biochemistry* **16**, 3838 (1977).
64. C. L. Borders, Jr. and J. F. Riordan, *Biochemistry* **14**, 4699 (1975).
65. T. L. James, *Biochemistry* **15**, 4724 (1976).
66. M. Vašák K., Nagayama, K. Wüthrich, M. L. Mertens, and J. H. R. Kagi, *Biochemistry* **18**, 5050 (1979).
67. M. A. Marletta and G. L. Kenyon, *J. Biol. Chem.* **254**, 1879 (1979).
68. M. I. Schimerlik and W. W. Cleland, *J. Biol. Chem.* **248**, 8418 (1973).
69. Y. Engelborghs, A. March, and H. Gutfreund, *Biochem. J.* **151**, 47 (1975).
70. E. J. Milner-White and D. C. Watts, *Biochem. J.* **122**, 727 (1971).
71. P. F. Cook, G. L. Kenyon, and W. W. Cleland, *Biochemistry* **20**, 1204 (1981).
72. G. G. Hammes and J. K. Hurst, *Biochemistry* **8**, 1083 (1969).

73. G. H. Reed, H. Diefenbach, and M. Cohn, *J. Biol. Chem.* **247**, 3066 (1972).
74. F. Travers, T. E. Barman, and R. Bertrand, *Eur. J. Biochem.* **100**, 149 (1979).
75. T. E. Barman, A. Brun, and F. Travers, *Eur. J. Biochem.* **110**, 405 (1980).
76. A. C. McLaughlin, J. S. Leigh, Jr., and M. Cohn, *J. Biol. Chem.* **251**, 2777 (1976).
77. G. H. Reed, C. H. Barlow, and R. A. Burns, Jr., *J. Biol. Chem.* **253**, 4153 (1978).
78. D. Dunaway-Mariano and W. W. Cleland, *Biochemistry* **19**, 1506 (1980).
79. G. H. Reed and T. S. Leigh, *Biochemistry* **19**, 5472 (1980).
80. B. D. Nageswara Rao, F. J. Kayne and M. Cohn, *J. Biol. Chem.* **254**, 2689 (1979).
81. D. E. Hansen and J. R. Knowles, *J. Biol. Chem.* **256**, 5967 (1981).
82. J. R. Knowles, *Annu. Rev. Biochem.* **49**, 877 (1980).

8

Regulation of Enzyme Activity

INTRODUCTION

Thus far only the catalytic properties of enzymes have been considered. However, in living systems hundreds of different enzyme catalyzed reactions occur simultaneously, and these reactions must be regulated for the proper functioning of a living system. This regulation is achieved by the modulation of key reactions that control metabolic fluxes. This can be done either by controlling the quantity of enzyme or by the control of the catalytic activity of a given amount of enzyme. The quantity of enzyme present is determined by the balance between enzyme synthesis and degradation. The synthesis of enzymes is regulated at the genetic level by induction and repression mechanisms. These regulatory processes, which have been extensively studied (cf. *1–4*), are not considered here. The direct modulation of the catalytic activity of enzymes occurs through several different types of mechanisms. A brief review of these mechanisms is given, followed by a discussion of the experimental manifestations of the regulation in the equilibrium and kinetic properties of the enzymes and an analysis of molecular models for the regulation. Conformational changes *within* a single polypeptide chain have been shown to be a central feature of enzyme catalysis. As will become clear, conformational changes that alter the interactions *between* polypeptide chains are of crucial importance for the regulation of enzymes. Thus, in terms of protein structure, many enzymes that are not regulated exist as single polypeptide chains, whereas almost all enzymes whose catalytic activity is regulated exist as structures containing multiple polypeptide chains. The actual regulation occurs through conformational changes altering interactions between polypeptide chains and/or through changes in aggregation state (polymerization–depolymerization).

The binding of small molecules such as substrates or effectors (activators and inhibitors) to enzymes is an important mode of regulation. *Feedback inhibition* is an important example. In this case, the first committed step in a biosynthetic pathway is inhibited by the ultimate end product of the pathway. A classic case is aspartate transcarbamoylase, which catalyzes the synthesis of carbamoyl aspartate from aspartate and carbamoyl phosphate (*5*). This enzyme is the first committed step in pyrimidine biosynthesis, and the ultimate end product of that pathway, cytidine 5′-triphosphate, is a strong inhibitor of aspartate transcarbamoylase although it is not structurally similar to the substrates. For multifunctional pathways, the feedback inhibition can be quite complex. For example, consider the branched pathway (*2*)

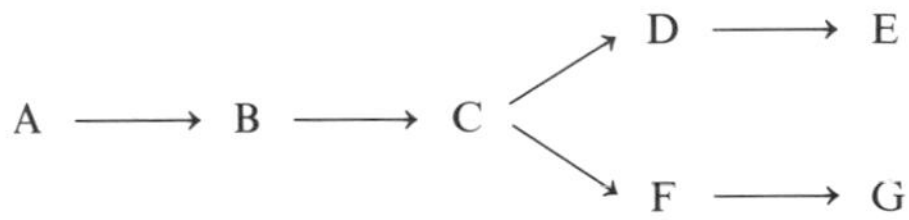

$$A \longrightarrow B \longrightarrow C \begin{array}{l} \nearrow D \longrightarrow E \\ \searrow F \longrightarrow G \end{array}$$

which can be regulated by a number of different mechanisms. *Enzyme multiplicity* is one of these mechanisms. If the first common step ($A \rightarrow B$) is catalyzed by two different enzymes, one can be feedback inhibited by E and the other by G. A second control is that G turns off the enzyme catalyzing $C \rightarrow F$ and E turns off the enzyme catalyzing $C \rightarrow D$. Thus, an excess of both products turns everything off, but if only one product is in excess, the other pathway can operate. An example is the biosynthesis of lysine, threonine, methionine, and isoleucine in *Escherichia coli*, all of which are derived from aspartate (*6*). Three different aspartokinases and two different homoserine dehydrogenases are involved in the biosynthetic pathway. In fact, enzyme multiplicity occurs whenever a given reaction has two or more functions as, for example, in a step common to biosynthetic and degradative pathways. Two different threonine deaminases are known in *E. coli*, one is inducible and one is subject to feedback control by isoleucine (*7*). *Sequential feedback control* occurs when E shuts off the utilization of the intermediate C ($C \rightarrow D$), which is still available for the other pathway. The product G also can shut off the utilization of C ($C \rightarrow F$). If both $C \rightarrow D$ and $C \rightarrow F$ are inhibited, the intermediate C accumulates and shuts down the whole pathway ($A \rightarrow B$). This type of regulation is found in the biosynthesis of aromatic amino acids in *Bacillus subtilis* (*8*). *Concerted feedback inhibition* occurs when two or more end products (E and G) inhibit the first step ($A \rightarrow B$), but neither alone works. For example, aspartokinase in *Bacillus polymxa* is insensitive to any one of the end products of aspartate metabolism, but threonine and lysine inhibit (*9*). Variants of this mechanism are *synergistic feedback inhibition*, *cumulative feedback inhibition*, and *heterogeneous metabolic pool inhibition*. In the first of these, each of the end products inhibits somewhat, but the presence of both yields a total inhibition greater than the sum of the independent inhibition. For example, in purine nucleotide biosynthesis, glutamine phosphoribosylphosphate amidotransferase is inhibited by 6-hydroxy and 6-amino purine nucleotides in this way (*10*). In the second case, a high concentration of each end product causes partial inhibition, and each end product acts independently of the others. Therefore, the simultaneous presence of end products produces a cumulative effect. A classic example of this occurs with glutamine synthetase from *E. coli* where the six end products of glutamine metabolism, L-alanine, and glycine are cumulative inhibitors (*11*). Finally with heterogeneous metabolic pool inhibition, a single inhibitory site for end products exists, but the binding is so weak that the site is never fully occupied. In this case, the effective inhibitor concentration is the sum of the concentrations of all of the end products. Such a mechanism appears to be important in the regulation of purine mononucleotide pyrophosphorylase in *B. subtilis* (*12*).

Many situations also exist where enzymes are activated by metabolites and other small molecules. *Precursor substrate activation* is when an enzyme catalyzing a key metabolic step is activated by a precursor metabolite. For example, in *Salmonella typhimurium* phosphoenolpyruvate carboxylase is activated by fructose diphosphate (*13*). Antagonists for feedback inhibition also exist: ATP reverses the inhibitory effect of CTP on aspartate transcarbamolyase (*5*). *Hormones* often stimulate enzymes. Probably the best known case is adenylate cyclase, which is stimulated by several hormones. The cyclic-AMP produced then serves as a regulator for other enzymes. Inorganic ions are well known activators for many enzymes. For example, K^+ activates pyruvate kinase, and both Na^+ and K^+ are required for the activity of the membrane-bound ATPase that transports the two ions. Divalent cations such as Mg^{2+}, Ca^{2+}, and Mn^{2+} are required for kinases and ATPases; often these enzymes are very specific in their metal requirements.

The regulation of enzymes by ligands can be viewed as one broad class of mechanisms. A second broad class is the energy-dependent covalent modification of regulatory enzymes. Phosphorylation–dephosphorylation, adenylation–deadenylation, and methylation–demethylation, all catalyzed by specific enzymes, have important regulatory functions. Although these mechanisms are not discussed in detail, a few examples suffice to illustrate the general principles. In *E. coli*, glutamine activates an enzyme that adenylates glutamine synthetase and deactivates a deadenylating enzyme. The reverse effects are caused by α-ketoglutarate and UTP. Adenylation results in a less active form of the enzyme which is more susceptible to inhibition; in addition, the divalent cation specificity is changed from Mg^{2+} to Mn^{2+} (*2*). The phosphorylation and dephosphorylation of glycogen phosphorylase is an important regulatory mechanism (cf. *2*). Glycogen phosphorylase *b* is a low activity dimer, which requires AMP. When it is phosphorylated by a kinase, phosphorylase *a* is formed; phosphorylase *a* is more active, does not require AMP and is tetrameric. A specific phosphatase exists to catalyze the conversion of phosphorylase *a* to phosphorylase *b*. The crystal structure of phosphorylase is known, and many of the regulatory aspects of its behavior can be understood in molecular terms (*14*). This enzyme is part of a relatively simply *cascade* mechanism illustrated in Fig. 8-1. Epinephrine binds to the membrane and activates adenylate cyclase; the cyclic-AMP formed activates phosphorylase kinase kinase by dissociating an inactive dimer to a regulatory polypeptide chain that binds cyclic-AMP and to an active catalytic polypeptide chain. The kinase kinase activates phosphoryl kinase, which in turn catalyzes the conversion of phosphorylase *b* to phosphorylase *a*. Cascades are important regulatory mechanisms and can be very complex (*15*). Another regulatory mechanism involving the breaking of covalent bonds is limited proteolysis. For example, many proteases are activated in this way (e.g., pepsinogen → pepsin, trypsinogen → trypsin). The synthesis of a protease in an inactive form protects it from proteolysis. Proteolysis is an important feature in blood coagulation, which proceeds by a complex cascade mechanism.

A third broad class of regulation is the use of special structures. Multienzyme complexes and membranes are examples of such structures. Multi-

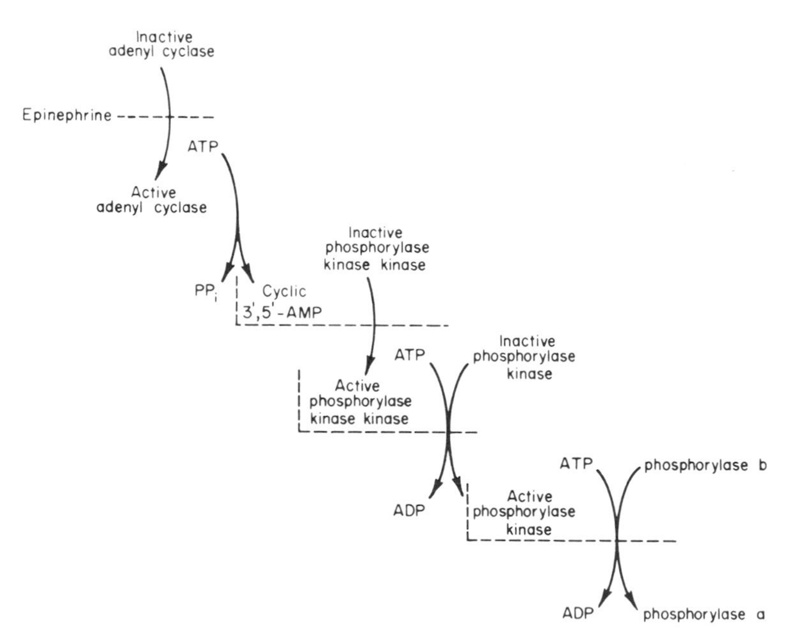

Fig. 8-1. Cascade effect in the conversion of phosphorylase b to phosphorylase a. The dashed lines indicate either an activation or a catalytic function. [From E. R. Stadtman, *in* "The Enzymes" (P. Boyer, ed.), 3rd Ed., Vol. **1**, p. 397. Academic Press, New York, 1970.]

enzyme complexes contain several enzyme activities that catalyze a sequence of reactions, with the product of the first enzyme activity being a substrate of the second enzyme activity, etc. Substrates can be passed from enzyme to enzyme without the dissociation of intermediates so that metabolites are channeled along a given metabolic pathway. An example of a multienzyme complex is pyruvate dehydrogenase from *E. coli* (*16*), which catalyzes the overall reaction

$$CH_3\overset{\overset{\displaystyle O}{\|}}{C}COOH + CoA + NAD^+ \longrightarrow \text{acetyl-CoA} + CO_2 + NADH + H^+$$

and contains three different enzymes in multiple copies: pyruvate decarboxylase decarboxylates pyruvate and forms a hydroxyethyl derivative with thiamin pyrophosphate; the acetyl group is transferred to a lipoic acid on dihydrolipoyl transacetylase; this acetyl group is transferred to CoA with concomitant formation of fully reduced lipoic acid; the lipoic acid is then oxidized by lipoamide dehydrogenase through a flavin, and finally the flavin is oxidized by NAD^+ to restore the system to its original state. This multienzyme complex also is subject to regulation by the binding of small molecules. Membranes provide a variety of regulatory functions. The simplest is compartmentalization. For example, the electron transport system is confined to mitochondria; ions are confined to specific structures; and concentration gradients sometimes are maintained. The specific transport of materials is an important function of membranes. Membranes also can markedly alter the properties of enzymes. For example, the membrane-bound enzyme β-hydroxybutyrate dehydrogenase has an absolute dependence on lecithin (*17*), and the steady-state kinetic parameters of mitochondrial ATPase are different on and off the membrane (*18*). Finally, mention should be made of the coupling between transport across the membrane and enzyme catalysis, which is an important method of physiological regulation. This is illustrated by the (Na^+, K^+) activated-ATPase which pumps Na^+ and K^+ and the Ca^{2+}activated-ATPase from sarcoplasm reticulum which pumps Ca^{2+} concomitant with ATP hydrolysis. Multienzyme complexes and membrane-bound enzymes are discussed in more detail in Chapters 10 and 11, respectively.

EXPERIMENTAL MANIFESTATIONS OF REGULATORY ENZYMES

Perhaps the simplest level of control of enzyme activity is exerted by the substrate concentration itself. Often a plot of the initial reaction velocity v and/or the number of the occupied substrate sites on the enzyme r versus the substrate concentration is a hyperbolic curve, as described, for example, by the Michaelis–Menten relationship (cf. Fig. 8-2a)

$$v = V_S[S]/(K_S + [S]) \quad (8\text{-}1)$$

When the substrate concentration [S] is smaller than or comparable to the Michaelis constant K_S, the rate of the enzymatic reaction is regulated by the substrate level. When the substrate concentration is much higher than K_S, the reaction rate reaches the maximal value V_S and is no longer sensitive to changes in substrate concentration.

Although some regulatory enzymes obey Michaelis–Menten kinetics, many of them exhibit nonhyperbolic kinetic and/or binding isotherms. Such behavior can be explained in terms of multiple substrate binding sites on the enzyme. If the binding of a substrate molecule to the first site facilitates binding to the second, etc. for successive sites, the binding or velocity isotherm is sigmoidal (Fig. 8-2a). This type of interaction is called *positive cooperativity*. On the other hand, if the binding of the substrate molecule to the first site inhibits the binding to the second, etc., the binding isotherm is said to exhibit *negative cooperativity* (Fig. 8-2a). In the case of positive cooperativity, a region in the velocity isotherm exists where the reaction rate is much more sensitive to the substrate concentration than is the case for a hyperbolic isotherm; for negative cooperativity, a concentration region exists where the reaction rate is less sensitive to the substrate concentration than for a hyperbolic isotherm. The interactions described above, which occur between binding sites for identical ligands, are termed *homotropic* interactions (*19*).

Heterotropic interactions (*19*) are those that occur between binding sites of dissimilar ligands and are typified by the effects of activators and inhibitors on enzyme activity. The effects of activators or inhibitors can be classified as being either *competitive* or *noncompetitive* with respect to substrate binding to the enzyme. Systems exhibiting competitive type kinetic behavior are called *K systems* since they involve changes in the apparent Michaelis constant of the substrate but no change in the maximum velocity. Similarly, noncompetitive systems are referred to as *V systems* since they involve changes only in the maximum velocity (*19*).

In both K and V systems, an effector can act as an activator (positive effector) or as an inhibitor (negative effector) of the enzyme, as illustrated in Fig. 8-3. In fact, all four possible classes of effects have been found among regulatory enzymes studied. In a K system, a positive effector reduces the positive homotropic interactions of substrates, whereas a negative effector enhances such interactions.

The hypothesis that effectors exert regulatory control over the catalysis by reacting at an *allosteric site*, distinct from the catalytic site is now generally

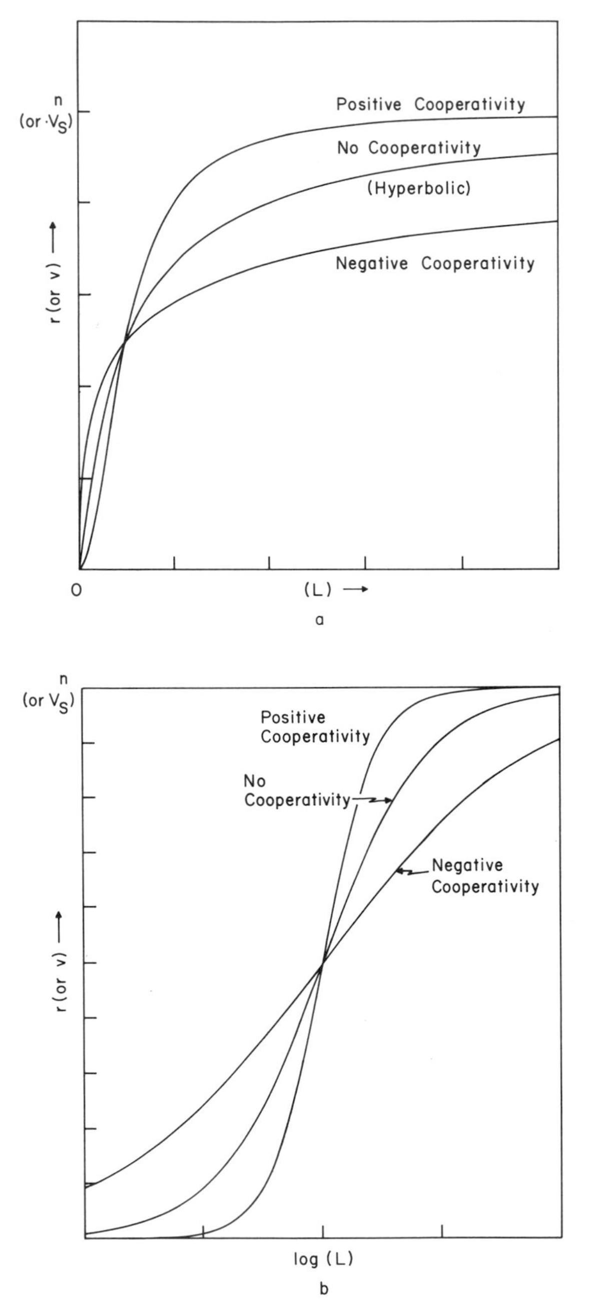

Fig. 8-2. Schematic illustrations of various representations of equilibrium binding and initial velocity data. In these figures r is the number of moles of ligand bound per mole of enzyme, v is the initial steady-state velocity of the enzymatic reaction, n is the total number of ligand binding sites per enzyme molecule, and V_S is the maximum velocity. Examples of binding and initial velocity–ligand isotherms displaying no cooperativity, positive cooperativity, and negative cooperativity are shown. [Reproduced, with permission, from G. G. Hammes and C.-W. Wu, *Annu. Rev. Biophys. Bioeng.*, **3**, 1. © 1974 by Annual Reviews, Inc.]

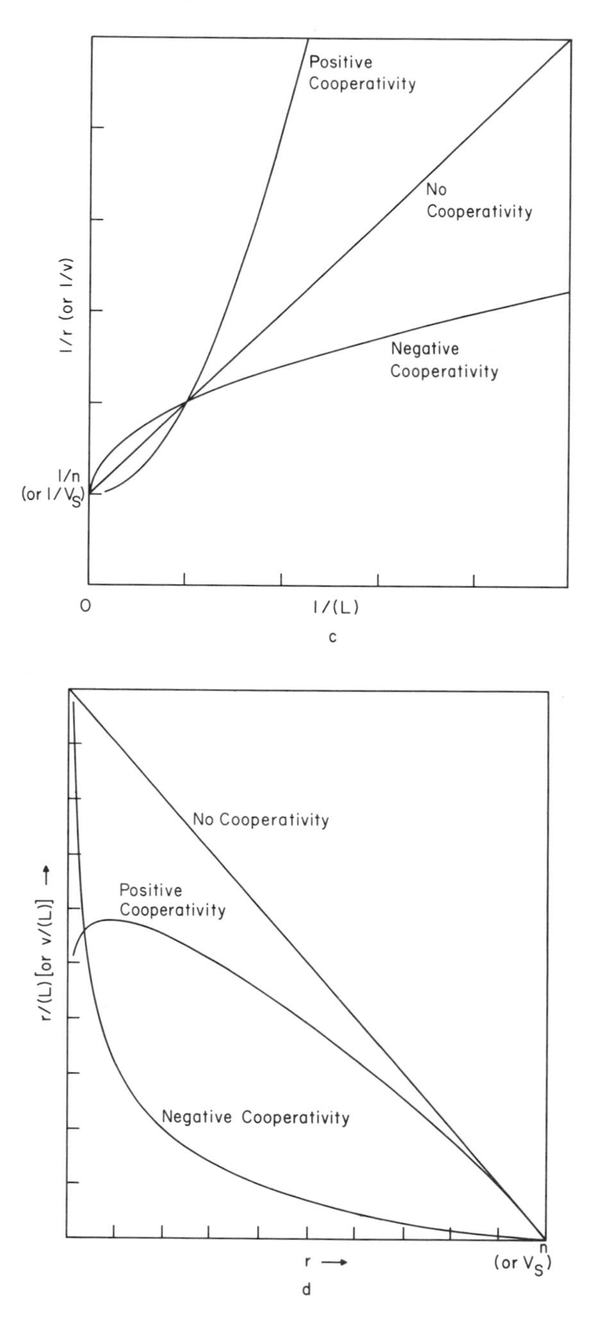

Fig. 8-2. (*Continued*)

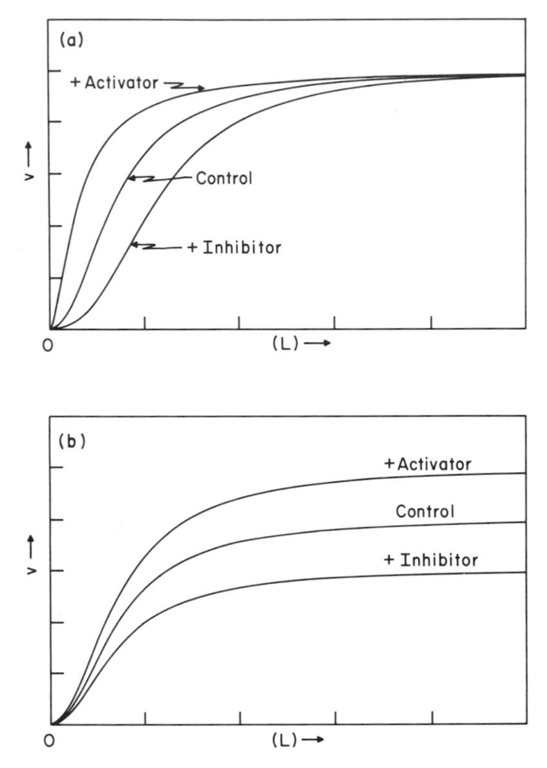

Fig. 8-3. Schematic illustrations of the behavior of allosteric enzymes: (a) K system and (b) V system. In these figures v is the initial velocity of the enzymatic reaction and [L] is the ligand concentration. [Reproduced, with permission, from G. G. Hammes and C.-W. Wu, *Annu. Rev. Biophys. Bioeng.*, **3**, 1. © 1974 by Annual Reviews, Inc.]

accepted. Some of the experimental facts that lead to this hypothesis are as follows.

1. Effectors are usually structurally dissimilar to substrates.
2. Regulatory enzymes often can be made insensitive to metabolic inhibition or activation without alteration of catalytic activity.
3. Feedback inhibitors sometimes protect an enzyme against denaturation, whereas substrates do not (and vice versa).
4. Gene mutations have been obtained that alter the susceptibility of some enzymes to feedback inhibition without altering catalytic activity.
5. Direct binding studies show the binding of feedback inhibitors is not influenced by saturating concentrations of substrates and vice versa.
6. Effector binding sites have been found to occur on subunits different than those containing the catalytic site.

Allosteric interactions are defined as indirect interactions between topographically distinct binding sites mediated by the protein molecules through conformational changes or alterations in subunit interactions. Such interactions stabilize conformational states having a different affinity for substrate (K system) or having a different catalytic potential (V system) or both.

EQUILIBRIUM BINDING ISOTHERMS

The study of substrate and effector binding to regulatory enzymes is a primary method for investigating enzyme regulation. A variety of experimental methods is available for determining binding isotherms. The direct methods include equilibrium dialysis, ultrafiltration, forced dialysis, and gel chromatography. Spectral methods such as difference spectroscopy and fluorescence often are used, but they are not as reliable since the assumption is made that some spectral parameter is directly proportional to the extent of binding. Such an assumption may not always be correct.

To illustrate the analysis of equilibrium binding isotherms, consider first the case of a protein P that has n binding sites for a ligand L. The binding equilibria and the corresponding association constants K_i can be written as

$$\begin{array}{ll} P + L \rightleftharpoons PL_1 & K_1 = [PL_1]/[P][L] \\ PL_1 + L \rightleftharpoons PL_2 & K_2 = [PL_2]/[PL_1][L] \\ \vdots \quad \vdots & \vdots \\ PL_{i-1} + L \rightleftharpoons PL_i & K_2 = [PL_i]/[PL_{i-1}][L] \\ \vdots \quad \vdots & \vdots \\ PL_{n-1} + L \rightleftharpoons PL_n & K_n = [PL_n]/[PL_{n-1}][L] \end{array} \tag{8-2}$$

The number of ligand molecules bound per protein molecule, r, is

$$r = \frac{\text{moles of L bound}}{\text{moles of protein}} = \frac{\sum_{i=1}^{n} i[PL_i]}{\sum_{i=0}^{n} [PL_i]}$$

$$r = \frac{[PL_1] + 2[PL_2] + \cdots + n[PL_n]}{[P] + [PL_1] + [PL_2] + \cdots + [PL_n]} \tag{8-3}$$

or

$$r = \frac{K_1[L] + 2K_1K_2[L]^2 + \cdots + nK_1K_2 \cdots K_n[L]^n}{1 + K_1[L] + K_1K_2[L]^2 + \cdots + K_1K_2 \cdots K_n[L]^n} \tag{8-4}$$

Equation (8-4), while quite complex, is quite general for describing complex binding and is used in the absence of auxiliary information about the binding sites. Some special cases that provide simpler equations are now described.

The first case is when all the sites are independent and identical so that they all have the same intrinsic association constant K. This intrinsic constant is not equal to the observed constant because of the statistical effect of multiple binding sites: the unoccupied molecule has n sites available for the first ligand; when one site is occupied $n - 1$ sites are available for the next ligand, etc.; a similar phenomenon exists for the dissociation process. As a result, the relationship between the macroscopic and intrinsic binding constant is

$$K_i = \frac{\#\text{ of free sites on P before binding}}{\#\text{ of occupied sites on P after binding}} K$$
$$= \frac{n - i + 1}{i} K \qquad i \geq 1 \tag{8-5}$$

Substitution of this result gives

$$r = \frac{nK[\mathrm{L}] + \frac{2n(n-1)}{2!} K^2[\mathrm{L}]^2 + \cdots + nK^n[\mathrm{L}]^n}{1 + nK[\mathrm{L}] + \frac{n(n-1)}{2!} K^2[\mathrm{L}]^2 + \cdots + K^n[\mathrm{L}]^n} \tag{8-6}$$
$$= \frac{nK[\mathrm{L}](1 + K[\mathrm{L}])^{n-1}}{(1 + K[\mathrm{L}])^n} = \frac{nK[\mathrm{L}]}{1 + K[\mathrm{L}]}$$

where the binomial theorem has been used to obtain the final result. This special case results in a hyperbolic binding isotherm indistinguishable in form from ligand binding to a single site.

A second limiting case is when a macromolecule possesses classes of different types of sites. If class 1 has n_1 equivalent sites with an association constant K_1, class 2 has n_2 equivalent sites with an association constant K_2, etc. and the binding to each class of sites occurs independently, then for m classes of binding sites

$$r = \sum_{i=1}^{m} n_i K_i[\mathrm{L}]/(1 + K_i[\mathrm{L}]) \tag{8-7}$$

The general case of identical interacting sites is, of course, represented by Eq. (8-4). The association constants used in this equation are sometimes replaced by the intrinsic association constants K_i' to take into account the statistical effect of identical sites. Equation (8-4) can then be rewritten as

$$r = \frac{\sum_{i=1}^{n} i[n!/(n-i)!i!]K_1'K_2' \cdots K_i'[\mathrm{L}]^i}{1 + \sum_{i=1}^{n} [n!/(n-i)!i!]K_1'K_2' \cdots K_i'[\mathrm{L}]^i} \tag{8-8}$$

For negative cooperativity $K'_1 > K'_2 > \cdots > K'_n$, whereas for positive cooperativity $K'_1 < K'_2 < \cdots < K'_n$. In principle, both positive and negative cooperativity can occur in a single binding isotherm. If the terms in Eq. (8-7) are put over a common denominator, an equation similar in form to Eq. (8-8) is obtained; however, the combinations of constants occurring permit only the possibility of negative cooperativity. Thus, negative cooperativity and multiple classes of binding sites cannot be distinguished from an equilibrium binding isotherm.

As a useful illustration of the relationships between the different types of binding isotherms, consider the binding of ligands to two sites on a protein according to the scheme

$$\begin{array}{ccccc} & \overset{K_a}{\rightleftharpoons} & (EL)_a + L & \overset{K_c}{\rightleftharpoons} & \\ E + 2L & & & & EL_2 \\ & \underset{K_b}{\rightleftharpoons} & (EL)_b + L & \underset{K_d}{\rightleftharpoons} & \end{array}$$

where the *microscopic binding constants* are

$$K_a = \frac{[EL]_a}{[E][L]} \qquad K_c = \frac{[EL_2]}{[EL]_a[L]}$$

$$K_b = \frac{[EL]_b}{[E][L]} \qquad K_d = \frac{[EL_2]}{[EL]_b[L]}$$

and $K_a K_c = K_b K_d$. The general binding isotherm is

$$r = \frac{[EL]_a + [EL]_b + 2[EL_2]}{[E] + [EL]_a + [EL]_b + [EL_2]} = \frac{(K_a + K_b)[L] + 2K_a K_c [L]^2}{1 + (K_a + K_b)[L] + K_a K_c [L]^2}$$

Comparison with Eq. (8-4) shows that the macroscopic and microscopic association constants are related by $K_1 = (K_a + K_b)$ and $K_1 K_2 = K_a K_c = K_b K_d [K_2 = K_a K_c/(K_a + K_b)]$. For identical, noninteracting sites, $K_a = K_b = K_c = K_d$, and as expected [Eq. (8-6)]

$$r = \frac{2K_a[L]}{1 + K_a[L]}$$

For nonidentical, noninteracting sites $K_a = K_d$ and $K_b = K_c$. Then

$$r = \frac{(K_a + K_b)[L] + 2K_a K_b [L]^2}{1 + (K_a + K_b)[L] + K_a K_b [L]^2}$$

$$= \frac{K_a[L]}{1 + K_a[L]} + \frac{K_b[L]}{1 + K_b[L]}$$

This is in agreement with Eq. (8-7); moreover, since $K_a K_b/(K_a + K_b) < (K_a + K_b)$, this isotherm has the same form as if negative cooperativity is occurring. For identical sites, with negative or positive cooperativity, $K_a = K_b$ and $K_c = K_d$ so that

$$r = \frac{2K_a[\mathrm{L}] + 2K_aK_c[\mathrm{L}]^2}{1 + 2K_a[\mathrm{L}] + K_aK_c[\mathrm{L}]^2}$$

which is identical in form to Eq. (8-8).

A careful analysis of binding data is essential to establish the nature of the cooperativity between binding sites. Positive and negative cooperativity can be easily distinguished. However, whereas positive cooperativity must be due to true cooperative interactions, negative cooperativity can arise from true cooperativity or from different classes of noninteracting sites (or equivalently from protein heterogeneity). Distinguishing between these two possible mechanisms for negative cooperativity usually requires experiments other than equilibrium binding; in some cases, binding experiments in the presence of competing ligands are useful (*25*). Several different methods of plotting binding data have been suggested. Thus far only plots of r versus [L] have been discussed (Fig. 8-2a). For noninteracting identical sites a hyperbolic curve is obtained. For positive cooperativity the curve is sigmoidal, whereas for negative cooperativity the final value of r is approached more slowly than in a hyperbolic curve. If the enzyme contains more than two binding sites, both negative and positive cooperativity are possible, and the plot of r versus [L] may display undulations or bumps.

A plot of r versus log [L] is often used to represent the data (Fig. 8-2b). Such a plot can cover a very wide range of ligand concentrations and the steepness of the binding isotherm gives a convenient measure of the cooperativity, with equal steepness for equal cooperativity (*20*). All binding isotherms appear sigmoidal and the steepness of the curve is in the order of positive cooperativity > no cooperativity > negative cooperativity.

A plot of $1/r$ versus $1/[\mathrm{L}]$ is linear for noninteracting identical sites, is concave upward for positive cooperativity, and is concave downward for negative cooperativity (Fig. 8-2c). However, a concave downward curve can also be obtained from polymorphic forms, electrostatic repulsion, and other causes (*21*). Thus, such a curve cannot be regarded as a proof of negative cooperativity. For noninteracting identical sites the slope of the plot is equal to $1/nK$ and the intercept is $1/n$ [Eq. (8-6)].

Equation (8-6), which describes ligand binding to n independent equivalent sites, can be rewritten as

$$r/[\mathrm{L}] = nK - Kr \qquad (8\text{-}9)$$

A plot of $r/[\mathrm{L}]$ versus r, usually called a Scatchard plot, is convenient for determining both n and K (*22*). When cooperativity is present the plot is no longer linear: it is concave downward and concave upward for positive and negative cooperativity, respectively (Fig. 8-2d).

In studying oxygen binding to hemoglobin, Hill showed that a plot of r versus $[\mathrm{L}]$ could be described by (*23*)

$$r = n[\mathrm{L}]^h/(K + [\mathrm{L}]^h) \tag{8-10}$$

This can be rewritten as

$$\mathrm{Log}\, r/(n - r) = h \log[\mathrm{L}] - \log K \tag{8-11}$$

so that a plot of $\log r/(n - r)$ versus $\log[\mathrm{L}]$, usually called a Hill plot, gives a straight line with an intercept equal to $\log K$ and a slope of h, which is defined as the Hill coefficient (Fig. 8-4). A Hill coefficient greater than one indicates positive cooperativity, less than one negative cooperativity, and

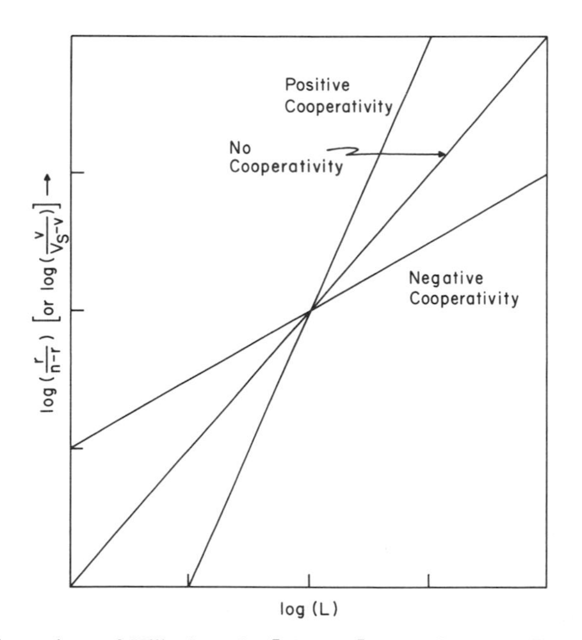

Fig. 8-4. Illustrations of Hill plots: $\log[r/(n - r)]$ or $\log[v/(V_S - v)]$ versus $\log[\mathrm{L}]$ for data conforming to Eq. (8-11) with $h = 1$ for no cooperativity, $h = 2$ for positive cooperativity and $h = 0.5$ for negative cooperativity, where r is the number of moles of ligand bound per mole of enzyme, n is the total number of ligand binding sites per enzyme molecule, v is the initial velocity of the enzymatic reaction and V_S is the maximum velocity. [Reproduced, with permission, from G. G. Hammes and C.-W. Wu, *Annu. Rev. Biophys. Bioeng.* **3**, 1, ©1974 by Annual Reviews, Inc.]

equal to one no cooperativity. In general, complex binding isotherms cannot be described by Eq. (8-10), and a Hill plot is a curve. The steepest slope is then usually used to determine the Hill coefficient, which is always less than the number of sites n. The Hill coefficient, therefore, is some measure of the extent of cooperativity in the binding process.

Each of the methods of treating binding data has some virtue, and it is useful to utilize more than one in carrying out a preliminary examination of the data. Ultimately, of course, binding data should be quantitatively fit using a least-squares analysis with proper weighting factors; the binding parameters obviously do not depend on how the data are plotted.

MOLECULAR MODELS FOR ALLOSTERISM

The molecular models proposed for allosteric control mechanisms are based on the hypothesis that control occurs through alterations in the interactions between protein polypeptide chains. These changes are caused by conformational changes within the protein molecule. Two limiting models have been proposed: one is due to Monod, Wyman, and Changeux (Monod–Wyman–Changeux model or MWC model, *19*), and the other is due to Koshland, Nemethy, and Filmer (*20*) who developed a molecular model for an empirical equation suggested by Adair for describing oxygen binding to hemoglobin (*24*, Adair–Koshland–Nemethy–Filmer model or AKNF model).

The MWC model is based on three assumptions: (a) the enzyme consists of two or more identical subunits, each containing a site for the substrate or effector; (b) at least two different conformational states (usually designated as R and T states) are in equilibrium and differ in their affinities for substate or effector; and (c) the conformational changes of all subunits occur in a concerted manner (conservation of structural symmetry). A schematic illustration of the MWC model for a four subunit enzyme is presented in Fig. 8-5. In the absence of ligand, the enzyme exists largely in the T states (the square conformation), but substrate binds preferentially to the R states (the circular conformation), so that the conformational equilibrium is shifted to the R states by the binding of substrate. This can lead to sigmoidal binding isotherms. The positive and negative effectors, by binding preferentially to the R and T states, respectively, can enhance or reduce the sigmoidicity of the substrate binding isotherms, as shown in Fig. 8-3a. Hybrid conformational states (i.e., states containing both square and circular conformations) are absent in the MWC model.

Quantitative analysis of this model is elegantly simple. For an n subunit enzyme (containing n equivalent ligand binding sites), the model can be

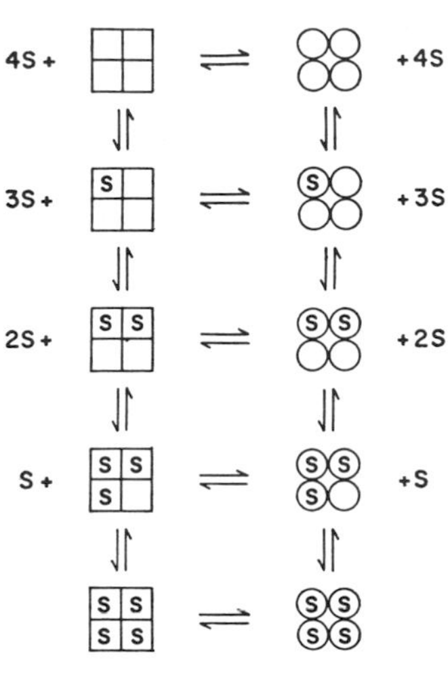

Fig. 8-5. Schematic representation of the Monod–Wyman–Changeux model for a four-subunit enzyme. The squares and circles designate different subunit conformations and S is the substrate.

written as

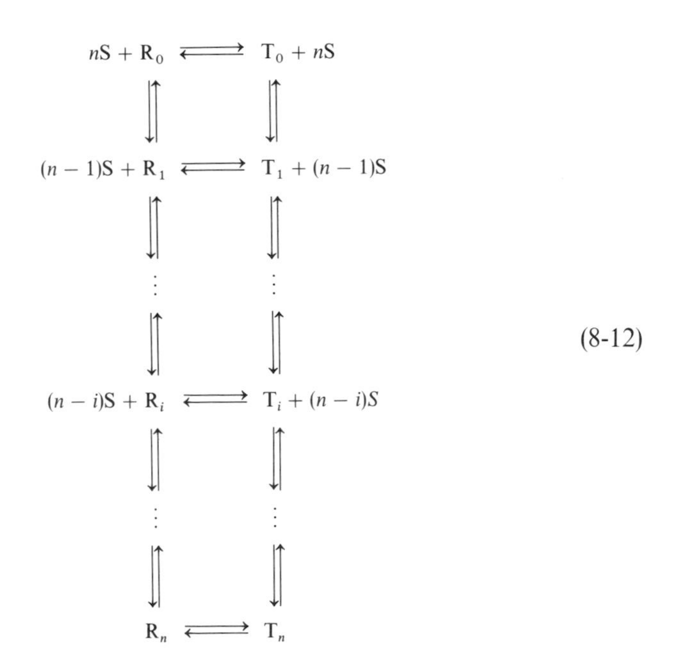

(8-12)

where R_0 and T_0 are two different conformational states in the absence of ligand, and R_i and T_i designate their complexes with i molecules of ligand S. Three constants are needed to specify the equilibrium binding isotherms: the intrinsic dissociation constants for ligand binding to the R and T states and the equilibrium constant for the ratio of the R_0 to T_0 states. These constants can be written as

$$\begin{aligned} L_0 &= [T_0]/[R_0] \\ K_R &= [(n-i+1)/i][R_{i-1}][S]/[R_i] \\ K_T &= [(n-i+1)/i][T_{i-1}][S]/[T_i] \end{aligned} \tag{8-13}$$

The saturation function (i.e., the fraction of sites bound by the ligand) $\bar{Y}$ can be expressed as

$$\begin{aligned} \bar{Y} = \frac{r}{n} &= \frac{([R_1]+2[R_2]+\cdots+n[R_n])+([T_1]+2[T_2]+\cdots+n[T_n])}{n\{([R_0]+[R_1]+\cdots+[R_n])+([T_0]+[T_1]+\cdots+[T_n])\}} \\ &= \frac{L_0 c\alpha(1+c\alpha)^{n-1}+\alpha(1+\alpha)^{n-1}}{L_0(1+c\alpha)^n+(1+\alpha)^n} \end{aligned} \tag{8-14}$$

where $\alpha = [S]/K_R$ and $c = K_R/K_T$. The nature of the binding isotherm depends on the values of L_0 and c. A hyperbolic isotherm is obtained when $c = 1$ or when L_0 and c are both either very large or very small. However, when L_0 is large and c is small, sigmoidal binding isotherms occur, as illustrated in Fig. 8-6. Binding isotherms displaying negative cooperativity cannot be obtained with the MWC model.

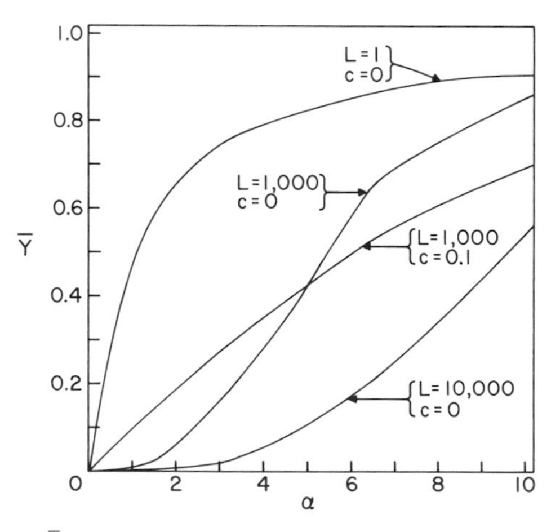

Fig. 8-6. A plot of $\bar{Y}$ versus α for the Monod–Wyman–Changeux model [Eq. (8-14)]. Plots for several values of c and L are shown.

When c is very small, Eq. (8-14) simplifies to

$$\bar{Y} = [\alpha(1+\alpha)^{n-1}]/[L_0 + (1+\alpha)^n] \tag{8-15}$$

With this limiting equation, it can be readily seen that a hyperbolic binding isotherm is found when L_0 is small, whereas a sigmoidal isotherm is predicted when L_0 is large.

In the MWC model, heterotropic effects are attributed to a shift in the equilibrium between R and T states. The limiting saturation function in the presence of an inhibitor I and an activator A may be written exactly as Eq. (8-14) except that L_0 is replaced by an apparent allosteric constant L'_0, defined as

$$L'_0 = \frac{\text{sum of different complexes of T states with I}}{\text{sum of different complexes of R states with A}}$$

where the inhibitor and activator are assumed to bind exclusively to the T and R states, respectively. If the same procedure as for the calculation of $\bar{Y}$ is used, it can be readily shown that

$$L'_0 = L_0[(1 + \beta)^n/(1 + \gamma)^n] \tag{8-16}$$

where $\beta = (\mathrm{I})/K_\mathrm{I}$, $\gamma = (\mathrm{A})/K_\mathrm{A}$, and K_A and K_I are the microscopic dissociation constants for A and I binding to the R and T states, respectively. This equation, coupled with Eq. (8-14), predicts that when the substrate binds cooperatively to the enzyme, the inhibitor increases the cooperativity of the substrate saturation curve, whereas the activator tends to abolish the cooperativity—exactly as shown earlier in Fig. 8-3a.

The basic assumptions of the AKNF theory are that (a) two conformational states A and B are available to each subunit, (b) only the subunit to which the ligand is bound changes its conformation, and (c) the ligand-induced conformational change in one subunit alters its interactions with neighboring subunits. The strength of the subunit interactions may be increased, decreased, or remain the same. Three parameters are defined in this model to calculate ligand binding isotherms—an association constant K_s characterizing the intrinsic affinity of a ligand for a given conformational state of the subunit (Eq. 8-17);

$$K_s = [\mathrm{BS}]/[\mathrm{B}][\mathrm{S}] \tag{8-17}$$

an equilibrium constant K_t characterizing the relative stability of subunit conformations A and B [Eq. (8-18)]; and constants expressing the relative

$$K_t = [\mathrm{B}]/[\mathrm{A}] \tag{8-18}$$

stability of subunit interactions between A and B conformations [Eq. (8-19)]

$$\begin{aligned} K_\mathrm{AA} &= 1 \\ K_\mathrm{AB} &= [\mathrm{AB}][\mathrm{A}]/[\mathrm{AA}][\mathrm{B}] \\ K_\mathrm{BB} &= [\mathrm{BB}][\mathrm{A}][\mathrm{A}]/[\mathrm{AA}][\mathrm{B}][\mathrm{B}] \end{aligned} \tag{8-19}$$

In Eq. (8-19), [AB] refers to interacting A and B subunits, whereas [A] and [B] refer to noninteracting subunits; setting K_{AA} equal to unity arbitrarily sets the standard free energy of interaction between two A subunits equal to zero. These molecular parameters can be used to characterize each individual molecular species present and can then be used to calculate the saturation function $\bar{Y}$ or binding isotherm r.

As an example, consider the calculation of the binding isotherm for a tetrameric enzyme in a square configuration as depicted in Fig. 8-7. To calculate the concentrations of each species, the statistical occurrence of each configuration and the subunit interactions must be taken into account.

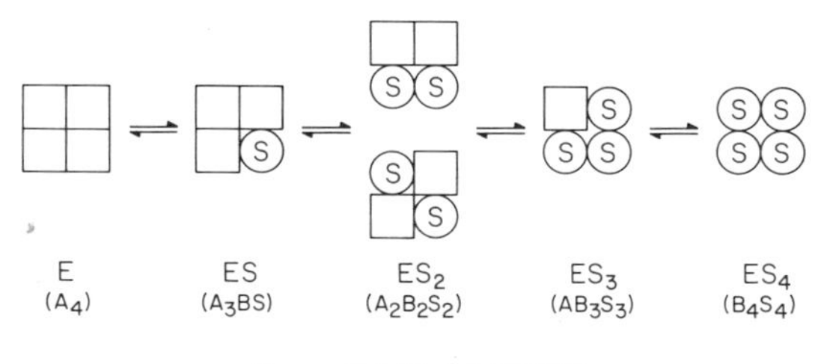

Fig. 8-7. Schematic representation of the Adair–Koshland–Nemethy–Filmer model for a four-subunit (square) enzyme. The squares and circles designate different subunit configurations, A and B, and S is the substrate. The free substrate has been omitted for the sake of simplicity.

This can be conveniently done by use of Table 8-1. The concentrations of the individual species can be written in terms of Eq. (8-17)–(8-19)

$$
\begin{aligned}
[\mathrm{ES}] &= 4K_{\mathrm{AB}}^{2}K_{s}K_{t}[\mathrm{S}][\mathrm{E}] \\
[\mathrm{ES}_2] &= ([4K_{\mathrm{AB}}^{2}K_{\mathrm{BB}} + 2K_{\mathrm{AB}}^{4})(K_{s}K_{t}[\mathrm{S}])^{2}[\mathrm{E}] \\
[\mathrm{ES}_3] &= 4K_{\mathrm{AB}}^{2}K_{\mathrm{BB}}^{2}(K_{s}K_{t}[\mathrm{S}])^{3}[\mathrm{E}] \\
[\mathrm{ES}_4] &= K_{\mathrm{BB}}^{4}(K_{s}K_{t}[\mathrm{S}])^{4}[\mathrm{E}]
\end{aligned}
\tag{8-20}
$$

Table 8-1

Characterization of the Square Tetramer in Terms of the AKNF Model[a]

Interactions: number of S bound	1	2	3	4
Formula	A_3BS	$A_2B_2S_2$	AB_3S_3	B_4S_4
Number of ways of arranging S	4	4 2	4	1
Number of AB pair interactions	2	2 4	2	0
Number of BB pair interactions	0	1 0	2	4

[a] Figure 8-7; note $A_2B_2S_2$ has two configurations.

The binding isotherm is

$$r = \frac{[ES_1] + 2[ES_2] + 3[ES_3] + 4[ES_4]}{[E] + [ES_1] + [ES_2] + [ES_3] + [ES_4]}$$

$$= \frac{4K_{AB}^2 K_s K_t[S] + 2(4K_{BB}K_{AB}^2 + 2K_{AB}^4)(K_s K_t[S])^2 + 12K_{AB}^2 K_{BB}^2 (K_s K_t[S])^3 + 4K_{BB}^4 (K_s K_t[S])^4}{1 + 4K_{AB}^2 K_s K_t[S] + (4K_{AB}^2 K_{BB} + 2K_{AB}^4)(K_s K_t[S])^2 + 4K_{AB}^2 K_{BB}^2 (K_s K_t[S])^3 + K_{BB}^4 (K_s K_t[S])^4} \quad (8\text{-}21)$$

This has the same form as Eq. (8-8) with $K_s K_t K_{AB}^2$ playing the role of the intrinsic association constant. Only two other independent constants exist for this model, K_{AB} and K_{BB}. If $K_{AB} = K_{BB} = 1$, no interactions between subunits occurs and Eq. (8-21) reduces to

$$r = \frac{4K_s K_t[S]}{1 + K_s K_t[S]}$$

(A simple hyperbolic curve also is obtained when $K_{AB}^{1/2} = K_{BB}$, even with subunit interactions occurring.)

Heterotropic interactions are explained by effector-induced conformational changes in each subunit which alter any of the constants K_s, K_t, K_{AB} and K_{BB}. Clearly, this model can display negative, positive, or mixed cooperativity, and any geometric arrangement of subunits can be interpreted in terms of this model.

Both the AKNF and MWC models are limiting cases of a more general scheme shown in Fig. 8-8. This scheme illustrates a general mechanism involving a tetrameric protein and only two conformational states for each subunit. Actually the situation is more complex than that shown since permutations of the ligand among the subunits for a given conformational state are not shown. The extreme right- and left-hand columns, enclosed by dashed lines, represent the MWC model, whereas the diagonal, enclosed by dotted lines, represents the AKNF model. Thus, these two limiting models can be viewed as limiting cases of a more complex mechanism. In practice, distinguishing between these two limiting models is very difficult. Both fit sigmoidal binding isotherms equally well. However, only the AKNF model or the assumption of several classes of independent binding sites is consistent with the observation of negative cooperativity. Binding studies in the presence of a second ligand competing for the same binding site can be useful in distinguishing between mechanistic alternatives (*25*).

The AKNF and MWC models both predict that conformational changes should accompany ligand binding. Experimental verification of the occurrence of conformational changes is consistent with these mechanisms. Many experimental techniques have been used to monitor conformational changes

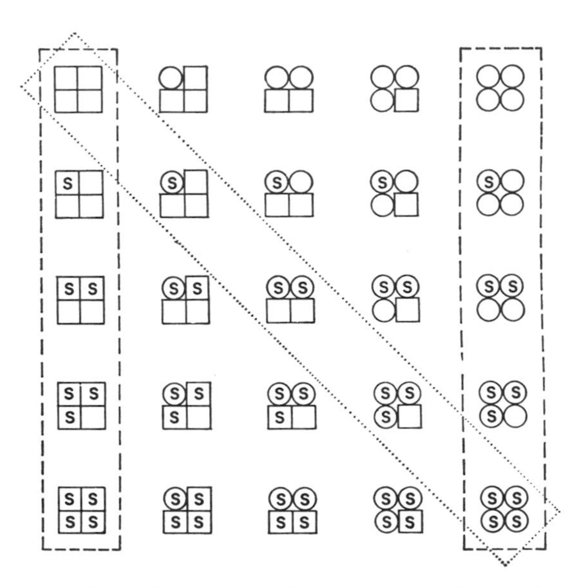

Fig. 8-8. Schematic illustration of a general allosteric model for the binding of a ligand, S, to a four-subunit enzyme. The squares and circles designate different subunit conformations. The portion enclosed by dashed lines is the MWC model, while that enclosed by dotted lines is the AKNF model. The free ligand, arrows connecting the various conformational states, and permutations of the ligand among the subunits of a given species are omitted for the sake of simplicity.

including fluorescence, ultraviolet absorption, circular dichroism and ultracentrifugation. However, conformational changes can occur even when cooperativity in ligand binding is not present. In the case of the MWC mechanism, the assumption often is made that the change in a given physical property (such as fluorescence, absorption etc.) is a direct measure of the change in conformation, and therefore can be used to measure the fraction of enzyme in the R and T state. The fraction of enzyme in the R state is

$$\bar{R} = \frac{\sum[R_i]}{\sum[R_i] + \sum[T_i]}$$

which can be written as [see the derivation of Eq. (8-14)]:

$$\bar{R} = \frac{(1 + \alpha)^n}{(1 + \alpha)^n + L_0(1 + c\alpha)^n}$$

Comparison of this result with Eq. (8-14) indicates that for a given value of α, $\bar{R}$ is always larger than $\bar{Y}$ (except when $\alpha = 1$). In other words, the fractional extent of the conformational change is always greater than the fraction of ligand saturation. For the AKNF model, the extent of the conformational change can be related to the extent of ligand saturation in almost any way, depending on how much of the total conformational change occurs as each subunit changes its conformation. Thus the correlation of conformational changes with ligand binding can indicate if a particular model is consistent with the data but is not generally a powerful discriminatory tool. Of course, if cooperativity is not associated with ligand binding, conformational changes may not occur.

STEADY-STATE KINETICS

Much of the information available on allosteric enzymes has been obtained by steady-state kinetic studies. This is because these studies require only catalytic concentrations of enzyme and because often they can be carried out with crude extracts. Moreover, the equilibrium binding isotherms for most enzymes are difficult to determine directly, but can be inferred from kinetic measurements by assuming that the number of ligand sites occupied, r, is proportional to the initial velocity, v. Various plots, exactly analogous to those described in the previous section for equilibrium binding isotherms, can then be used to analyze the kinetic data. In these plots, r is simply replaced by v. For the Hill plot, $\log[v/(V_S - v)]$ is plotted against $\log[L]$. However, the validity of all these steady-state kinetic analyses is based on the assumption that v is proportional to r. Although this assumption may be valid in many cases, it is not a necessary feature for all enzymes, and some of the alternatives will be considered.

For a number of enzymes the velocity plots are, in fact, essentially identical with the binding isotherms. This indicates that the binding steps prior to the rate determining step are rapid and reversible, and that the turnover numbers for all the binding sites are identical. Under these conditions the relationship between v and r is given by Eq. (8-22)

$$v = V_S r \tag{8-22}$$

(The maximum velocity is assumed to be expressed as the product of the molar concentration of enzyme and the turnover number per site; if the enzyme concentration is expressed in terms of the molar concentration of sites, then $v = V_S r/n$.)

If the turnover numbers of various enzyme forms or binding sites differ, the binding and kinetic data may not parallel each other even if the binding equilibria are adjusted rapidly relative to the rate-determining step. By assigning different maximal velocities for various molecular species or binding sites, the initial velocities for the MWC and AKNF models can be written as in Eqs. (8-23) and (8-24), respectively. In these equations, V_{SR} and

V_{ST} are the maximum velocities for the R and T conformational states

$$v = \frac{V_{SR}n(1 + \alpha)^{n-1} + V_{ST}Lc\alpha n(1 + c\alpha)^{n-1}}{(1 + \alpha)^n + L(1 + c\alpha)^n} \tag{8-23}$$

$$\begin{aligned} v &= \sum_{i=1}^{n} \frac{nV_{Si}[EL_i]}{[E] + [EL] + \cdots + [EL_n]} \\ &= \sum_{i=1}^{n} \frac{nV_{Si}K_1K_2 \cdots K_i[L]^i}{1 + K_1[L] + \cdots + K_1K_2 \cdots K_n[L]^n} \end{aligned} \tag{8-24}$$

respectively, and V_{Si} is the maximum velocity for the species EL_i. These equations are quite complex, but some applications to the analysis of kinetic data have been made (*26*). Even for hyperbolic velocity isotherms, the Michaelis constant, of course, is not necessarily a true equilibrium constant and, therefore, may differ significantly from binding constants obtained from thermodynamic measurements.

The substrate saturation curves (or v versus [L] plots) of allosteric enzymes usually deviate from normal Michaelis–Menten kinetics; consequently the usual plots of $1/v$ versus $1/[L]$ are nonlinear (Fig. 8-2c). They can either be concave upward (apparent positive cooperativity) or concave downward (apparent negative cooperativity). Several types of models that can account for such behavior are now discussed.

Most unisubstrate enzymes follow Michaelis–Menten kinetics. Nonlinear reciprocal plots of $1/v$ versus $1/[L]$ may arise from the following two situations. If two or more enzymes catalyze the same reaction—for example, if the enzyme exists as isozymes—the initial velocity is the sum of two or more individual velocities, as expressed by Eq. (8-25) for n enzymes. Formally, this corresponds to the case of two or more classes

$$v = \sum_{i=1}^{n} V_{Si}[L]/(K_{Si} + [L]) \tag{8-25}$$

of independent binding sites so that the double reciprocal plots for this case are always either concave downward (apparent negative cooperativity) or linear. A second possibility is that the substrate can exist in two or more different forms—for example, substrate isomers—which are active to a varying degree.

Although cooperative models provide an explanation for sigmoidal relationships between velocity and substrate concentration, cooperativity is not essential for sigmoidicity. Models having only a single independent active site but containing alternative kinetic pathways for the reaction of two or more substrates can also predict sigmoidal velocity isotherms (*27*). However, it has not yet been shown that such an explanation adequately describes the regulatory properties of enzymes.

Rate equations for enzymes containing interacting sites can be readily obtained; however, the mathematical operations become very complex for enzymes containing more than two subunits, and a useful general treatment has not yet been possible.

A serious drawback of the steady-state kinetic approach for studying regulatory enzymes is that sigmoidal saturation curves cannot be reliably interpreted in terms of binding stoichiometry and interacting binding sites. Therefore, steady-state kinetic data should not be regarded as an adequate substitute for thermodynamic binding data, which are essential for an understanding of ligand–enzyme interactions.

PROTEIN POLYMERIZATION–DEPOLYMERIZATION

Prior to the current emphasis attributing virtually all allosteric phenomena to conformational changes of proteins, sigmoidal binding curves obtained with the oxygen–hemoglobin system were suggested to be due to oxygen binding to a dissociated form of hemoglobin, which existed in equilibrium with a polymeric nonbinding form (*28*). In fact, many enzymes are involved in polymerization–depolymerization equilibria, and if a ligand can influence the state of polymerization of the protein, cooperative binding of the ligand can occur (*29–31*).

Consider a system in which a monomeric enzyme E is in equilibrium with a higher polymer E_x

$$xE \underset{}{\overset{K}{\rightleftarrows}} E_x \tag{8-26}$$

Here $K = ([E_x]/[E]^x)$ is the equilibrium constant for the polymerization reaction. The binding of ligand to the protein can be described by the equilibria constants given in Eq. (8-27)

$$\begin{aligned}
E + L &\rightleftarrows EL & K_1 &= [EL]/[E][L] \\
&\vdots & &\vdots \\
EL_{n-1} + L &\rightleftarrows EL_n & K_n &= [EL_n]/[EL_{n-1}][L] \\
E_x + L &\rightleftarrows E_xL & Q_1 &= [E_xL]/[E_x][L] \\
&\vdots & &\vdots \\
E_xL_{m-1} + L &\rightleftarrows E_xL_m & Q_m &= [E_xL_m]/[E_xL_{m-1}][L]
\end{aligned} \tag{8-27}$$

In these equations, n binding sites per monomer and m binding sites per polymer are present. If the assumption is made that all the binding sites on the monomer are independent and equivalent and all the binding sites on the polymer are independent and equivalent, intrinsic association constants K_E and K_{EX} for monomer and polymer, respectively, can be defined by expressions analogous to Eq. (8-5). The moles of ligand bound per mole of enzyme (expressed in terms of the monomer molecular weight), r, can be obtained as in previous cases

$$r = \frac{nK_E[E][L](1 + K_E[L])^{n-1} + mK_{EX}[E_x][L](1 + K_{EX}[L])^{m-1}}{[E](1 + K_E[L])^n + x[E_x](1 + K_{EX}[L])^m} \tag{8-28}$$

$$r = \frac{n\alpha(1 + \alpha)^{n-1} + K[E]^{x-1}mc\alpha(1 + c\alpha)^{m-1}}{(1 + \alpha)^n + xK[E]^{x-1}(1 + c\alpha)^m} \tag{8-29}$$

In Eq. (8-29), $\alpha = K_E[L]$, $c = K_{EX}/K_E$, and the polymeric species E_x has been eliminated by the use of the polymerization equilibrium constant K. When [E] or $[E_x]$ equals zero, the binding isotherm for the binding of ligand to independent, equivalent sites on a single protein molecule is obtained [Eq. (8-6)]. When $x = 1$, the model simplifies to an isomerizing system governed by a dimensionless equilibrium constant; if, in addition, $n = m$, the binding isotherm of the MWC model is obtained [Eq. (8-14)] with $L_0 \equiv K$). For $x > 1$, cooperative binding isotherms can be readily generated. Obviously, more complex polymerization and ligand binding equilibria, including interacting sites, are possible to treat in principle, although difficult in practice. However, the important point to note is that cooperative binding isotherms can be readily obtained with polymerizing systems. The intrinsic activity or turnover number of the monomeric and polymeric form of the enzyme may or may not be the same. For the present, the assumption is made that the polymerization equilibria are adjusted rapidly relative to the turnover number. If E and E_x have the same turnover number, the polymerization equilibria can produce a greater degree of sensitivity of the enzyme activity on the ligand concentration, which corresponds to a K system. However, the quantitative behavior also would depend on the enzyme concentration. In some cases, this may be equivalent to converting a kinetically normal enzyme to an allosteric one. If the turnover numbers of E and E_x are different, the situation becomes more complex. The expression for the initial velocity would be Eq. (8-29) with each term in the numerator multiplied by the maximum velocity of the appropriate enzyme species. Extension of these considerations to nonequilibrium (steady-state) systems is possible, but of little practical use.

HYSTERETIC ENZYMES

Thus far the response of allosteric enzymes to changes in ligand concentrations has been assumed to be rapid. However, many systems are known where ligands can induce changes in enzyme activity that occur much more slowly than the rate of the overall catalytic reaction, resulting in a time lag in the response of the enzyme to changes in the ligand level. Such slowly responding enzymes have been termed *hysteretic* (*32*). Several mechanisms can account for slow responses including ligand-induced conformational changes of the enzyme, displacement of a tightly bound ligand by a different ligand, and enzyme polymerization–depolymerization (*32–34*).

As mentioned previously, an important characteristic attributed to regulatory enzymes is that the enzyme may exist in two or more conformational states that differ in their catalytic or binding properties. The rate of interconversion of conformational states can be very slow; a simple mechanism incorporating a slow conformational transition is (*32*)

$$\begin{array}{ccccccc} E + S & \underset{}{\overset{K_1}{\rightleftharpoons}} & ES & \underset{k_{-3}}{\overset{k_3}{\rightleftharpoons}} & E'S & \underset{}{\overset{K_4}{\rightleftharpoons}} & E' + S \\ & & \downarrow k_5 & & \downarrow k_6 & & \\ & & P + E & \underset{k_{-2}}{\overset{k_2}{\rightleftharpoons}} & E' + P & & \end{array} \tag{8-30}$$

In this equation, the K_is are equilibrium constants, the k_is are rate constants, X, P, and E are substrate, product, and enzyme, respectively, and the prime denotes a change in enzyme conformation. The following assumptions are now made: (a) the substrate binding steps are in rapid equilibrium; (b) the overall reaction is irreversible; (c) the substrate concentration remains constant; (d) no product inhibition occurs; and (e) the rates of the conformational changes are slow (that is, k_2, k_{-2}, k_3, and k_{-3} are small relative to k_5 and/or k_6). Under these conditions, the rate of the reaction can be written as

$$d[\mathrm{P}]/dt = v_t = v_\infty + (v_0 - v_\infty)e^{-k't} \tag{8-31}$$

where v_t, v_∞, and v_0 are the velocities at time t, at infinite time, and at zero time, respectively, and k' is a complex rate constant given by Eq. (8-32). The detailed balance condition $k_{-2}k_3K_4 = k_2k_{-3}K_1$ and the relationship $K_2 = k_2/k_{-2}$ have been used to obtain Eq. (8-32). If the enzyme

$$k' = k_2[1 + k_3[\mathrm{S}]/k_2K_1]\left\{\frac{1}{1 + [\mathrm{S}]/K_1} + \frac{1}{K_2(1 + [\mathrm{S}]/K_4)}\right\} \tag{8-32}$$

exists in the E form only before substrate addition (that is, $[\mathrm{E}'] = 0$ at $t = 0$), the initial velocity v_0 is defined by Eq. (8-33) and the final velocity v_∞ by Eq. (8-34) with $[\mathrm{E}_0]$ being the total enzyme concentration

$$\frac{v_0}{[\mathrm{E}_0]} = \frac{k_5}{1 + K_1/[\mathrm{S}]} \tag{8-33}$$

$$\frac{v_\infty}{[\mathrm{E}_0]} = \frac{k_5(1 + K_1K_2k_6/k_5K_4)/(1 + K_1K_3/K_4)}{1 + (K_1/[\mathrm{S}])[(1 + K_2)/(1 + K_1K_2/K_4)]} \tag{8-34}$$

Equation (8-31) can be readily integrated, if the substrate concentration is assumed to remain constant, to give

$$[\mathrm{P}]_t = v_\infty t - (1/k')(v_\infty - v_0)(1 - e^{-k't}) \tag{8-35}$$

The second term in Eq. (8-35) represents a lag function: the extent of the lag before reaching a constant rate of product formation depends on k', which in turn is a function of the substrate concentration. An example of an enzyme whose behavior is consistent with Eq. (8-31) is frog phosphorylase: the increase in activity during arsenolysis after adding glycogen can be quantitatively described by this equation (*32*).

The simple mechanism given in Eq. (8-30) can be readily extended to include effectors. The lag time then becomes a function of both substrate and effector concentrations. In any case, v_0 and v_∞ could differ markedly so that hysteretic responses to changes in substrate and effector concentrations could occur. The further extension of this mechanism to include interacting ligand binding sites also is possible.

The fact that nonhyperbolic binding of ligands may occur when there are rapidly adjusted enzyme polymerization–depolymerization equilibria has been discussed earlier. If the different molecular weight forms have different kinetic characteristics, and if ligand-induced interconversion of polymeric forms is slow, then slow time-dependent changes of the enzymatic activity can occur in response to effectors. Rate equations for ligand-induced slow polymerization–depolymerization reactions can be derived and they are analogous to those presented above for slow conformational changes. However, the rate equations become very complex and generalization is difficult.

Slow ligand-induced transitions also can give rise to cooperative phenomena. An analysis of steady-state kinetics and simulation studies for the slow isomerization mechanism, assuming the binding steps are in a steady state but not at equilibrium, have demonstrated that both positive and negative cooperativity can occur with either lags or bursts of enzyme activity (*33*). Thus, for those cases in which a slow transition occurs, the transition may be associated with a nonlinear double reciprocal plot of the steady-state kinetic data without requiring interactions between sites. The combination of slow transitions and site–site interactions could generate even more complex types of cooperative kinetics.

Hysteretic enzyme systems may be relevant in enzyme regulation. They could provide a time-dependent buffering that would prevent immediate changes in the concentration levels of metabolites, but would allow a slow approach to new metabolic conditions. At this time, however, the physiological significance of hysteretic enzymes is uncertain, although a number of regulatory enzymes appear to be associated with slow ligand-induced changes in activity (*31–34*).

TRANSIENT KINETICS

Thus far the analysis of equilibrium binding isotherms and steady-state kinetics for allosteric enzymes has been considered. Equilibrium binding studies are a necessary prelude to the understanding of allosteric enzymes. However, such studies detect only overall changes of complex systems, and it is very difficult, if possible at all, to establish mechanistic features from thermodynamic data. Dynamic information is a necessity for the elucidation of molecular mechanisms. Steady-state kinetic studies can be of some assistance, but, as has been emphasized, a clear-cut interpretation of steady-state kinetic data is difficult. In addition, the very low concentrations of enzyme employed in steady-state kinetic studies preclude the observation of reaction intermediates. Ideally, the elucidation of an allosteric mechanism requires that the overall changes be broken down to their elementary steps by a direct study of the kinetic behavior of all reaction intermediates. This corresponds to spreading out the reaction mechanism on the entire time axis, as well as the concentration axis, and requires the use of fast reaction techniques.

As an illustration of the kinetic analysis of systems near equilibrium, the two simple mechanisms presented in Eqs. (8-36) and (8-37) are now considered

$$\mathrm{E} + \mathrm{L} \underset{k_{-1}}{\overset{k_1}{\rightleftharpoons}} \mathrm{EL} \underset{k_{-2}}{\overset{k_2}{\rightleftharpoons}} \mathrm{E'L} \qquad (8\text{-}36)$$

$$\begin{array}{ccc} \mathrm{E} + \mathrm{L} & \underset{k_{-2}}{\overset{k_2}{\rightleftharpoons}} & \mathrm{EL} \\ k_1 \big\downarrow\uparrow k_{-1} & & \\ \mathrm{E'} & & \end{array} \qquad (8\text{-}37)$$

The first mechanism represents a conformational change of the enzyme induced by ligand binding, as proposed in the AKNF model, whereas the second represents a pre-equilibration between two conformations of the enzyme with the ligand binding preferentially to one conformation, as proposed in the MWC model. These examples do not involve the question of sequential versus concerted conformational changes, which is a fundamental difference between the AKNF and MWC models. Two relaxation processes are associated with each of these mechanisms. If the assumption is made that the bimolecular binding steps are much faster than the isomerization steps, the reciprocals of the two relaxation times τ_1 and τ_2 associated with the mechanism of Eq. (8-36) can be written as

$$1/\tau_1 = k_{-1} + k_1([\bar{\mathrm{E}}] + [\bar{\mathrm{L}}])$$
$$1/\tau_2 = k_{-2} + \frac{k_2}{1 + (k_{-1}/k_1)/([\bar{\mathrm{E}}] + [\bar{\mathrm{L}}])} \tag{8-38}$$

and those associated with the mechanism of Eq. (8-37) can be written as

$$1/\tau_1 = k_{-2} + k_2([\bar{\mathrm{E}}] + [\bar{\mathrm{L}}])$$
$$1/\tau_2 = k_1 + \frac{k_{-1}}{1 + [\bar{\mathrm{L}}]/\{(k_{-2}/k_2) + [\bar{\mathrm{E}}]\}} \tag{8-39}$$

In these equations the overbars designate equilibrium concentrations. These two mechanisms can be distinguished by the different dependence of the relaxation times on substrate concentration. In the first case, $1/\tau_2$ increases as $[\bar{\mathrm{L}}]$ increases, whereas in the second case, $1/\tau_2$ decreases as $[\bar{\mathrm{L}}]$ increases. In both cases $1/\tau_2$ approaches a limiting value at high substrate concentrations. Thus kinetic measurements can distinguish between these two mechanisms. Allosteric models can be analyzed similarly.

If in the MWC model [Eq. (8-12)] the vertical ligand binding steps are assumed to equilibrate much more rapidly than the horizontal isomerization steps, the relaxation spectrum associated with the mechanism then can be readily calculated. The vertical steps can be viewed separately as coupled bimolecular reactions

$$\mathrm{R} + \mathrm{S} \underset{k_{\mathrm{D}}^{\mathrm{R}}}{\overset{k_{\mathrm{A}}^{\mathrm{R}}}{\rightleftharpoons}} \mathrm{RS}$$
$$\mathrm{T} + \mathrm{S} \underset{k_{\mathrm{D}}^{\mathrm{T}}}{\overset{k_{\mathrm{A}}^{\mathrm{T}}}{\rightleftharpoons}} \mathrm{TS} \tag{8-40}$$

Here $k_{\mathrm{D}}^{\mathrm{R}}/k_{\mathrm{A}}^{\mathrm{R}} = K_{\mathrm{R}}$, $k_{\mathrm{D}}^{\mathrm{T}}/k_{\mathrm{A}}^{\mathrm{T}} = K_{\mathrm{T}}$, and R and T represent free binding sites in the R and T states, respectively, that is, $[\mathrm{R}] = n[\mathrm{R}_0] + (n-1)[\mathrm{R}_1] + \cdots + [\mathrm{R}_{n-1}]$ and $[\mathrm{T}] = n[\mathrm{T}_0] + (n-1)[\mathrm{T}_1] + \cdots + [\mathrm{T}_{n-1}]$. These reactions are coupled since they share the same ligand as a reactant. The rate equations for this mechanism in the neighborhood of equilibrium are

$$-\frac{d\Delta[\mathrm{R}]}{dt} = \{k_{\mathrm{A}}^{\mathrm{R}}([\bar{\mathrm{R}}] + [\bar{\mathrm{S}}]) + k_{\mathrm{D}}^{\mathrm{R}}\}\,\Delta[\mathrm{R}] + k_{\mathrm{A}}^{\mathrm{R}}[\bar{\mathrm{R}}]\,\Delta[\mathrm{T}]$$

$$-\frac{d\Delta[\mathrm{T}]}{dt} = k_{\mathrm{A}}^{\mathrm{T}}[\bar{\mathrm{T}}]\,\Delta[\mathrm{R}] + \{k_{\mathrm{A}}^{\mathrm{T}}([\bar{\mathrm{T}}] + [\bar{\mathrm{S}}]) + k_{\mathrm{D}}^{\mathrm{T}}\}\,\Delta[\mathrm{T}]$$

The relaxation times associated with this mechanism are

$$1/\tau_{1,2} = \{(a_{11} + a_{22}) \pm [(a_{11} + a_{22})^2 - 4(a_{11}a_{22} - a_{12}a_{21})]^{1/2}\}/2 \tag{8-41}$$

where

$$a_{11} = k_A^R([\bar{R}] + [\bar{S}]) + k_D^R \qquad a_{12} = k_A^R[\bar{R}]$$
$$a_{21} = k_A^T[\bar{T}] \qquad a_{22} = k_A^T([\bar{T}] + [\bar{S}]) + k_D^T$$

Both of these relaxation times become shorter as the substrate concentration is raised since the binding steps become faster. Under certain limiting conditions—for example, if the R state reacts much faster than the T state or vice versa—the reciprocal relaxation times become linear functions of simple concentration functions (*35, 36*).

The third relaxation time characterizing the MWC mechanism is associated with the relatively slow equilibration between the R and T states, i.e., the concerted conformational change

$$\sum R_i \underset{k_{-i}}{\overset{k_i}{\rightleftharpoons}} \sum T_i \tag{8-42}$$

The relaxation time for this process can be derived as follows. The rate law is

$$-\frac{d\Delta \sum_{i=0}^{n} [R_i]}{dt} = k_0\Delta[R_0] + k_1\Delta[R_1] + \cdots + k_n\Delta[R_n] - k_{-0}\Delta[T_0] - \cdots - k_{-n}\Delta[T_n] \tag{8-43}$$

$$= \left[\sum_{i=0}^{n} \Delta[R_i]\right] \Big/ \tau_3 \tag{8-44}$$

Mass conservation requires that

$$\sum_{i=0}^{n} \Delta[R_i] = \Delta[R_0]\left(1 + \sum_{i=1}^{n} \frac{\Delta[R_i]}{\Delta[R_0]}\right)$$

or

$$\Delta[R_0] = \frac{\sum_{i=0}^{n} \Delta[R_i]}{1 + \sum_{i=1}^{n} \Delta[R_i]/\Delta[R_0]}$$

and

$$\Delta[R_i] = (\Delta[R_i]/\Delta[R_0])\Delta[R_0]$$

Insertion of these relationships and similar ones for the T state into Eqs. (8-43) and (8-44) gives

$$\frac{1}{\tau_3} = \frac{k_0 + k_1\Delta[R_1]/\Delta[R_0] + \cdots + k_n\Delta[R_n]/\Delta[R_0]}{1 + \Delta[R_1]/\Delta[R_0] + \cdots + \Delta[R_n]/\Delta[R_0]} + \frac{k_{-0} + k_{-1}\Delta[T_1]/\Delta[T_0] + \cdots + k_{-n}\Delta[T_n]/\Delta[T_0]}{1 + \Delta[T_1]/\Delta[T_0] + \cdots + \Delta[T_n]/\Delta[T_0]} \tag{8-45}$$

If $\Delta[\text{L}] \approx 0$, that is the ligand concentration is buffered because it is much larger than the enzyme concentration, then from Eq. (8-13)

$$K_{\text{R}} = [(n - i + 1)/i][\bar{\text{S}}]\,\Delta[\text{R}_{i-1}]/\Delta[\text{R}_i]$$
$$K_{\text{T}} = [(n - i + 1)/i][\bar{\text{S}}]\,\Delta[\text{T}_{i-1}]/\Delta[\text{T}_i]$$

Substitution of these equations into Eq. (8-45) yields

$$1/\tau_3 = \frac{\sum_0^n k_j[n!/(n-j)!j!]([\bar{\text{S}}]/K_{\text{R}})^j}{(1 + [\bar{\text{S}}]/K_{\text{R}})^n} + \frac{\sum_0^n k_{-j}[n!/(n-j)!j!]([\bar{\text{S}}]/K_{\text{T}})^j}{(1 + [\bar{\text{S}}]/K_{\text{T}})^n} \tag{8-46}$$

Qualitatively, if $k_i + k_{-i} > k_{i-1} + k_{-(i-1)}$ for all i, $1/\tau_3$ increases with increasing ligand concentrations, whereas if $k_i + k_{-i} < k_{i-1} + k_{-(i-1)}$ for all i, $1/\tau_3$ decreases with increasing ligand concentrations. The reciprocal relaxation time also can go through a maximum or minimum depending on the relative values of the rate constants. If the further assumption is made that $k_{-0} = k_{-1} = \cdots k_{-n}$, then by detailed balance, $k_0 = k_1/c = k_2/c^2 = \cdots k_n/c^n$ and Eq. (8-46) can be simplified to give

$$1/\tau_3 = k_{-0} + k_0\left(\frac{1 + [\bar{\text{S}}]/K_{\text{T}}}{1 + [\bar{\text{S}}]/K_{\text{R}}}\right)^n \tag{8-47}$$

As shown in Fig. 8-9, when $K_{\text{R}} < K_{\text{T}}$, $1/\tau_3$ decreases as $[\bar{\text{S}}]$ increases, whereas when $K_{\text{R}} > K_{\text{T}}$, $1/\tau_3$ increases as $[\bar{\text{S}}]$ increases. An alternative simplifying assumption is that $k_0 = k_1 = \cdots k_n$; detailed balance then requires that $k_{-0} = ck_{-1} = c^2k_{-2} = \cdots c^nk_n$. Because of the symmetric nature of the MWC mechanism, the expression obtained for $1/\tau_3$ is the same as Eq. (8-47) except that the subscript R is interchanged with T and -0 with 0. Now when $K_{\text{R}} < K_{\text{T}}$, $1/\tau_3$ increases as $[\bar{\text{S}}]$ increases and when $K_{\text{R}} > K_{\text{T}}$, $1/\tau_3$ decreases as $[\bar{\text{S}}]$ increases. In terms of the original MWC mechanism, the only meaningful cases are when $K_{\text{R}} < K_{\text{T}}$. Thus this limiting form of the MWC mechanism predicts three relaxation times each with its own characteristic concentration dependence.

The AKNF model proposes a sequence of ligand binding steps coupled to conformational changes, which can be written as

$$\begin{array}{lllll} \text{E} & + \text{S} \rightleftarrows & \text{ES}_1 & \rightleftarrows & \text{ES}'_1 \\ \text{ES}'_1 & + \text{S} \rightleftarrows & \text{ES}'_2 & \rightleftarrows & \text{ES}''_2 \\ \vdots & & & & \vdots \\ \text{ES}''^{\cdots\prime}_{n-1} & + \text{S} \rightleftarrows & \text{ES}''^{\cdots\prime\prime}_n & \rightleftarrows & \text{ES}''^{\cdots\prime\prime\prime}_n \end{array} \tag{8-48}$$

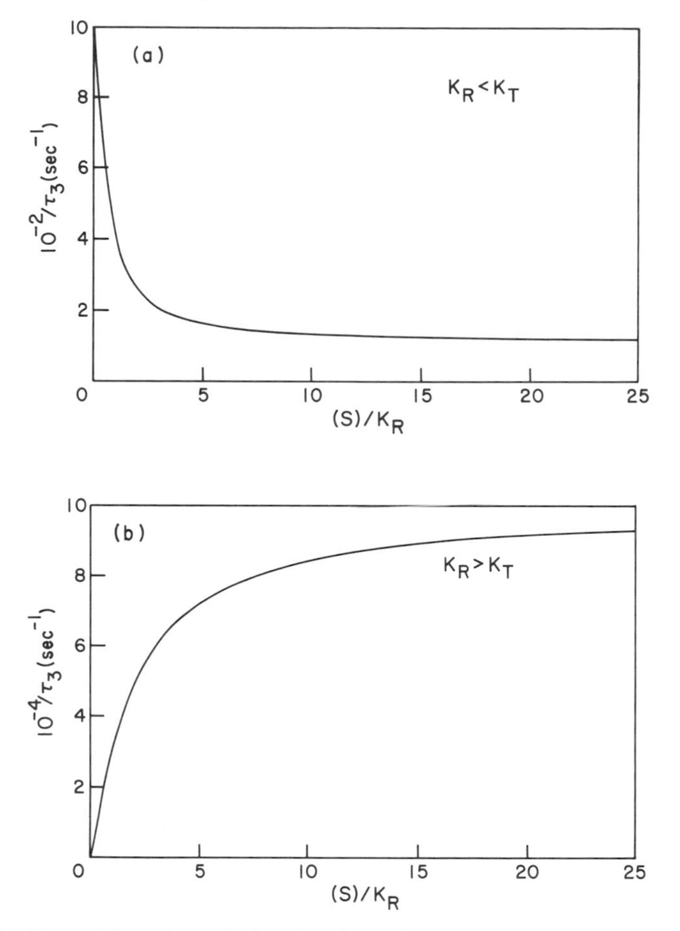

Fig. 8-9. Plots of the reciprocal relaxation time, $1/\tau_3$, associated with concerted conformational changes in the MWC model versus the ratio of the substrate concentration [S] to the intrinsic dissociation constant K_R for binding to the R state. The reciprocal relaxation times are calculated according to Eq. (8-47) with $n = 2$, $k_{-0} = 100 \sec^{-1}$, $k_0 = 1000 \sec^{-1}$ and (a) $K_R = 0.1\ K_T$ or (b) $K_R = 10K_T$. When $[S] = 0$, $1/\tau_3 = k_{-0} + k_0$ and when $[S] = \infty$, $1/\tau_3 = k_{-0} + k_0(K_R/K_T)^n$ in both cases. [Reproduced, with permission, from G. G. Hammes and C.-W. Wu, *Annu. Rev. Biophys. Bioeng.* **3**, 1. © 1974 by Annual Reviews, Inc.]

A spectrum of $2n$ relaxation times is predicted by this mechanism. In practice, the experimental resolution of such a complex relaxation spectrum is very difficult, and in general not all relaxation processes would be expected to be observed under all conditions. Their observation depends on the relative population of each state, the specific rate of each elementary step, and the amplitude of the relaxation effects. For example, if one conformational transition is assumed to be much slower than all other steps, as shown in Eq. (8-49),

$$
\begin{aligned}
E + S &\underset{}{\overset{K_1}{\rightleftharpoons}} X_1 \\
X_1 + S &\underset{}{\overset{K_2}{\rightleftharpoons}} X_2 \\
&\vdots \\
X_i &\underset{k_{-2}}{\overset{k_2}{\rightleftharpoons}} X_i' \\
X_i' + S &\underset{}{\overset{K_j}{\rightleftharpoons}} X_{i+1} \\
&\vdots \\
X_{n-1} + S &\underset{}{\overset{K_n}{\rightleftharpoons}} X_n
\end{aligned}
\tag{8-49}
$$

where the transition $X_i \rightleftharpoons X_i'$ is assumed to be much slower than all other reactions, then the smallest (slowest) reciprocal relaxation time is given by

$$1/\tau = \frac{k_2}{1 + [1/\sum_{m=1}^{i} \prod_{l=1}^{m} K_l[\bar{S}]^m]} + \frac{k_{-2}}{1 + \sum_{m=i+1}^{n} \prod_{l=i+1}^{m} K_l[\bar{S}]^{m-i}} \tag{8-50}$$

Again the substrate concentration has been assumed to be buffered. The other fast conformational changes associated with ligand binding can be included in the K_is and need not be explicitly written. The first term of Eq. (8-50) increases as $[\bar{S}]$ increases, whereas the second term decreases as $[\bar{S}]$ increases. This reciprocal relaxation time can increase or decrease or go through a maximum or minimum, similar to the slowest relaxation process associated with the MWC model discussed earlier.

In principle, determination of the relaxation spectrum associated with ligand binding to an allosteric enzyme should distinguish between the MWC and AKNF models. However, the practical limitation is that often the complete relaxation spectrum cannot be determined. Some situations are relatively clear-cut. If the three relaxation times associated with the MWC mechanism discussed above are all observed, then it seems unlikely that the AKNF mechanism is operative, although a devout skeptic might argue that this is not the case and instead all the expected relaxation times associated with the AKNF model simply are not seen. On the other hand, if a complex spectrum of relaxation times is seen that can be interpreted in terms of sequential binding and conformational change relations, the AKNF model is the clear-cut choice. (The fact that the MWC model does not predict negative cooperativity also is of importance.)

A less clear-cut case occurs when only a single relaxation process is associated with ligand binding and is characteristic of a conformational change. In principle, this is consistent with both the MWC and AKNF models, but in the former case the assumption must be made that the bimolecular ligand binding steps cannot be detected, whereas in the latter case the assumption must be made that $2n - 1$ relaxation processes are not observed, including conformational transitions presumably very similar to that detected. Bimolecular enzyme–ligand reactions are usually very rapid, and, therefore, difficult to study so that it is not too surprising that the relaxation times associated with such reactions are not detected. However, this does not explain the unseen relaxation times associated with conformational transitions in the AKNF model. On the basis of these considerations, and in the absence of any conflicting data, the simplest interpretation of a single relaxation process associated with a conformational change accompanying cooperative ligand binding is in terms of the MWC model. While this is not as definitive as determination of the complete relaxation spectrum, it does permit the establishment of a working hypothesis on which future experiments can be designed. The more general allosteric model illustrated in Figure 8-8 may be required in situations where neither of the two limiting models is sufficient to account for the experimental observations.

CONCLUSION

A number of explanations for the regulatory properties of enzymes has been presented. From the data already available it is apparent that physiological systems rely on a variety of regulatory mechanisms. Perhaps the differences between the two limiting mechanisms proposed, the MWC and AKNF models, have been overemphasized: on a molecular basis their similarities seem more important, namely, the postulation of conformational transitions and changing subunit interactions leading to regulation. However, regulation through kinetic effects and enzyme polymerization–depolymerization should not be overlooked. Regulation need not be rapid, and hysteretic enzymes may be simply a logical extension of other models. Finally, it seems very likely that physiological systems will employ regulatory schemes more complex than any single limiting model discussed; this may even involve combinations of regulatory mechanisms such as conformational changes and polymerization–depolymerization. A combination of equilibrium ligand binding and transient kinetic measurements is a powerful tool for the elucidation of regulatory mechanisms, but must be used carefully and critically since ambiguities in interpretation still can occur. Obviously, structural studies also are of great importance, and only a correlation of thermodynamic, kinetic, and structural information can lead to a complete molecular mechanism.

REFERENCES

1. A. L. Lehninger, "Biochemistry" 2nd ed. Worth, New York, 1975.
2. E. R. Stadtman, *in* "The Enzymes" (P. Boyer, ed.), 3rd ed., Vol. 1, p. 397. Academic Press, New York, 1970.
3. R. T. Schimke, *Adv. Enzymol.* **37**, 135 (1973).
4. R. Revel and Y. Groner, *Annu. Rev. Biochem.* **47**, 1079 (1978).
5. J. C. Gerhart and A. B. Pardee, *J. Biol. Chem.* **237**, 891 (1962).
6. G. N. Cohen, *in* "Current Topics in Cellular Regulation" (B. L. Horecker and E. R. Stadtman, eds.), Vol. 1, p. 183. Academic Press, New York, 1969.
7. H. E. Umbarger, *Annu. Rev. Biochem.* **38**, 323 (1969).
8. E. W. Nester and R. A. Jensen, *J. Bacteriol.* **91**, 1591 (1966).
9. H. Paulus and E. Gray, *J. Biol. Chem.* **239**, PC4008 (1964).
10. C. T. Casky, D. M. Ashton, and J. B. Wyngaarden, *J. Biol. Chem.* **239**, 2570 (1964).
11. C. A. Woolfolk and E. R. Stadtman, *J. Bacteriol.* **118**, 736 (1967).
12. R. D. Berlin and E. R. Stadtman, *J. Biol. Chem.* **241**, 2679 (1966).
13. P. Maeba and B. D. Sanual, *Biochem. Biophys. Res. Commun.* **22**, 194 (1966).
14. R. J. Fletterick and N. B. Madsen, *Annu. Rev. Biochem.* **48**, 31 (1980).
15. P. B. Chock, S. G. Rhee, and E. R. Stadtman, *Annu. Rev. Biochem.* **48**, 813 (1980).
16. L. J. Reed, *Acc. Chem. Res.* **7**, 40 (1974).
17. P. Jurtshuk, I. Sekuzu and D. E. Green, *Biochem. Biophys. Res. Commun.* **6**, 76 (1961).
18. G. G. Hammes and D. A. Hilborn, *Biochim. Biophys. Acta* **233**, 580 (1971).
19. J. Monod, J. Wyman and J.-P. Changeux, *J. Mol. Biol.* **12**, 88 (1965).
20. D. E. Koshland, Jr., G. Nemethy and G. Filmer, *Biochemistry* **5**, 365 (1966).
21. A. Levitzki and D. E. Koshland, Jr., *Proc. Natl. Acad. Sci. U.S.A.* **62**, 1121 (1969).
22. G. Scatchard, *Ann. N.Y. Acad. Sci.* **51**, 660 (1949).
23. A. V. Hill, *J. Physiol.* (*London*) **40**, iv–viii (1910).
24. G. S. Adair, *J. Biol. Chem.* **63**, 529 (1925).
25. Y. I. Henis and A. Levitzki, *Eur. J. Biochem.* **102**, 449 (1979).
26. M. E. Kirtly and D. E. Koshland, Jr., *J. Biol. Chem.* **242**, 4192 (1967).
27. J. R. Sweeny and J. R. Fisher, *Biochemistry* **7**, 561 (1968).
28. R. W. Briehl, *J. Biol. Chem.* **238**, 2361 (1963).
29. C. Frieden and R. Colman, *J. Biol. Chem.* **242**, 1705 (1967).
30. L. W. Nichol, W. J. H. Jackson and D. J. Winzor, *Biochemistry* **6**, 2449 (1967).
31. K. E. Neet, *Bull. Mol. Biol. Med.* **4**, 100 (1979).
32. C. Frieden, *J. Biol. Chem.* **245**, 5788 (1970).
33. G. R. Ainslie, Jr., J. P. Shill and K. E. Neet, *J. Biol. Chem.* **247**, 7088 (1972).
34. C. Frieden, *Annu. Rev. Biochem.* **48**, 471 (1979).
35. M. Eigen, *in* "Fast Reactions and Primary Processes in Chemical Kinetics" (S. Claesson, ed.), p. 333. Wiley (Interscience), New York, 1979.
36. K. Kirschner, *Curr. Top. Cell. Regul.* **3**, 167 (1971).

9

Case Studies of Selected Regulatory Enzymes

INTRODUCTION

Because regulatory enzymes generally are more complex structurally than enzymes having only a catalytic function, an understanding of regulatory mechanisms in molecular detail is not at the level found for the hydrolytic mechanisms discussed in Chapter 7. Nevertheless, a large number of regulatory enzymes are available in reasonably pure form and in sufficiently large amounts for extensive study. Three-dimensional structures are being determined, and a few already are available, although not yet at atomic resolution. In this chapter, two specific regulatory enzymes are discussed, which have been extensively studied by a variety of different methods.

ASPARTATE TRANSCARBAMOYLASE

Aspartate transcarbamoylase catalyzes the first committed step in pyrimidine biosynthesis, namely, the carbamoylation of aspartic acid

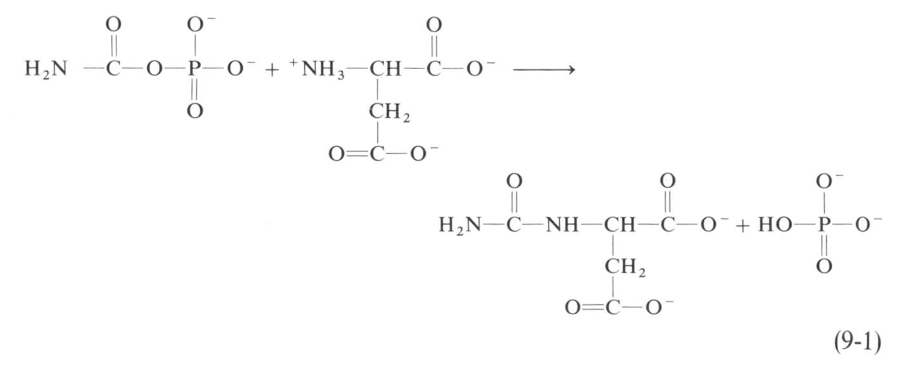

(9-1)

The enzyme is inhibited by the ultimate end product of the biosynthetic pathway, CTP. In fact, this is one of the first enzymes for which the mechanism of feedback inhibition was established (*1*). The purine nucleotide ATP activates the enzyme; this may be a regulatory mechanism for controlling the relative amounts of purines and pyrimidines. The discussion here is confined to the enzyme found in *Escherichia coli*. This enzyme has been probed with a large variety of different methods by many different investigators, and extensive reviews are available (cf. *2*,*3*). An endearing feature of the *E. coli* enzyme is that a pyrimidine auxotroph has been found in which 8% of the total protein of the bacteria is aspartate transcarbamoylase. Therefore, large amounts of pure enzyme can be obtained relatively easily. A kinetic isotherm in which the initial steady-state velocity is plotted versus the concentration of aspartate at saturating concentrations of carbamoyl phosphate is shown schematically in Fig. 9-1. When CTP is added, the isotherm becomes more sigmoidal and when ATP is added less sigmoidal, with the same maximum velocity being reached in all cases. Thus this is a classical K system. Furthermore, treatment of the enzyme with mercurials, heat, urea, trypsin, or other reagents, eliminates the effects of CTP and ATP, suggesting aspartate transcarbamoylase is an allosteric enzyme.

The enzyme has a molecular weight of 300,000. When it is treated with parahydroxymercuribenzoate (or many other reagents that react reversibly

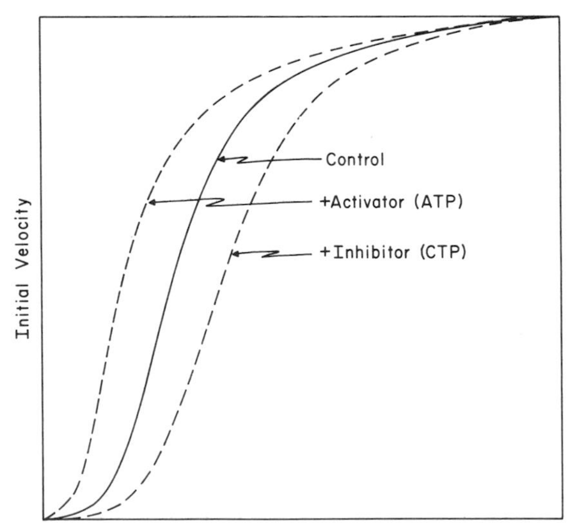

Fig. 9-1. Schematic representation of a plot of the initial steady state velocity for the reaction catalyzed by aspartate transcarbamoylase versus aspartate at a saturating concentration of carbamoyl phosphate with no effectors (—), ATP (– – –), and CTP (– – –) present.

with sulfhydrl groups) and passed through a DEAE-Sephadex column, two proteins are obtained: one is catalytically active, but the activity is not regulated by CTP and ATP; the other has no catalytic activity, but binds CTP tightly. If the two proteins are mixed together and the resulting solution dialyzed, the native enzyme is obtained with the catalytic activity regulated by CTP and ATP. This result establishes clearly the allosteric nature of the regulation since the catalytic sites and CTP binding sites are on different proteins. The catalytic subunit, as the protein with catalytic activity is called, has a molecular weight of 100,000; two are present in the native enzyme. The catalytic subunit contains three identical polypeptide chains of molecular weight 33,000. The catalytic subunit has a turnover number approximately twice as great as the native enzyme. The finding that regulated enzymes have less than maximum catalytic efficiency is quite general and not unexpected. The regulatory subunit has a molecular weight of 34,000; three are present in the native enzyme. The regulatory subunit contains two identical polypeptide chains of molecular weight 17,000. The amino acid sequence of the regulatory polypeptide has been published (*4*) and that of the catalytic polypeptide is known but not yet published. While the stoichiometry of six identical catalytic polypeptide chains and six identical regulatory polypeptide chains per enzyme molecule is now firmly established, it was the center of considerable controversy for several years. A single zinc ion is found per regulatory polypeptide chain. The zinc is not involved in CTP binding, but is essential for reconstitution of the enzyme, suggesting the zinc may be at the interface of a regulatory and catalytic polypeptide. Several other metals can be substituted for zinc (e.g., Co and Mn).

A number of chemical modification studies have been carried out with aspartate transcarbamoylase and its subunits (*2*). The catalytic polypeptide chain has a single sulfhydryl residue and modification of it by a variety of reagents results in inactivation of the enzyme; structural studies indicate the sulfhydryl group is very near the active site. The regulatory polypeptide chain has four sulfhydryl residues which may be involved in binding zinc. A single amino group on the catalytic polypeptide chain reacts with pyridoxal phosphate, which competitively inhibits carbamoyl phosphate binding (*5,6*). If the resultant Schiff base is reduced with borohydride, the polypeptide is covalently labeled with pyridoxamine phosphate (at lysine 80 in the amino acid sequence), and the enzyme is inactive. The amino group clearly is near the active site. Photooxidation experiments indicate two histidines per catalytic chain are oxidized, resulting in enzyme inactivation (*5*). Tetranitromethane specifically nitrates tyrosine (cf. *7*). If one tyrosine per catalytic polypeptide chain is nitrated (tyrosine-213), the enzyme is active, and the nitrotyrosine group serves as a spectral probe for monitoring conformational changes. Nitration of a second tyrosine (160) inactivates the enzyme. Several

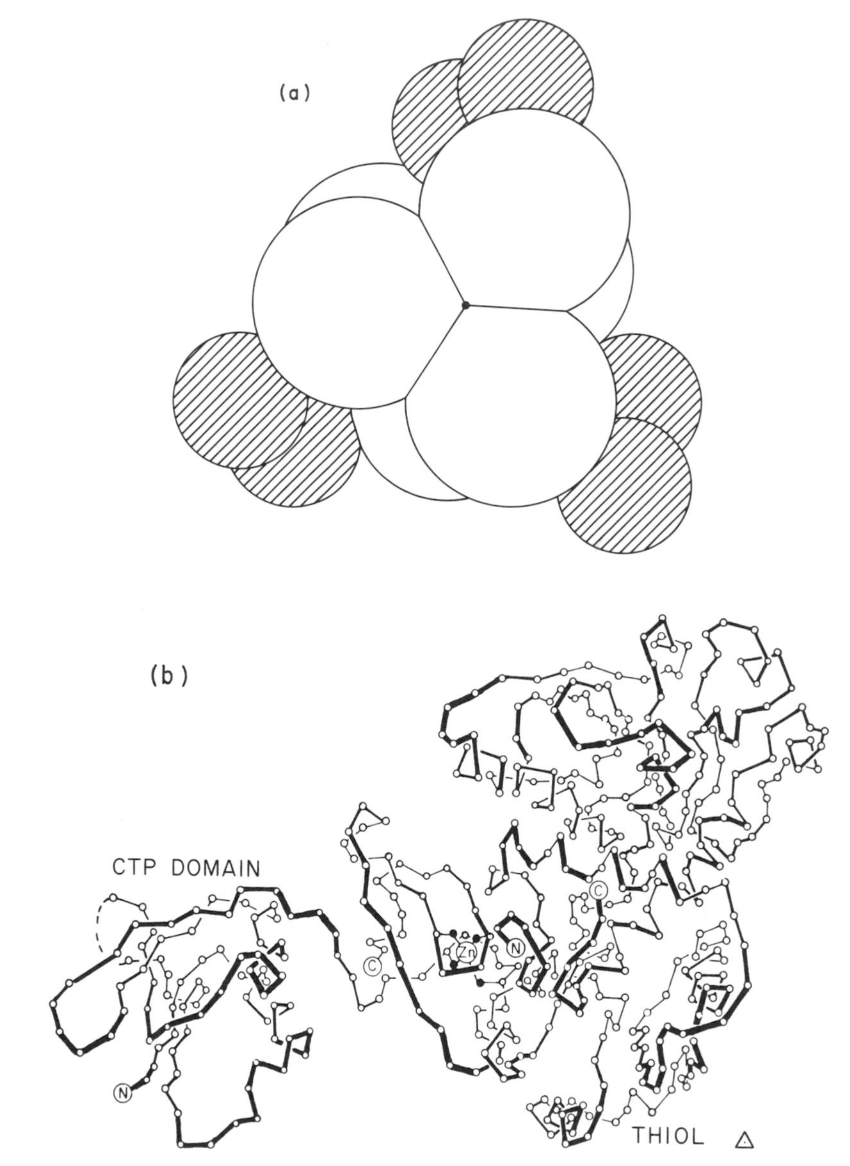

Fig. 9-2. Representations of the structure of aspartate transcarbamoylase. (a) A ball model looking down the threefold rotation axis. The striped balls are regulatory subunits and the white balls catalytic subunits. Note that the regulatory subunits are connected to the catalytic subunits in a helical-like fashion. (b) Structure of the polypeptide chains of a regulatory (*left*) and catalytic (*right*) polypeptide as derived from x-ray crystallographic studies. The thiol pocket is near the catalytic site, and the triangle represents the threefold rotation axis. [From H. L. Monaco, J. L. Crawford, and W. N. Lipscomb, *Proc. Natl. Acad. Sci. U.S.A.* **75**, 5276 (1978).]

studies have been carried out in which arginine is modified by phenylglyoxal (cf. *8*). Because several different arginines are modified, the results cannot be interpreted unambiguously. Apparently one arginine per catalytic polypeptide is important for catalytic activity, and two arginines per regulatory polypeptide are important for regulation. Finally a rather unique chemical modification has been made by isolating aspartate transcarbamoylase from a mutant bacterium (*9*). A single amino acid substitution has been made in the catalytic polypeptide, glycine has been substituted for aspartate at position 125 in the sequence, and the enzyme is inactive.

The overall morphology of the enzyme was deduced from the experiments demonstrating that the native molecule contains two catalytic trimers and three regulatory dimers. An obvious structure is to have the catalytic trimers held together by the regulatory dimers with each polypeptide of the dimer interacting with a catalytic polypeptide of a different trimer. Electron micrographs suggest such a structure (*10*), and the three-dimensional structure of the native enzyme and the native enzyme plus CTP has been determined to 2.8 Å resolution (cf. *11*). Both structures are essentially identical. The overall dimensions of the molecule are 110 × 110 × 90 Å, and a large central aqueous cavity (25 × 50 × 50 Å) is present. The polypeptides in the two catalytic subunits are staggered; no direct interaction between the two subunits occurs; and they are connected by regulatory subunits. The structure has both a threefold and twofold rotational axis. Pictorial representations of the structure are shown in Fig. 9-2. Because of intrinsic disorder in the crystals, the individual amino acid residues cannot be located unambiguously in the structure. However, a representation of the structure of catalytic and regulatory polypeptides is included in Fig. 9-2. The sulfhydryl groups on the catalytic polypeptides can be located by heavy atom labeling. As shown in Fig. 9-3, they form a small equilateral triangle, 22 Å on each side, near the center of the catalytic trimer and the aqueous cavity. The planes of the triangles on different catalytic trimers are 42 Å apart. The zinc, which is located at an interface between catalytic and regulatory polypeptides, is 24 Å from the nearest thiol, and CTP is 43 Å from the nearest thiol. The catalytic and regulatory binding sites are far apart, clearly demonstrating that regulation occurs through conformational changes.

The relative location of specific sites on the enzyme also has been determined using fluorescence resonance energy transfer and nuclear magnetic resonance (*12–14*). The fluorescence of covalently attached pyridoxamine phosphate can be completely quenched by suitable chromaphores on the catalytic polypeptide sulfhydryl groups. This indicates the catalytic site and thiol are adjacent (within a few Ångstroms). Furthermore, the distance between active sites and between an active site and a nonadjacent thiol is 26 Å, in good agreement with x-ray results. The distance between the closest active

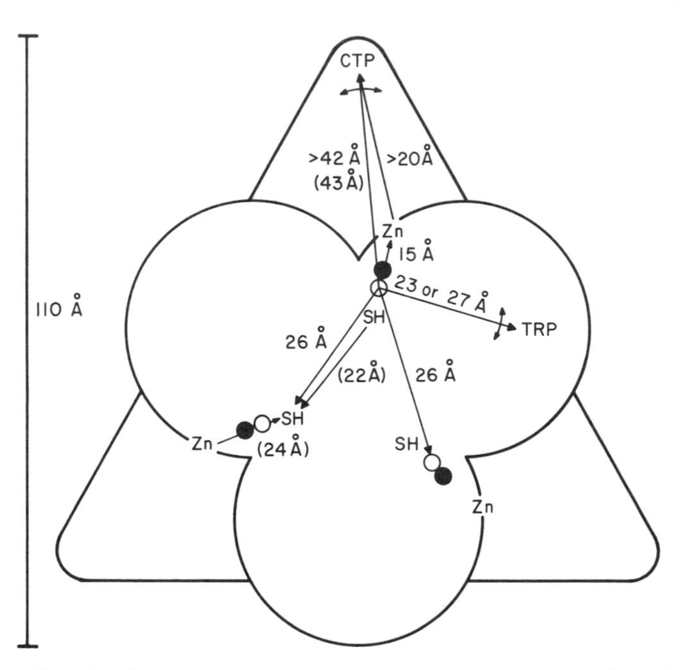

Fig. 9-3. Site–site distances in aspartate transcarbamoylase determined by fluorescence resonance energy transfer, nuclear magnetic resonance, and x-ray crystallography (numbers in parentheses). The open circle represents bound pyridoxamine phosphate and the filled circle bound succinate. The distance between the planes defining the catalytic sites on *different* subunits was reported to be 42 Å from x-ray crystallographic measurements (Hg on thiols) and 30 Å from fluorescence resonance energy transfer measurements (pyridoxal phosphate on an amino group near the active site).

sites on different catalytic trimers is 30 Å. The discrepancy with the x-ray results may be due to whether the probe molecules are oriented into the central aqueous cavity or into the bulk solution. The distance between the active site and CTP binding site was too long to measure with available probes; only a lower bound of 42 Å could be assessed. Tryptophan fluorescence can be quenched by molecules bound to the active site. Since only two tryptophans are present per catalytic polypeptide, the distance between the catalytic site and nearest tryptophan can be calculated to lie in the range 23–27 Å. If Mn^{2+} is substituted for Zn^{2+} in the enzyme, the nuclear magnetic resonance relaxation times of succinate (an aspartate analog) protons bound to the enzyme are altered. As discussed in Chapter 2, this paramagnetic perturbation can be utilized to calculate the distance between the succinate protons and Mn^{2+}. The site–site distances are summarized in Fig. 9-3. Altogether a quite good picture of the essential features of the native enzyme is available, although the atomic details of individual amino acids remain to be determined.

The catalytic mechanism has been studied with the isolated catalytic subunit. No cooperativity is observed so apparently all three catalytic sites function independently. Although some controversy exists, an ordered binding mechanism appears to occur with carbamoyl phosphate binding first (*2*). A ternary complex is formed, and direct transfer of the carbamoyl moiety is demonstrated by the lack of exchange between [$^{32}P_i$] and carbamoyl phosphate in the presence of the enzyme. A very interesting and useful inhibitor of the catalytic activity is *N*-phosphonacetyl-L-aspartate (PALA) (*15*).

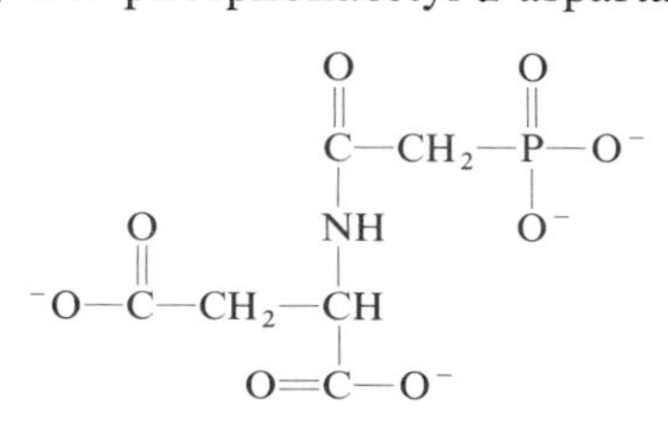

This compound is a *transition state analog* and binds very tightly to the enzyme. (This dissociation constant is 10^{-7}–10^{-8} *M*.) It is a noncompetitive inhibitor of L-aspartate, as expected for an ordered binding mechanism. (It would be a competitive inhibitor if the binding mechanism were random.) Nuclear magnetic resonance and kinetic studies with carbamoyl phosphate and its analogs indicate the following (*2*): the carbonyl group and phosphate are tightly bound; the amino group is unnecessary (acetyl phosphate is a substrate); and bulky groups cannot be attached to the amino group nor can the carbon skeleton be lengthened. When succinate is bound in the presence of carbamoyl phosphate, the temperature dependence of the nuclear magnetic resonance relaxation times of the succinate protons indicates slow exchange is occurring. Temperature jump measurements under similar conditions show that a conformational change following the initial reaction of succinate and enzyme is rate determining in the binding (*3*). The rate constants derived from the nuclear magnetic resonance measurements are consistant with the temperature jump results. The pH dependence of the nuclear magnetic relaxation times suggests an ionizable group with a p*K* of about 8.3 is important in the initial binding process and one with a p*K* of 6.8 is important for the conformational change (*16*). Direct measurements of proton absorption accompanying succinate binding suggest the involvement of additional ionizable groups (*17*). The binding of [^{13}C]carbamoyl phosphate to the catalytic subunit has been studied with nuclear magnetic resonance (*18*). Only a small chemical shift (2 Hz) is observed when the carbonyl carbon is bound to the enzyme. When succinate is added, a large down field shift (18 Hz) occurs; a similar shift is found with PALA. This is probably due to protonation of the carbonyl; the extent of protonation corresponds to lowering the carbonyl p*K* 9–10 units. This protonation evidently occurs because of the conformational change accompanying binding, and the protonation makes the carbonyl carbon more susceptible to reaction with the aspartate nitrogen. The activated form of carbamoyl phosphate must be protected from the solvent since hydrolysis does not occur. This is consistent with the formation of a hydrophobic pocket, which was previously postulated to be of general importance in enzyme catalysis (Chapter 6). Succinate binds more tightly to the enzyme than aspartate or another substrate analog, L-malate. In the latter two cases, the conformational change may not be able to occur fully because of steric hinderance. The enzyme could enhance catalysis by compressing carbamoyl phosphate and aspartate together via the conformational change (*2*).

The regulatory mechanism for aspartate transcarbamoylase is still incompletely understood although considerable information is available. Equilibrium and kinetic studies of ligand binding have been particularly useful tools. Early studies of nucleotide binding were quite confusing and stoichiometries of 3–8 CTP bound per enzyme can be found in the literature (cf. *19*, *20*). The reason for this dilemma can be appreciated by the binding isotherms in Fig. 9-4. In the presence of saturating carbamoyl phosphate, a limiting stoichiometry of 6 CTP per enzyme is obtained, but negative cooperativity occurs. The structure of the enzyme suggests that negative cooperativity occurs within a regulatory dimer. The results can be analyzed in terms of three tight and three loose binding sites (preexisting asymmetry in a dimer) or in terms of ligand-induced negative cooperativity. If carbamoyl phosphate is not present, CTP binds weakly to the catalytic site and is an active site inhibitor. Thus, the complete binding isotherm can be written as

$$r = \frac{3K_1[\mathrm{CTP}]}{1 + K_1[\mathrm{CTP}]} + \frac{3K_2[\mathrm{CTP}]}{1 + K_2[\mathrm{CTP}]} + \frac{6K_3[\mathrm{CTP}]}{1 + K_3[\mathrm{CTP}]} \qquad (9\text{-}2)$$

where r is the moles of bound ligand per mole of enzyme and the K_is are binding constants. The binding constants K_1 and K_2 become larger by comparable amounts as the temperature is lowered indicating a negative enthalpy for binding and a primarily entropic difference between binding at the two sites. A very surprising result, not yet well understood, is that the values of K_1 and K_2 are not markedly changed when carbamoyl phosphate and succinate are bound to the enzyme. The binding of ATP follows a similar pattern, except that binding to the regulatory sites is about an order of magnitude weaker than for CTP. An important question to consider is whether all six regulatory sites are equally important for regulation. In fact, inhibition and activation correlate linearly with binding to all six regulatory sites (*20*). The nucleotide specificity has been examined in considerable detail, e.g., CTP, cytidine, and cytosine all inhibit; UTP, ITP, and GTP are very poor inhibitors. Modifications at the 5 position of the ring have no effect on binding (even very bulky modifications), but all other ring modifications greatly

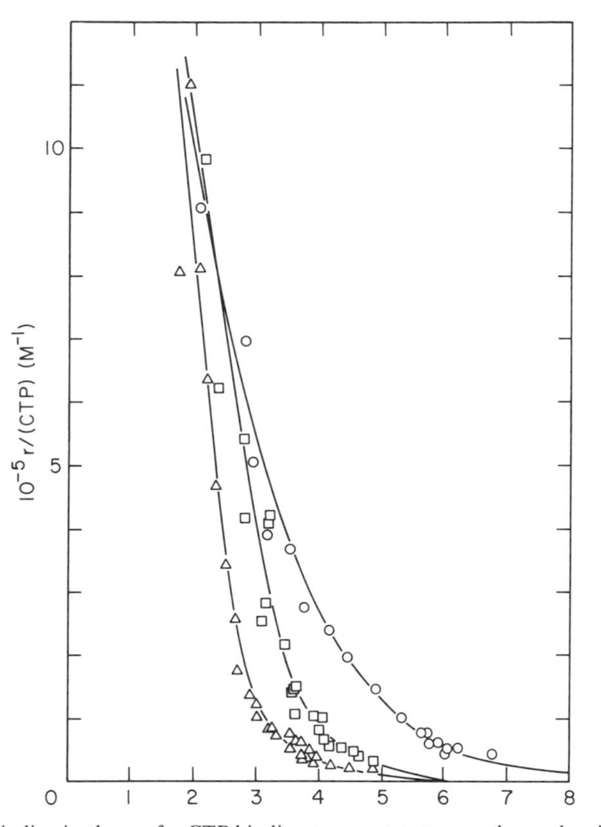

Fig. 9-4. Binding isotherms for CTP binding to aspartate transcarbamoylase in the absence of ligands (○), plus saturating concentrations of carbamoyl phosphate (□), and plus saturating concentrations of carbamoyl phosphate and succinate (△). Here *r* is the moles of CTP bound per mole of protein. The lines have been calculated assuming the model in the text. [Adapted with permission from S. Matsumoto and G. G. Hammes, *Biochemistry* **12**, 1388 (1973). Copyright (1973) American Chemical Society.]

weaken binding. The binding of ATP also is quite specific; a sulfhydryl group can be substituted for the amino group on the ring, but other ring modifications have dire effects.

The kinetics of nucleotide binding to the regulatory sites has been studied with the temperature jump method using pH changes and ultraviolet difference spectra accompanying binding to monitor the process (*3*). In all cases, a single relaxation process is observed; the reciprocal relaxation time increases with increasing nucleotide concentration to a limiting value of about 10^3–10^4 sec^{-1}. A binding mechanism qualitatively consistant with the results is

$$E + N \underset{}{\overset{K_1}{\rightleftharpoons}} X_1 \underset{k_{-2}}{\overset{k_2}{\rightleftharpoons}} X_2 \tag{9-3}$$

where N is the nucleotide, and X_1 and X_2 are different conformations of the nucleotide–enzyme complex. To quantitatively fit the data, the following assumptions are made: (1) the same conformational change occurs independently at each site and (2) the negative cooperativity arises entirely from changes in the equilibrium dissociation constant K_1. The values of K_1, k_2, and k_{-2} have been determined for several nucleotides under a variety of conditions. The most important conclusions reached are that (1) the same conformational change occurs with ATP and CTP, (i.e., a single relaxation process is observed in the presence of both nucleotides); (2) the nucleotide binding sites are equivalent before ligand binding (*21*); and (3) CTP favors the formation of one conformation (X_2), whereas ATP and carbamoyl phosphate plus succinate favor the formation of the other conformation (X_1). However, the changes in the equilibrium constant k_2/k_{-2} caused by CTP, ATP, and carbamoyl phosphate plus succinate are not large so that this cannot be the primary regulatory mechanism. The conformational change in the regulatory subunit accompanying nucleotide binding has been monitored by nuclear magnetic resonance with ^{13}C amino acids incorporated into the polypeptide (*22*). The results suggest that the conformational change occurring involves only small rearrangements of tertiary structure (consistant with the results deduced from the kinetic studies) and that at least one phenylalanine and one histidine residue are present in the region of the CTP binding site.

The binding of carbamoyl phosphate to the native enzyme is not well understood. In the presence of succinate, six carbamoyl phosphates per enzyme molecule are tightly bound, as expected. However, in the absence of succinate the binding is complex, with different results being reported from different laboratories (*23, 24*). The suggestion has been made that the heterogeneity observed is due to incomplete removal of formyl methionine from the N terminus of the polypeptide chain (*23*). The binding to the isolated catalytic subunit is characterized by a single relaxation time with a concentration dependence characteristic of a bimolecular reaction (*3*). Since the reaction mechanism is ordered, the binding of carbamoyl phosphate must cause a conformational change, but this has not been observed with relaxation kinetics.

Aspartate binds very weakly to the enzyme in the absence of carbamoyl phosphate, and obviously its binding cannot be measured in the presence of carbamoyl phosphate. Instead the binding of the aspartate analog succinate in the presence of carbamoyl phosphate has been extensively studied. The binding isotherm obtained with the catalytic subunit is hyperbolic, but a sigmoidal binding isotherm is obtained with the native enzyme (*25*). A tryptophan difference spectrum is observed indicating a perturbation of the protein structure. The temperature jump method has been used to study the binding process (*3*). As previously indicated, binding to the catalytic subunit is a two-step process, a bimolecular reaction followed by a conformational change. The identical conformational change is found with the native enzyme, and the associated rate constants are not altered by CTP. Clearly, this conformational change is restricted to the catalytic subunit. A slower relaxation process also is observed in the presence of saturating carbamoyl phosphate: the dependence of the relaxation time on succinate concentration is presented in Fig. 9-5. As shown in Fig. 9-5, the regulatory nucleotides, ATP and CTP, alter the concentration dependence in opposite manners. These data can be quantitatively analyzed in terms of the Monod–Wyman–Changeux model for allosterism discussed in Chapter 8. With this model, the enzyme exists primarily in the T state and is converted to the R state by succinate binding, i.e., succinate binds preferentially to the R state. As expected, CTP binds preferentially to the T state and ATP preferentially to the R state. This conformational change is quite distinct from that associated with nucleotide binding since relaxation processes for both transitions can be observed simultaneously. A slow relaxation process also can be detected at high concentrations of succinate and low concentrations of carbamoyl phosphate: the dependence of the relaxation time on carbamoyl phosphate concentration is included in Fig. 9-5. Again the relaxation time is altered in the presence of CTP. These results also can be interpreted in terms of the Monod–Wyman–Changeux model, with carbamoyl phosphate binding preferentially to the R state. The two concerted conformational changes observed at low succinate and carbamoyl phosphate concentrations appear to be related (the same limiting relaxation time is

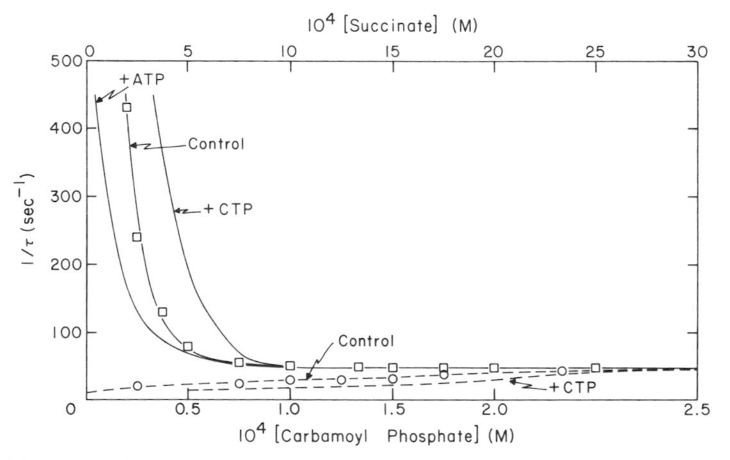

Fig. 9-5. Plots of the reciprocal relaxation time versus the concentration of succinate (1 mM carbamoyl phosphate; upper abscissa —, □) and the concentration of carbamoyl phosphate (10 mM succinate; lower abscissa ---, ○). The effects of nucleotides are shown schematically. The lines are consistant with the Monod-Wyman-Changeux mechanism for allosterism and have been derived from data in G. G. Hammes and C.-W. Wu, *Biochemistry* **10**, 1051, 2150 (1971).

Table 9-1

Detection of Conformational Changes in Aspartate Transcarbamoylase[a]

Method	Ultraviolet difference spectrum	Change in sedimentation coefficient	Change in optical rotation (233 nm)	Rate of reaction with *p*-hydroxy mercuribenzoate	Rate of denaturation
Catalytic subunit					
+ Carbamoyl phosphate	None	+(?)	—		
+ Carbamoyl phosphate and succinate	Yes	+ +	—		
Native enzyme					
+ Carbamoyl phosphate	None	−(?)	−(?)	+	+
+ Carbamoyl phosphate and succinate	Yes	—	+	+ +	+
+ ATP or CTP	Yes	?	—	—	—

[a] Tabulated from Refs. *2* and *26*. An increase or decrease is designated by + or −, respectively. A question mark indicates a very small change of questionable significance. More than one plus or minus indicates relatively larger changes.

reached), but not identical. (CTP decreases the reciprocal relaxation time in one case and increases it in the other; increasing succinate decreases the reciprocal relaxation time and carbamoyl phosphate increases it.)

Thus, the binding of ligands to aspartate transcarbamoylase is associated with at least three and probably four conformational changes. Many different physical measurements have been used to detect the conformational changes. A qualitative summary of some of the results obtained is given in Table 9-1 (cf. *2, 26*). Different conformational states are stabilized by the following conditions: (1) carbamoyl phosphate plus succinate binding to the isolated catalytic subunit; (2) carbamoyl phosphate plus succinate binding to the native enzyme; (3) nucleotide binding to the regulatory subunit on the native enzyme; and (4) perhaps carbamoyl phosphate binding to the native enzyme and isolated catalytic subunit. The transition state analog PALA causes essentially the same changes found with saturating carbamoyl phosphate and succinate. A quantitative analysis of a variety of different results has been made assuming that only a single concerted conformational change occurs (*26*). The results analyzed include the initial steady-state velocity–aspartate isotherm (with saturating carbamoyl phosphate), changes in sedimentation coefficient caused by substrates, nucleotides and their analogs, the reactivity of sulfhydryl groups as a function of PALA binding, stimulation of activity by low concentrations of succinate, and spectral and activity measurements with hybrid molecules containing modified and unmodified catalytic polypeptides. The values of the ratio [T]/[R] calculated for some limiting conditions are given in Table 9-2. The use of hybrid enzyme molecules merits special mention since this approach is useful for studying

Table 9-2

Calculated Ratios of [T]/[R] for Aspartate Transcarbamoylase[a]

Ligands	[T]/[R]
None	250
Carbamoyl phosphate	7
ATP	70
CTP	1250

[a] Calculated from the experimental data discussed in the text assuming only a single concerted conformational change occurs (Monod–Wyman–Changeux model) and saturation of all sites by the ligand (*26*).

the communication between polypeptides. For example, an enzyme containing native regulatory subunits and catalytic subunits with one tyrosine per polypeptide nitrated has been prepared (*27*). When CTP is added to active and regulated enzyme, the light absorption of the nitrotyrosine increases, whereas when ATP is added it decreases. This demonstrates that ATP and CTP binding cause conformational effects that are promulgated to catalytic polypeptides. Note that all of the analyses suggest that the unliganded enzyme is largely in the T state and that CTP binds preferentially to this state. This explains why the crystal structures of the native enzyme and the native enzyme plus CTP are essentially identical.

The final model emerging from the above results is that a major concerted conformational transition is associated with ligand binding. This is largely responsible for the positive cooperativity observed in the initial velocity–aspartate isotherm. The nature of the conformational change is unknown but a possibility is rotation about the threefold axis, and/or a change in the separation between catalytic subunits (expansion–contraction), which would alter interactions between regulatory and catalytic polypeptide chains (*28*). In addition, local conformational changes in the regulatory subunits that are coupled to the rest of the molecule accompany nucleotide binding. The local changes are associated with the negative cooperativity found with nucleotide binding. Also, a local conformational change in the catalytic subunit accompanies substrate binding and probably is associated with catalysis. Regulation then involves *multiple conformational transitions*. This provides a sensitive switching mechanism for a variety of metabolic conditions.

PHOSPHOFRUCTOKINASE

Phosphofructokinase catalyzes the phosphorylation of fructose 6-phosphate to fructose 1,6-diphosphate.

$$\text{Fructose 6-phosphate} + \text{ATP} \rightleftharpoons \text{fructose 1,6-diphosphate} + \text{ADP} + \text{H}^+ \quad (9\text{-}4)$$

A divalent metal ion is required: Mg^{2+} is utilized physiologically, but Mn^{2+} and Co^{2+} also work quite well; Ca^{2+} is not an activator. Phosphofructokinase also is activated by low concentrations of NH_4^+ and K^+. The reaction catalyzed is the first unique step in glycolysis and, as might be expected, is subject to metabolic regulation. The enzyme has been isolated from several different sources (e.g., *E. coli*, plants, yeast, liver, and muscle); many of the enzymes have different regulatory properties. We will consider the enzyme isolated from rabbit muscle, which has been extensively studied. Several recent reviews of phosphofructokinase are available (*29–32*).

The kinetic behavior of rabbit muscle phosphofructokinase is shown schematically in Fig. 9-6. The initial steady-state velocity increases sigmoidally with increasing fructose 6-phosphate concentration at a constant MgATP concentration. This curve is shifted to higher sugar concentrations as the MgATP concentration is raised. On the other hand, the initial velocity first increases hyperbolically with increasing MgATP concentration at constant fructose 6-phosphate concentration, but then decreases dramatically at high concentrations of MgATP. The MgATP concentration at which inhibition occurs increases as the fructose 6-phosphate concentration is raised. Thus the two substrates interact strongly in modulating enzyme activity. Positive effectors of the enzyme raise the activity by relieving the MgATP inhibition and shifting the initial velocity–fructose 6-phosphate isotherm to the left. Adenine nucleotides and phosphate compounds are strong activators, e.g., ADP, AMP, cyclic-AMP, phosphate, fructose 6-phosphate, and fructose 1,6-diphosphate. While activators apparently function primarily by altering the binding properties of the enzyme, regulation by the negative effector citrate is more complex. This will be discussed further below, but the binding of ATP and citrate are strongly coupled and citrate can alter the turnover number of the enzyme. Protons also are allosteric effectors. Above about pH 7.5, the regulatory properties of the enzyme are not observed in steady-state kinetic studies; the regulatory properties become more accentuated as the pH is lowered into the range 6–7. Finally phosphofructokinase has been isolated with its polypeptide chains phosphorylated. The evidence that

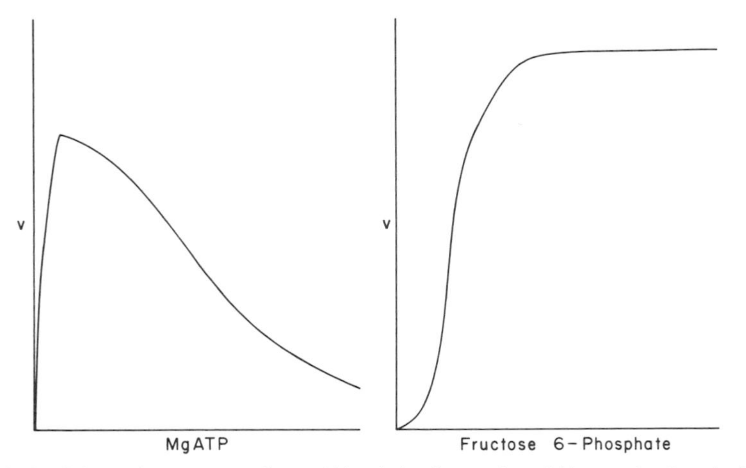

Fig. 9-6. Schematic representations of kinetic isotherms for rabbit muscle phosphofructokinase at pH 7. The initial steady-state velocity, v, is plotted versus the concentration of fructose 6-phosphate (at constant MgATP concentration) and MgATP (at constant fructose 6-phosphate concentration).

this is a manifestation of regulation by phosphorylation–dephosphorylation is quite weak, but this matter merits further attention (*29*).

Rabbit muscle phosphofructokinase contains a single type of polypeptide chain of molecular weight 80,000 (*33–35*). Under physiological conditions, the enzyme is primarily dimers and tetramers. The structure of the enzyme has been investigated by electron microscopy, low angle x-ray scattering, and light scattering (*32, 36*). The monomer is a prolate ellipsoid about 65 Å in length, and the tetramer has an effective Stoke's radius of approximately 67 Å. The structure of phosphofructokinase from *Bacillus stearothermophilus* has been determined to 2.4 Å resolution by x-ray crystallography (*37*). A schematic representation of the tetramer is shown in Fig. 9-7. This enzyme is much smaller than the rabbit muscle enzyme as the monomer has a molecular weight of 33,900. A fascinating aspect of this structure is the fructose 6-phosphate binding sites (A in Fig. 9-7) are formed by two adjacent subunits, as is the effector site for phosphate compounds (C in Fig. 9-7). The ADP binding site primarily resides on a single polypeptide chain (B in Fig. 9-7). Of course, the relevance of this structure for rabbit muscle phosphofructokinase is unknown. However, a dimer is the fundamental unit of

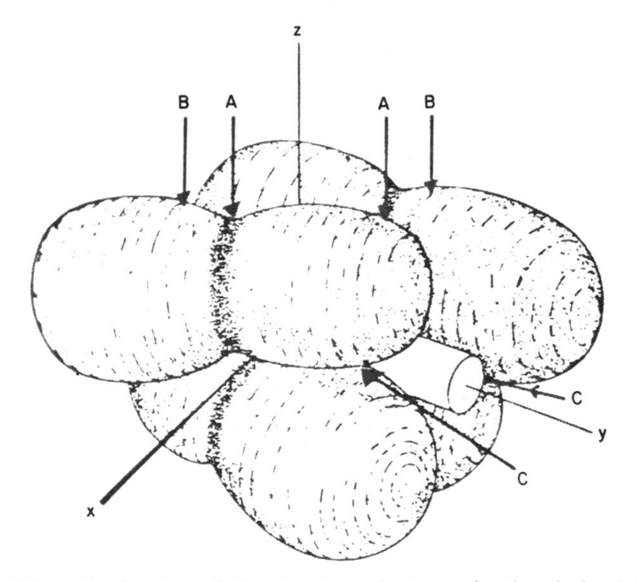

Fig. 9-7. Schematic drawing of the structure of tetrameric phosphofructokinase from *Bacillus stearothermophilus* as determined by x-ray crystallography. The cylinder in the center represents a solvent filled hole of 7 Å diameter. The catalytic site for fructose 6-phosphate is designated by A and that for MgADP by B. The effector site, which binds P_i, MgADP, and phosphoenolypyruvate, is designated by C. Both sites A and C lie between subunits. [Reprinted by permission from P. R. Evans and P. J. Hudson, *Nature* (*London*) **279**, 500. Copyright © 1979 Macmillan Journals Limited.]

this enzyme and also appears to be a fundamental structural unit of the rabbit muscle enzyme. Large aggregates of the rabbit muscle enzyme also can be observed by electron microscopy including highly symmetric filamentous structures; the physiological significance of these aggregates is unknown.

The aggregation behavior of phosphofructokinase has been studied by ultracentrifugation and gel filtration (*33, 34, 38*). The enzyme can exist in a number of interconvertible polymeric forms. Some high molecular weight aggregates not in equilibrium with other species also are observed. At physiological concentrations of a few tenths of a milligram per milliliter, the enzyme exists as a tetramer at high pH (>7) but at low pH ($\sim 6-7$), a dimer–tetramer equilibrium exists. The correlation between the aggregation state and the specific activity at pH 7 is shown in Fig. 9-8 (*39*); the dimer and monomer possess very little, if any, activity and the tetramer and higher

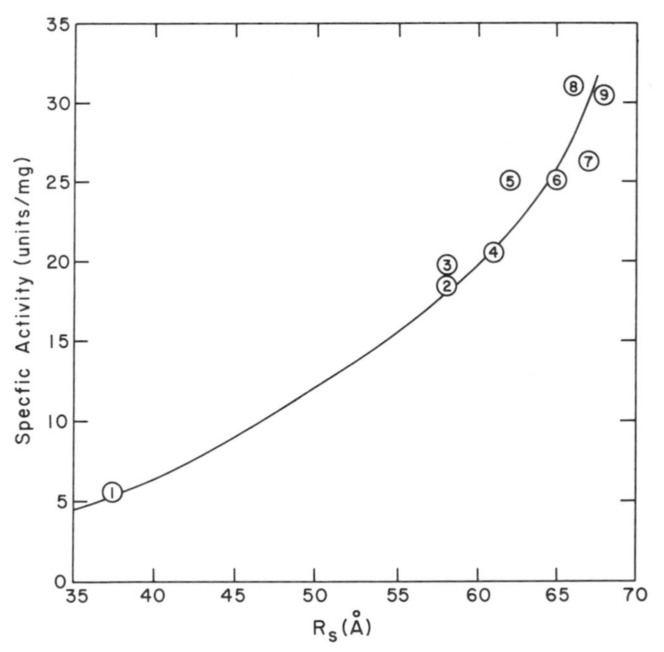

Fig. 9-8. Correlation of the Stokes' radius, R_s (determined by gel chromatography), with specific activity at pH 7 (unless otherwise indicated) for rabbit muscle phosphofructokinase. The dimer has a Stokes' radius of about 37 Å and the tetramer one of about 67 Å. The numbers indicate the following ligands: 1, 5 m*M* citrate; 2, 0.1 *M* phosphate; 3, 10 m*M* Mg^{2+}; 4, 5 m*M* MgAMP; 5, 5 m*M* ATP-0.5 m*M* Mg^{2+}; 6, 5 m*M* MgATP; 7, 0.1 *M* phosphate, pH 8; 8, 10 m*M* fructose 6-phosphate; and 9, 5 m*M* fructose 1,6-diphosphate. [Adapted with permission from P. M. Lad, D. E. Hill, and G. G. Hammes, *Biochemistry* **12**, 4303 (1973). Copyright (1973) American Chemical Society.]

aggregates have the same specific activity. Moreover, activators of the enzyme stabilize the tetramer, whereas the inhibitor citrate stabilizes low molecular weight species, primarily the dimer (Fig. 9-8). Thus at least part of the regulation of phosphofructokinase involves the aggregation state of the protein.

However, allosteric regulation also occurs within the tetrameric enzyme. For example, the inhibitor MgATP stabilizes the tetramer. Chemical cross-linking can be used to obtain a catalytically active tetrameric enzyme that cannot dissociate (*40*). Homotropic fructose 6-phosphate interactions are not observed with the cross-linked tetramer, but allosteric regulation occurs: the Michaelis constant for fructose 6-phosphate is decreased by AMP, cyclic-AMP, and other substances; it is increased by MgATP; and citrate decreases the maximum velocity. Phosphofructokinase is cold labile, and an investigation of its inactivation by lowering the temperature and/or pH has led to the model for regulation shown in Fig. 9-9a (*41*). In this mechanism, protonation of the enzyme plays a key role in regulation: under most physiological conditions the equilibria between E_4, HE_4, and HE'_4 (a different conformational state of HE_4) dominates the regulatory behavior because the dissociation of HE'_4 to dimers occurs relatively slowly. The inhibitors ATP and citrate bind preferentially to the protonated form, whereas fructose 6-phosphate, fructose 1,6-diphosphate, and AMP bind preferentially to the unprotonated forms.

Many studies of ligand binding to phosphofructokinase have been carried out. The characteristics for fructose 6-phosphate binding are particu-

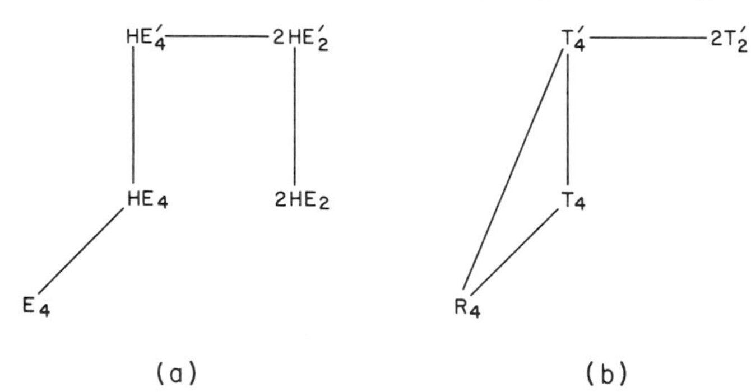

Fig. 9-9. Pathways for the regulatory behavior of rabbit muscle phosphofructokinase. (a) The mechanism derived by Bock and Frieden (*41*), primarily from studies of inactivation by low pH and temperature. The mechanism involves protonated and unprotonated dimers and tetramers, and changes in conformation. More than a single proton is involved in the transitions, although only one is shown. A prime indicates a different conformation. (b) The mechanism derived by Goldhammer and Hammes (*52*), primarily from steady-state studies. The active state is represented by R_4 and inactive states by T_4, T'_4 and T'_2. The correspondence between the two mechanisms is evident.

larly interesting (*42*). One binding site per polypeptide chain is found, but the cooperativity observed depends on the aggregation state. With the dimer, extreme negative cooperativity is found; less negative cooperativity is found with the tetramer and higher aggregates display virtually no cooperativity in the binding process. If the ATP analog adenylyl imidodiphosphate (AMP-PNP) is present with the tetrameric enzyme, positive cooperativity is found. Such behavior requires a model such as that of Adair, Koshland, Nemethy, and Filmer. Kinetic studies of fructose 6-phosphate binding have not been carried out.

Studies of the binding of cyclic-AMP and other activating molecules indicate one site per polypeptide chain (*43*). This site can be covalently modified with the affinity label 5′-[*p*-fluorosulfonylbenzoyl] adenosine (*44*), and the modified enzyme is much less susceptible to ATP inhibition. Both AMP-PNP and ATP bind to the enzyme in a similar manner (*45,46*): two binding sites per polypeptide chain are found, one with a dissociation constant of about 1 μM and the other with a dissociation constant of about 100 μM. The former can be identified with the catalytic site and the latter with the regulatory site. The binding isotherms for a given type of site are hyperbolic or exhibit some negative cooperativity. The protein fluorescence is slightly enhanced when ATP binds to the catalytic site and greatly quenched when ATP binds to the regulatory site. A model in which ATP binds at the regulatory site exclusively to a protonated form of the enzyme, with two protons being associated with the binding, is consistant with the data. A carlorimetric investigation of ATP binding also suggests at least two protons are associated with binding (*47*). This is consistant with the model in Fig. 9-9a. The physiological inhibitor citrate strengthens the binding of ATP to the enzyme, and ATP enhances the binding of citrate (*48*). Modification of a single lysine residue per polypeptide with pyridoxal phosphate, under the proper conditions, blocks citrate binding (*48*). If borohydride is used to give a covalently linked pyridoxamine phosphate, a tetrameric enzyme can be obtained, which is largely inactive and depolymerizes more readily than the native enzyme. This indicates that citrate can act as an inhibitor both by inhibiting the activity of the tetramer and by depolymerizing the enzyme. Thus, the following functional sites have been identified in phosphofructokinase: (1) a catalytic site that binds ATP and fructose 6-phosphate; (2) a regulatory site that binds citrate; (3) a regulatory site that binds ATP; and (4) a regulatory site that binds cyclic-AMP and other adenine nucleotides. Sites 3 and 4 may be identical; this question has not yet been resolved.

The affinity labeling of the cyclic-AMP and citrate sites represents two important chemical modifications of phosphofructokinase. A single sulfhydryl group per polypeptide also can be modified specifically because it reacts with sulfhydryl group reagents much more rapidly than other sulfhydryl groups on the enzyme (*29–32*). This sulfhydryl group is not essential for activity, but obviously is of considerable structural importance. The enzyme with a single modified sulfhydryl group has greatly reduced catalytic activity and its regulatory properties are altered. Fluorescence resonance energy transfer has been used to measure the distance between this sulfhydryl group and the citrate and cyclic-AMP binding sites; the distances found are 38 and 28 Å, respectively (*49, 50*). A nitroxide spin label has been used to label the sulfhydryl group and to monitor ATP binding to the enzyme (*51*). The results suggest that a MnATP site is quite near to the sulfhydryl group; the weakness of the binding suggests the regulatory MnATP site is being monitored.

Steady-state kinetic studies of the enzyme at pH values greater than 8, where the allosteric properties of the enzyme are not observed, indicate that a ternary complex is formed with a random addition of substrates (*29*). However, an ordered binding mechanism has not yet been unequivocally excluded. Steady-state kinetic studies at pH values where phosphofructokinase is tetrameric and exhibits allosteric behavior can be explained by assuming a Monod–Wyman–Changeux model in which the enzyme exists in the two conformational states, R_4 and T_4 (*52,53*). For two substrates A and B which bind independently to the tetrameric enzyme, the fraction of enzyme in the R_4 state, f_{R_4}, can be written as

$$f_{R_4} = \frac{\alpha\beta(1+\alpha)^3(1+\beta)^3}{(1+\alpha)^4(1+\beta)^4 + L(1+c\alpha)^4(1+d\beta)^4} \tag{9-5}$$

In this equation $\alpha = [A]/K_{RA}$, $\beta = [B]/K_{RB}$, $c = K_{RA}/K_{TA}$, $d = K_{RB}/K_{TB}$, K_{RA} is the intrinsic dissociation constant for A binding to the R_4 state, K_{TA} is the intrinsic dissociation constant for A binding to the T_4 state, K_{RB} and K_{TB} are the corresponding constants for B binding to the enzyme, and L is the ratio of the concentration of T to R states in the absence of ligands A and B. [If the binding of A and B is not independent, a more complex equation is obtained (*53*).] The binding of activators and inhibitors changes the constant L as discussed in Chapter 8. For example, if MgATP binds exclusively to the T state at the regulatory site with a dissociation constant K_{MgATP}, $L = L_0(1 + [MgATP]/K_{MgATP})^4$ where L_0 is the value of L in the absence of MgATP. The kinetic results indicate that fructose 6-phosphate binds preferentially to the R_4 state; MgATP shows little preference for the R_4 and T_4 states when binding at the catalytic site, but binds exclusively to the T_4 state regulatory site; and cyclic-AMP binds preferentially to the R_4 state regulatory site. The inhibition by citrate requires modification of the model, namely, a new conformation T'_4, which dissociates into dimers more readily than the native enzyme. This more complete model is shown schematically in Fig. 9-9b. The correspondence between this model and that in Fig. 9-9a is obvious if $R_4 \equiv E_4$, $T_4 \equiv HE_4$, $T'_4 \equiv HE'_4$, and $T'_2 \equiv HE'_2$. Thus, quite diverse experiments have led to a similar model. A relaxation process, which has been found to accompany the binding of ATP and AMP-PNP to the enzyme can be attributed to the interconversion of the R_4 and T_4 states (*54*). The temperature jump method with fluorescence detection was used in this study. The rate constants for interconversion are about 10^3 sec^{-1}. The parameters obtained are similar to, but not identical with, those found from the steady-state studies. The kinetics of the binding of the fluorescent ATP analog, Mg^{2+}-1,N^6-etheno-ATP to the catalytic site of the enzyme has been studied with the stopped flow (*55*). An interconversion between two conformations of the enzyme can be detected which can be identified with the transition $R_4 \rightleftharpoons T_4$. The fluorescence of the etheno-ATP bound to the catalytic site is quite different for the R_4 and T_4 states.

The regulatory mechanism for phosphofructokinase obviously is very complex. A concerted conformational change and depolymerization of the tetramer account for much of the results, but even here two different conformations of the inactive (T_4) state are required. Different experiments and different experimental conditions give somewhat different parameters for the allosteric model (*52–55*), also suggesting the occurrence of multiple R_4 and T_4 conformational states. In addition the negative cooperativity associated with sugar-phosphate and AMP-PNP binding require other conformational changes to be of importance. Some of these conformational changes can be quite local in nature. As with aspartate transcarbamoylase, *multiple conformational changes* clearly are of importance in regulation. While a limiting allosteric model such as that of Monod, Wyman, and Changeux is a useful simplification for interpreting and understanding regulatory processes, the "real thing" clearly involves the coupling of several conformational transitions. With the intensive study of regulatory enzymes currently occurring, the development of more sophisticated and exact molecular models should not be long in coming.

REFERENCES

1. J. C. Gerhart and A. B. Pardee, *J. Biol. Chem.* **237**, 819 (1962).
2. G. R. Jacobson and G. R. Stark, *in* "The Enzymes" (P. Boyer, ed.), 3rd ed. Vol. **9**, p. 225. Academic Press, New York, 1973.
3. G. G. Hammes and C.-W. Wu, *Annu. Rev. Biophys. Bioeng.* **3**, 1 (1974).
4. K. Weber, *Nature* (*London*) **218**, 1116 (1968).
5. P. Greenwell, S. L. Jewett, and G. R. Stark, *J. Biol. Chem.* **248**, 5994 (1973).
6. T. D. Kempe and G. R. Stark, *J. Biol. Chem.* **250**, 6861 (1975).
7. A. M. Lauritzen, S. M. Landfear, and W. N. Lipscomb, *J. Biol. Chem.* **255**, 602 (1980).
8. E. R. Kantrowitz and W. N. Lipscomb, *J. Biol. Chem.* **252**, 2873 (1977).
9. K. A. Wall and H. K. Schachman, *J. Biol. Chem.* **254**, 11917 (1979).

10. K. E. Richards and R. C. Williams, *Biochemistry* **11**, 3393 (1972).
11. H. L. Monaco, J. L. Crawford and W. N. Lipscomb, *Proc. Natl. Acad. Sci. U.S.A.* **75**, 5276 (1978).
12. L. E. Hahn and G. G. Hammes, *Biochemistry* **17**, 2423 (1978).
13. S. Matsumoto and G. G. Hammes, *Biochemistry* **14**, 214 (1975).
14. S. Fan, L. W. Harrison and G. G. Hammes, *Biochemistry* **14**, 2219 (1975).
15. K. D. Collins and G. R. Stark, *J. Biol. Chem.* **246**, 6599 (1971).
16. H. I. Mosberg, C. B. Beard and P. G. Schmidt, *Biophys. Chem.* **6**, 1 (1977).
17. N. M. Allewell, G. E. Hofmann, A. Zaug and M. Lennick, *Biochemistry* **18**, 3008 (1979).
18. M. F. Roberts, S. J. Opella, H. M. Schaffer, H. M. Phillips and G. R. Stark, *J. Biol. Chem.* **251**, 5976 (1976).
19. C. C. Winland and M. J. Chamberlin, *Biochem. Biophys. Res. Commun.* **40**, 43 (1970).
20. S. Matsumoto and G. G. Hammes, *Biochemistry* **12**, 1388 (1973).
21. C. Tondre and G. G. Hammes, *Biochemistry* **13**, 3131 (1974).
22. A. C. Moore and D. T. Browne, *Biochemistry* **19**, 5768 (1980).
23. J. A. Ridge, M. F. Roberts, M. H. Schaffer, and G. P. Stark, *J. Biol. Chem.* **251**, 5966 (1976).
24. P. Suter and J. P. Rosenbusch, *J. Biol. Chem.* **251**, 5986 (1976).
25. J. -P. Changeux, J. C. Gerhart, and H. K. Schachman, *Biochemistry* **7**, 531 (1968).
26. G. J. Howlett, M. N. Blackburn, J. G. Compton, and H. K. Schachman, *Biochemistry* **16**, 5091 (1977).
27. P. Hensley and H. K. Schachman, *Proc. Natl. Acad. Sci. U.S.A.* **76**, 3732 (1979).
28. J. C. Gerhart, *Curr. Top. Cell. Regul.* **2**, 275 (1970).
29. K. Uyeda, *Adv. Enzymol.* **48**, 193 (1979).
30. D. P. Bloxham and H. A. Lardy, *in* "The Enzymes" (P. Boyer, ed.), 3rd ed., Vol. 8, p. 240. Academic Press, New York, 1973.
31. H. W. Hofer, *in* "Reaction Mechanisms and Control Properties of Phosphotransferases" (E. Hofman, and H. J. Bohme, eds.) p. 367. Akademic Verlag, Berlin. 1973.
32. A. R. Goldhammer and H. H. Paradies, *Curr. Top. Cell. Regul.* **15**, 109 (1979).
33. K. R. Leonard and I. O. Walker, *Eur. J. Biochem.* **26**, 442 (1972).
34. M. J. Pavelich and G. G. Hammes, *Biochemistry* **12**, 1408 (1973).
35. C. J. Coffee, R. B. Aaronson, and C. Frieden, *J. Biol. Chem.* **248**, 1381 (1973).
36. J. N. Telford, P. M. Lad, and G. G. Hammes, *Proc. Natl. Acad. Sci. U.S.A.* **72**, 3054 (1975).
37. P. R. Evans and P. J. Hudson, *Nature* (*London*) **279**, 500 (1979).
38. R. B. Aaronson and C. Frieden, *J. Biol. Chem.* **247**, 7505 (1972).
39. P. M. Lad, D. E. Hill, and G. G. Hammes, *Biochemistry* **12**, 4303 (1973).
40. P. M. Lad and G. G. Hammes, *Biochemistry* **13**, 4530 (1974).
41. P. E. Bock and C. Frieden, *J. Biol. Chem.* **251**, 5630 (1976).
42. D. E. Hill and G. G. Hammes, *Biochemistry* **14**, 203 (1975).
43. R. G. Kemp and E. G. Krebs, *Biochemistry* **6**, 423 (1967).
44. D. W. Pettigrew and C. Frieden, *J. Biol. Chem.* **253**, 3623 (1978).
45. N. M. Wolfman, W. R. Thompson, and G. G. Hammes, *Biochemistry* **17**, 1813 (1978).
46. D. W. Pettigrew and C. Frieden, *J. Biol. Chem.* **254**, 1887 (1979).
47. N. M. Wolfman and G. G. Hammes, *J. Biol. Chem.* **254**, 12289 (1979).
48. G. Colombo and R. G. Kemp. *Biochemistry* **15**, 1774 (1976).
49. N. W. Wolfman and G. G. Hammes, *Biochemistry* **16**, 4806 (1977).
50. D. W. Craig and G. G. Hammes, *Biochemistry* **19**, 330 (1980).
51. R. Jones, R. A. Dwek, and I. O. Walker, *Eur. J. Biochem.* **34**, 28 (1973).
52. A. R. Goldhammer and G. G. Hammes, *Biochemistry* **17**, 1818 (1978).
53. D. W. Pettigrew and C. Frieden, *J. Biol. Chem.* **254**, 1896 (1979).
54. N. M. Wolfman, A. C. Storer, and G. G. Hammes, *Biochemistry* **18**, 245 (1979).
55. D. Roberts and G. L. Kellett, *Biochem. J.* **189**, 561, 569 (1980).

10

Multienzyme Complexes

INTRODUCTION

An important aspect of physiological regulation not yet discussed is the existence of multienzyme complexes. As methods of protein purification have improved, a number of large, but well-defined, molecular aggregates have been isolated which catalyze more than a single reaction. Generally a sequence of reactions is catalyzed in which the product of the first enzyme activity is the substrate of the second enzyme activity and the product of the second enzyme activity is the substrate of the third enzyme, etc. For a sequence of $n - 1$ reactions of substrates S_i and enzyme activities E_i, this can be represented as

$$S_1 \xrightarrow{E_1} S_2 \xrightarrow{E_2} \cdots \xrightarrow{E_{n-1}} S_n \qquad (10\text{-}1)$$

In some cases, well-defined molecular aggregates catalyze two or more reactions that are not sequential, but are part of the same metabolic pathway. The biological significance of such multienzyme complexes is not clear; perhaps they represent parts of multienzyme complexes that have been partially disintegrated during isolation.

Multienzyme complexes can contain polypeptide chains either with a *single* enzyme activity or with *multiple* enzyme activities. The latter are often referred to as *multifunctional* proteins (*1*). However, the most important feature of multienzyme complexes is that each catalytic site has its own structural domain. Whether or not these domains are linked by a peptide bond is not of major importance. After discussion of some of the general principles of the operation of multienzyme complexes, a few examples are considered briefly. A number of reviews on this general subject are available (*1–5*).

GENERAL MECHANISTIC PRINCIPLES

The physiological advantages of multienzyme complexes appear to be threefold: (1) enhancement of intrinsic catalytic activity; (2) sequestering of reactive intermediates; and (3) regulation of metabolic fluxes. The enhancement of intrinsic catalytic activity can come about by the mutual restriction of the conformations of interacting polypeptide chains and/or by the spatial location of catalytic sites. Such a mechanism appears to be operative in the case of tryptophan synthase discussed below. The sequestering of reaction intermediates probably is the most common function of multienzyme complexes. Confining intermediates to the complex prevents their destruction by chemical reactions which compete with the desired reaction and in some cases increases the specificity of catalysis. This "channeling" effect by which the reactions of a series of intermediates are restricted to the desired ones is important in the pyruvate dehydrogenase, ketoglutarate dehydrogenase, and fatty acid synthase complexes to be discussed. The sequestering also gives rise to locally high concentrations of substrates; this is advantageous only if the combination of enzyme and substrate is rate limiting in catalysis, which is not often the case. The enhancement of intrinsic catalytic activity and sequestering of reaction intermediates are mechanisms for regulating metabolic fluxes. In addition, many of the multienzyme complexes also are regulated by classical allosteric mechanisms such as feedback inhibition and antagonism of specific metabolites. For optimal efficiency, the multienzyme complex should be regulated at a reaction early in the sequence catalyzed; one of the products of the reactions catalyzed by the complex may itself be a feedback inhibitor.

Some of the most probable mechanisms for the transfer of reaction intermediates within a multienzyme complex are (1) juxtaposition of binding sites, (2) diffusion, (3) macromolecular conformational changes, and (4) rotating arms. These mechanisms are depicted schematically in Fig. 10-1.

The simplest mechanism to envisage for the channeling of intermediates within a multienzyme complex is the juxtaposition of two or more catalytic sites located on different structural domains. This sharing of sites allows intermediates to be protected readily from competing reactions, and intermediates need to be transferred only over very short distances. In principle, such a structural arrangement could transform a multistep reaction sequence into a single concerted reaction.

The most obvious mechanism for transferring reaction intermediates between catalytic sites is diffusion. Bulk diffusion does not seem to be a likely candidate; this affords no protection to the intermediates and does not provide specificity for the transfer of intermediates. Surface diffusion is a

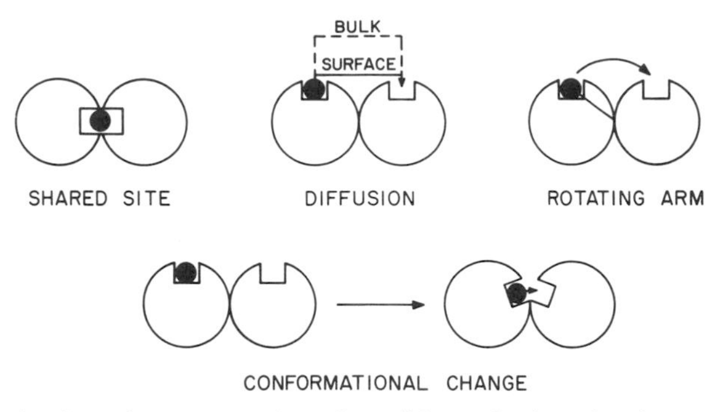

Fig. 10-1. A schematic representation of possible mechanisms for the transfer of intermediates in multienzyme complexes. [From G. G. Hammes, *Biochem. Soc. Symp.* **46**, 73 (1981).]

more attractive mechanism since intermediates can be sequestered, but again specificity is lacking. No compelling evidence exists for surface diffusion being involved in multienzyme complexes with small metabolites as substrates, but some type of one-dimensional diffusion along polynucleotide chains might be of importance for enzymes acting on nucleic acids.

A third potential mechanism for moving reaction intermediates is conformational changes. The triggering of conformational changes by ligand binding is, of course, a well-known and essential ingredient of catalysis and allosteric regulation. However, thus far conformational changes of 5–10 Å are considered large, and this will not carry an intermediate far. The possibility of utilizing large conformational changes for transferring intermediates in multienzyme complexes must be regarded as a potential, but unproven, mechanism.

The use of long chain coenzymes to transport intermediates in multienzyme complexes appears to be the most utilized mechanism. Some examples are the coenzymes lipoic acid,

$$\underset{\text{S—S}}{\text{(ring)}}\text{—}CH_2\text{—}CH_2\text{—}CH_2\text{—}CH_2\text{—}COOH$$

found in the pyruvate and ketoglutarate dehydrogenase multienzyme complexes and 4′-phosphopantetheine,

$$HS\text{—}(CH_2)_2\text{—}NH\overset{O}{\overset{\|}{C}}\text{—}(CH_2)_2\text{—}NH\overset{O}{\overset{\|}{C}}\text{—}\overset{OH}{\overset{|}{C}H}\text{—}\underset{CH_3}{\underset{|}{\overset{CH_3}{\overset{|}{C}}}}\text{—}CH_2\text{—}\underset{OH}{\underset{|}{\overset{O}{\overset{\|}{O P}}}}\text{—}OH$$

in fatty acid synthases. The former arm is 14 Å long when coupled to a lysine side chain of a protein so that a maximum distance of 28 Å can be spanned, whereas the latter coenzyme is 20 Å in length. In both instances, good evidence exists that intermediates are transported by substrates forming thiol adducts with the coenzymes although direct evidence for rotation of the arms and an assessment of the distances traversed have proved to be elusive. The free energy required for rotation between specific sites could be derived from the binding of successive intermediates on the multienzyme complex. A mechanism also has been proposed in which a polyglutamate chain on the substrate rotates the substrate between two catalytic sites (*6*); this is yet another variation of a rotating arm mechanism.

TRYPTOPHAN SYNTHASE

Tryptophan synthase from *Escherchia coli* is a simple multienzyme complex consisting of two types of subunits, α and β, with molecular weights of 29,000 and 45,000, respectively. The α subunit catalyzes the reaction

$$\text{Indole-3-glycerol phosphate} \rightleftharpoons \text{indole} + \text{glyceraldehyde 3-phosphate} \quad (10\text{-}2)$$

and the dimer β_2 catalyzes the reaction

$$\text{Indole} + \text{serine} \rightleftharpoons \text{tryptophan} \quad (10\text{-}3)$$

with pyridoxal 5′ phosphate being a required coenzyme. The β_2 dimer has two catalytic sites, and the pyridoxal phosphate is attached to the protein through a Schiff base linkage with an ε amino group. The intact enzyme, which has the structure $\alpha_2\beta_2$, catalyzes both of the above reactions [Eqs. (10-2) and (10-3)] much more efficiently than the individual subunits and also catalyzes the sum of the two reactions

$$\text{Indole-3-glycerol phosphate} + \text{serine} \rightleftharpoons \text{tryptophan} + \text{glyceraldehyde 3-phosphate} \quad (10\text{-}4)$$

Numerous structural, equilibrium, and kinetic studies have been carried out to elucidate the reaction pathway and reaction intermediates, but primarily aspects relevant to the multienzyme character of the enzyme are considered here (cf. *7, 8*). A similar enzyme is found in *Neurospora* except that the activities of α and β_2 are found on a single polypeptide chain, and the enzyme is a homodimer (*8, 9*); however, the structure and mechanism of the enzyme appear to be quite similar to those of the *E. coli* enzyme.

The interaction of the α and β_2 subunits is associated with an ultraviolet difference spectrum, and an alteration in the spectrum of bound pyridoxal phosphate indicates conformational changes occur when the $\alpha_2\beta_2$ enzyme is formed. Studies of coenzyme, substrate, and substrate analog binding to individual subunits and the intact enzyme have been carried out (cf. *10–13*). Temperature jump experiments have suggested that the isolated β_2 subunit can exist in at least two different conformations (*10*), and the α subunit in at least three different conformations (*12*). However, when the $\alpha_2\beta_2$ enzyme is formed only a single conformation can be detected. The overall process of multienzyme complex formation can be represented as

$$2(\alpha \rightleftharpoons \alpha' \rightleftharpoons \alpha'') + (\beta_2 \rightleftharpoons \beta_2') \rightleftharpoons \alpha_2\beta_2 \qquad (10\text{-}5)$$

Formation of the $\alpha_2\beta_2$ enzyme thus restricts the conformations of both the α and β polypeptides to those with the highest catalytic efficiency (*11*). The subunits are held together quite tightly; the equilibrium constant for the above reaction is about $10^{10}\ M^{-2}$ (*11*). The α polypeptide also restricts the occurrence of several side reactions catalyzed by β_2. Therefore, both intrinsic catalytic activity and specificity are enhanced by formation of the multienzyme complex.

Several steady-state kinetic studies have been carried out with the $\alpha_2\beta_2$ enzyme (*8*). A random addition of substrates apparently occurs although Schiff base formation of serine with pyridoxal phosphate clearly must occur before the reaction can proceed. Although the rates of individual reaction steps are different in the isolated enzymes and intact complex, the basic catalytic mechanism seems unchanged. Free indole is not found as a reaction intermediate with the *E. coli* enzyme so that the catalytic sites of the α and β subunits are presumed to be juxtaposed and a concerted reaction mechanism has been proposed (*14*). However, indole binding sites apparently are retained by both subunits in the complex since indole, serine, and indole propanol phosphate all can bind simultaneously to a catalytic center (*12*). Some incorporation of indole from the bulk solvent into tryptophan and equilibration of indole from the bulk solvent with radioactively labeled indole from indole glycerol phosphate occurs during tryptophan synthesis catalyzed by the *Neurospora* enzyme (*15*). (A similar experiment has not been done with the *E. coli* enzyme.) This suggests a stepwise mechanism through Eqs. (10-3) and (10-4) with free indole trapped in the site and rapidly moving a few Ångstroms to the serine through diffusion or a conformational change as an alternative mechanistic possibility. A schematic representation of a composite site between the α and β subunits is shown in Fig. 10-2. In any event a restriction of polypeptide conformation with concomitant enhanced catalytic activity and juxtaposition of catalytic sites are important factors in the function of the multienzyme complex.

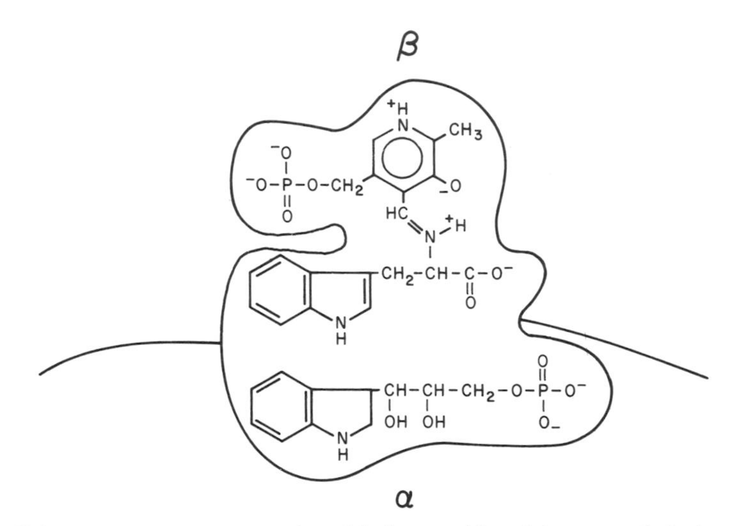

Fig. 10-2. A schematic representation of the juxtaposition of the two catalytic sites on the α and β subunits of tryptophan synthase from *Escherichia coli*. Serine is shown as a Schiff base with pyridoxal phosphate. [From G. G. Hammes, *Biochem. Soc. Symp.* **46**, 73 (1981).]

PYRUVATE AND α-KETOGLUTARATE DEHYDROGENASE

The pyruvate and α-ketoglutarate dehydrogenase multienzyme complexes are found in organisms ranging from bacteria to man (*3, 4*). They catalyze the overall reaction

$$\mathrm{R{-}\overset{O}{\overset{\|}{C}}{-}COOH + CoA + NAD^+ \longrightarrow R{-}\overset{O}{\overset{\|}{C}}{-}CoA + CO_2 + NADH + H^+} \tag{10-6}$$

where $R = CH_3$ for the pyruvate enzyme and $R = (CH_2)_2{-}COOH$ for the α-ketoglutarate enzyme. This overall reaction is carried out by three enzymes that are part of the complex: a decarboxylase (E_1), lipoate acetyltransferase or succinyltransferase (E_2), and lipoamide dehydrogenase (E_3). The sequence of reactions postulated to occur is

$$\mathrm{R\overset{O}{\overset{\|}{C}}{-}COOH + E_1[TPP] \xrightarrow{Mg^{2+}} CO_2 + E_1[R\overset{OH}{\overset{|}{C}}{-}TPP]} \tag{10-7}$$

$$\mathrm{E_1[R\overset{OH}{\overset{|}{C}}{-}TPP] + E_2[Lip\text{-}S_2] \rightleftharpoons E_1[TPP] + E_2[HS\text{-}Lip\text{-}S{-}\overset{O}{\overset{\|}{C}}{-}R]} \tag{10-8}$$

$$\mathrm{E_2[HS\text{-}Lip\text{-}S{-}\overset{O}{\overset{\|}{C}}{-}R] + CoA \rightleftharpoons E_2[Lip\text{-}(SH)_2] + R\overset{O}{\overset{\|}{C}}{-}CoA} \tag{10-9}$$

$$E_2[\text{Lip-(SH)}_2] + E_3[\text{FAD}] \rightleftharpoons E_2[\text{Lip-S}_2] + E_3[\text{FAD(red)}] \tag{10-10}$$

$$E_3[\text{FAD(red)}] + \text{NAD}^+ \rightleftharpoons E_3[\text{FAD}] + \text{NADH} + \text{H}^+ \tag{10-11}$$

where TPP, FAD, and Lip-S_2 represent the cofactors thiamin pyrophosphate, flavin adenine dinucleotide, and lipoic acid, respectively.

The pyruvate dehydrogenase enzyme from *E. coli* has been extensively studied and is now considered in detail. The structure of the enzyme still is somewhat controversial, but the most probable polypeptide chain stoichiometry is 24:24:12 (E_1:E_2:E_3) (*4, 16*). The symmetry requirements imposed by electron microscopy appear to require 24 acetyltransferase polypeptide chains as the core of the enzyme, with E_1 and E_3 combining with this core. This structure is in agreement with reconstitution experiments in which the subcomplexes E_1–E_2 and E_2–E_3 are formed but not E_1–E_3. The overall molecular weight of the complex is 4.8×10^6, and the molecular weights of the individual polypeptide chains are approximately 94,000, 76,000, and 54,000 (E_1, E_2, and E_3; *16*). A model of the octahedral structure suggested by the data is shown in Fig. 10-3, although the positions of E_1 and E_3 in the complex have not been definitively established. In this model, three E_2 polypeptide chains are located at the eight corners of a cube, a dimer of E_3 polypeptide chains is on each face of the cube, and an E_1 polypeptide chain is matched with each E_2 polypeptide chain. The subunit stoichiometry and molecular weight of the complex are consistent with the flavin content of the complex and the number of thiamin pyrophosphate, etheno-CoA, and pyruvate binding sites (*16*). Surprisingly 48 lipoic acids/complex have been found from *N*-ethyl maleimide modification and acetylation of the lipoic acids (*17–21*). This indicates two functional lipoic acids are present per polypeptide chain of E_2.

The regulation of pyruvate and acetyl-CoA concentrations is of great importance in metabolism. Therefore, as might be expected, the pyruvate

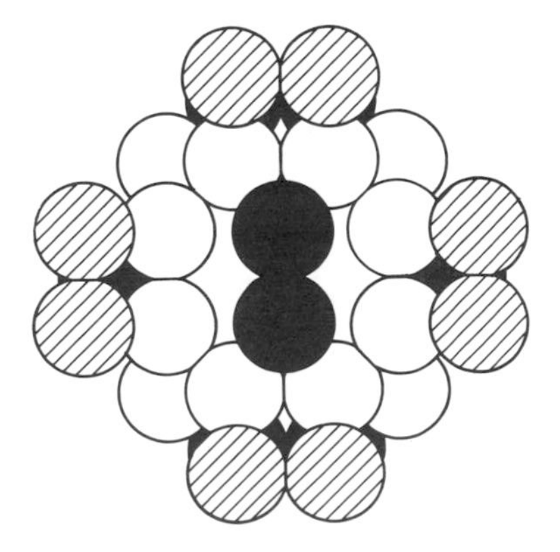

Fig. 10-3. A model of the pyruvate dehydrogenase multienzyme complex from *Escherichia coli*. The view is one face of the octahedral structure. The white balls are the lipoate acetyl transferase with three polypeptide chains on each corner of a cube. The dark balls on the faces of the cube are dimers of lipoamide dehydrogenase. A striped ball representing pyruvate decarboxylase should be on the edge of each lipoate acetyl transferase, but not all of these are shown so as not to obscure the model.

dehydrogenase multienzyme complex is subject to metabolite regulation. Since the reaction catalyzed by E_1 is irreversible, most of the regulation occurs with this enzyme. In fact, E_1 is a classic allosteric enzyme; it is feedback inhibited by acetyl-CoA and also is inhibited by GTP; the acetyl-CoA inhibition is reversed by nucleotide monophosphates, and the GTP inhibition is reversed by GDP; in addition, phosphoenolpyruvate and oxaloacetate are activators of E_1 (*4, 22*). The complex also is inhibited by NADH through binding to E_3. The mammalian pyruvate dehydrogenase complex contains two additional regulatory enzymes, a kinase and a phosphatase, and further regulation occurs by phosphorylation–dephosphorylation of E_1 (*3*).

In view of the results found with tryptophan synthase, an important question is whether formation of the multienzyme complex alters the conformations of the individual enzymes and/or enhances their catalytic efficiency. When E_1 and E_3 are dissociated from the enzyme, they are fully active dimers. Furthermore, if E_2 and E_3 are inactivated, E_1 remains catalytically active, and the binding of ligands to E_1 is unaltered. Similarly, if E_1 and E_2 are inactivated, E_3 is catalytically active. Also the binding of etheno-CoA to E_2 is altered very little when E_1 is removed from the complex entirely. All of these results indicate that mechanistically significant conformational coupling between the three different enzymes does not occur. The primary function of the multienzyme complex structure appears to be the correct geometrical placement of subunits, rather than conformational coupling, and the sequestering or channeling of reaction intermediates. The only clear-cut conformational coupling occurs *within* an E_1 dimer where conformational changes play a role in allosteric regulation.

The classical mechanism proposed for the pyruvate dehydrogenase complex is that lipoic acid transports reaction intermediates by rotating between the three catalytic sites of E_1, E_2, and E_3 (*4*). This is shown schematically in Fig. 10-4a. This mechanism requires that the catalytic sites lie within about 30 Å of each other. The distances between catalytic sites were measured by fluorescence resonance energy transfer (cf. *23*). Thiochrome diphosphate was utilized as a fluorescent analog of thiamin pyrophosphate, etheno-CoA as a fluorescent analog of CoA, and FAD on E_3 is a convenient energy acceptor. The catalytic sites of both E_1 and E_2, on the average, are 50 Å or more from FAD suggesting this simple mechanism may not be adequate. Fluorescent maleimides can be attached to an SH moiety of lipoic acid. Not all of the lipoic acids are chemically equivalent, and the distance between the end of the lipoic acid and thiochrome diphosphate or FAD varies depending on which chemical class of lipoic acids is labeled. The distances measured by resonance energy transfer between specific sites on the pyruvate dehydrogenase complex are summarized in Table 10-1. The experimental uncertainty in the measured distances is 5–10 Å so that unless a substantial fraction of

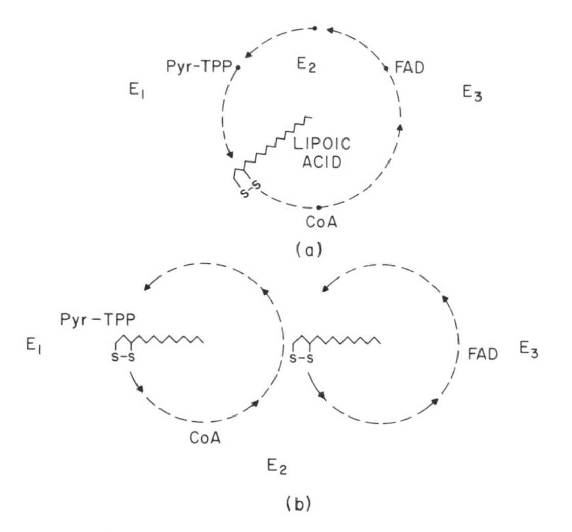

Fig. 10-4. (a) A schematic representation of the mechanism for the pyruvate dehydrogenase multienzyme complex in which a single lipoic acid transports intermediates by rotating between the catalytic sites. (b) A schematic representation of the mechanism for the pyruvate dehydrogenase multienzyme complex in which intermediates are transferred between adjacent lipoic acids and between the lipoic acids and the catalytic sites.

Table 10-1

Average Distances between Catalytic Centers in the Pyruvate Dehydrogenase Multienzyme Complex from *Escherichia coli*[a]

Sites	R (Å)
TPP (E_1)–FAD (E_3)	~50
CoA (E_2)–FAD (E_3)	>50
TPP (E_1)–lipoic acid (E_2)	37–>45
Lipoic acid (E_2)–FAD (E_3)	21–>41
Lipoic acid (E_2)–lipoic acid (E_2)	23–34

[a] Ref. *19*, *23*, *38*.

the catalytic sites and lipoic acids does not participate in the catalytic cycle, i.e., the average distances are not characteristic of those traversed in a catalytic cycle, a single lipoic acid is insufficient to span the required distances.

An alternative mechanism is shown in Fig. 10-4b. Here the acetyl group and electrons are passed between adjacent lipoic acids (*19*, *20*, *24*); as a result the maximum distance traversed by intermediates is 28 Å (twice the extension of a lipoic acid from the polypeptide backbone) multiplied by the number of lipoic acids participating in a catalytic cycle. Adjacent lipoic acids can be shown to approach very closely by energy transfer measurements in which the SH moieties of lipoic acids are simultaneously labeled with energy donors and acceptors and by the observation of excimers in the fluorescence emission when the lipoic acids are completely labeled with *N*-pyrene maleimide (*19, 23*). The distances of closest approach estimated in the first instance are included in Table 10-1, and excimer formation suggests a distance of about 10 Å. More direct support of this mechanism comes from three types of experiments. If the E_2–E_3 subcomplex is reconstituted with very few E_1 molecules, the entire complex can still be acetylated (*24*). This requires either transacetylation or rapid diffusion of E_1 on the surface, which is unlikely. Similarly if E_1 is almost completely inactivated by a reversible inhibitor, the complex can be completely acetylated (*20*). Either rapid dissociation or diffusion of the inhibitor occurs, or transacetylation takes place. On the other end of the reaction, the fact that only 12 flavins are present while up to 48 lipoic acids are acetylated suggests "trans" oxidation–reduction is a possibility. This could be explained by spatial proximity, but the finding that only one-half of the flavins are required for full activity makes this unlikely (*25*). Thus strong evidence exists that interlipoic acid transport of acetyl groups and electrons can occur. However, this does not prove that such transfers are part of a normal catalytic cycle.

Dissecting the normal catalytic cycle into its elementary steps has proved to be difficult. A fundamental question is how many catalytic sites of each type and how many lipoic acids participate in a catalytic cycle. The rate of loss of the overall activity of the complex is directly proportional to the rate of loss of E_1 activity when E_1 is titrated with an inhibitor (*25*); this suggests one E_1 is involved in each catalytic cycle. Titration of lipoic acids with *N*-ethyl maleimide indicates that a substantial fraction, up to one-half, of the lipoic acids is not required for the overall activity of the complex (*25,26*). The dependence of the decrease in activity on the extent of lipoic acid modification is complex, and its interpretation is unclear. All 48 lipoic acids can be acetylated and deacetylated, but only one-half of the lipoic acids are reoxidized by the flavin as judged by the number of equivalents of NADH produced from 48 acetylated lipoic acids (*27*). Direct kinetic studies of the acetylation and deacetylation and of flavin reduction have been carried out (*21, 22*). The results obtained indicate that one-half of the acetylation events occur too slowly to participate in the catalytic cycle. The first order rate constant for acetylation of those lipoic acids that participate in the catalytic cycle is 40–65 sec^{-1} at 4°C. The range in the rate constants is due to the fact that mechanisms assuming *series* acetylation, that is a transfer of acetyl groups between lipoic acids, and *parallel* acetylation, that is servicing of two lipoic acids by a single E_1 in parallel reactions, are both consistent with the data. Comparison of the first-order rate constant with the turnover number of the multienzyme complex suggests that only one-half of the E_1 molecules may participate in a catalytic cycle. To summarize, all experiments assessing the role of lipoic acids in catalysis suggest one-half of them are not necessary for catalysis, although they might play a physiologically relevant role as reserves. The evidence still is inconclusive as to whether transacetylation is part of a normal catalytic cycle and also as to whether all of the E_1 molecules participate in a catalytic cycle.

The rate determining step in overall catalysis is the first irreversible step, which is concerned with decarboxylation of pyruvate and formation of hydroxyethyl-thiamin pyrophosphate (*21*). The rate of deacetylation of the lipoic acids has not been measured directly, but is much faster than acetylation ($>40\ \text{sec}^{-1}$). The rate constant for reduction of flavin is similar to that for acetylation. Therefore, electron transfer reactions and the transport of intermediates are both rapid relative to the rate determining step in catalysis. Direct measurement of the rate of rotation of lipoic acids and presumably of the rate of transfer of intermediates has not been possible. Measurements of the fluorescence polarization of fluorescence probes attached to lipoic acid indicate considerable rotational mobility, with rotational correlation times of approximately nanoseconds (*19, 23*). A similar conclusion can be reached from experiments in which a spin label is attached to the lipoic acid (*28*). Unfortunately in neither case can a decision be reached as to whether local rotational mobility is being measured or actual rotation of the lipoic acid between catalytic centers.

Obviously, the mechanism utilized by the pyruvate dehydrogenase complex is not yet clear, primarily because many of the lipoic acids and perhaps some of the E_1 molecules are not utilized in a normal catalytic cycle. Transfer of acetyl groups and electrons between lipoic acids and between catalytic sites clearly can occur. Additional possibilities that have not been ruled out are that intermediates are passed to an intervening group on the protein and/or large protein conformational changes occur.

The α-ketoglutarate dehydrogenase multienzyme complex from *E. coli* has not been studied as extensively as the pyruvate dehydrogenase complex because it is more difficult to obtain in large quantities. However, 24 dihydrolipoyl transsuccinylase polypeptides appear to form an octahedral core as with the dihydrolipoyl transacetylase (*3, 4*). The molecular weight of the complex is $2.5-2.8 \times 10^6$, and the polypeptide chain stoichiometry proposed is 12:24:12 (E_1:E_2:E_3; *4*). Only one lipoic acid is found per E_2 polypeptide chain. In this case, the distance between the catalytic sites of E_1 and E_3 is ~ 32 Å, and the lipoic acids are close to both FAD on E_3 and the catalytic site on E_1 ($\lesssim 24$ Å; *29*). Moreover, fluorescence measurements do not reveal close interactions between lipoic acids as with the pyruvate dehydrogenase complex. Also in contrast, the overall activity falls off directly proportional to the number of lipoic acids labeled with *N*-ethyl maleimide. These results are consistent with a single lipoic acid rotating between the catalytic sites during a catalytic cycle. Perhaps the structure of the α-ketoglutarate dehydrogenase complex is a more fundamental unit than that of the pyruvate dehydrogenase complex. However, as a precautionary note, mention should be made that transsuccinylation and "trans" oxidation–reduction can occur if E_1 is blocked with an inhibitor or flavin is removed (*20, 29*).

FATTY ACID SYNTHASE

Fatty acid synthases are multienzyme complexes found in a wide range of organisms from microbes to mammals (cf. *30–32*). Those most extensively studied are from yeast and avian liver (pigeon or chicken). The overall reaction catalyzed is

$$\text{Acetyl-CoA} + n\text{malonyl-CoA} + 2n\text{NADPH} + 2n\text{H}^+ \longrightarrow \text{CH}_3\text{—(CH}_2\text{CH}_2)_n\text{—COOH} + (n+1)\text{CoA} + n\text{CO}_2 + 2n\text{NADP}^+ + n\text{H}_2\text{O} \quad (10\text{-}12)$$

In yeast and some other microbes, an acyl-CoA thioester of the fatty acid is the final product rather than the free acid found in bacteria and animals. As suggested by the reaction stoichiometry, the saturated fatty acid is formed by priming the chain with an acetyl group and then adding two carbons at a time with malonyl-CoA until a hydrocarbon chain of the desired length is obtained, namely, palmitic acid. In *E. coli* the required sequence of reactions is carried out by seven different enzymes rather than by a multienzyme complex (*30*). Elucidation of the sequence of reactions for *E. coli* was an important factor in understanding the multienzyme systems, which utilize a similar sequence. The central component in the reaction sequence is the acyl carrier protein (ACP) that contains 4′-phosphopantetheine bound to the protein as a phosphodiester with the hydroxyl of a serine residue. The acyl groups are bound in thioester linkages to the SH group of the coenzyme. The sequence of reactions can be represented as follows:

Acetyl transacylase (E_1)

$$\text{Acetyl-CoA} + \text{E}_1 + \text{ACP} \rightleftharpoons \text{acetyl-ACP} + \text{E}_1 + \text{CoA} \quad (10\text{-}13)$$

Malonyl transacylase (E_2)

$$\text{Malonyl-CoA} + \text{E}_2 + \text{ACP} \rightleftharpoons \text{malonyl-ACP} + \text{E}_2 + \text{CoA} \quad (10\text{-}14)$$

β-Ketoacyl synthase (E_3)

$$\text{Acetyl-ACP} + \text{malonyl-ACP} + E_3 \longrightarrow CH_3{-}CO{-}CH_2{-}CO{-}ACP + E_3 + ACP + CO_2 \tag{10-15}$$

β-Ketoacyl reductase (E_4)

$$CH_3{-}CO{-}CH_2{-}CO{-}ACP + E_4 + NADPH + H^+ \rightleftharpoons CH_3{-}CHOH{-}CH_2{-}CO{-}ACP + NADP^+ + E_4 \tag{10-16}$$

Dehydratase (E_5)

$$CH_3{-}CHOH{-}CH_2{-}CO{-}ACP + E_5 \rightleftharpoons CH_3{-}CH{=}CH{-}CO{-}ACP + H_2O + E_5 \tag{10-17}$$

Enoyl reductase (E_6)

$$CH_3{-}CH{=}CH{-}CO{-}ACP + NADPH + H^+ + E_6 \rightleftharpoons CH_3{-}CH_2{-}CH_2{-}CO{-}ACP + NADP^+ + E_6 \tag{10-18}$$

The chain obtained in the last step can combine with E_3 and then with malonyl-ACP to add another two carbons, etc. until palmityl-ACP is formed. A palmitoyl thioesterase hydrolyzes palmityl-ACP to give the free acid. The above reactions only indicate the overall processes; the mechanism for the multienzyme systems is discussed below.

The mammalian and avian multienzyme complexes all have a molecular weight of about 500,000 and consist of two polypeptide chains of molecular weight 250,000 (*31*). Whether or not these polypeptide chains are identical and contain all seven enzyme activities in a single polypeptide chain or whether they are nonidentical is controversial. Early analysis of the 4′-phosphopantetheine content of the complexes indicated one per molecule (*30*) suggesting nonidentity, but more recent analyses find two per molecule (*33*) which is consistent with the presence of two identical polypeptide chains per molecule. Separation of the two chains by affinity chromatography and identification of specific activities with each chain have been reported (*34*), but further substantiation is needed. The structure of the yeast enzyme is more clear-cut. It has a molecular weight of 2.2×10^6 and contains two different types of polypeptide chains with molecular weights of 185,000 and 180,000. The polypeptide chain stoichiometry is $(\alpha\beta)_6$, with each $\alpha\beta$ dimer forming a distinct structural and catalytic entity (*30, 32, 35*). Genetic analysis of yeast fatty acid auxotrophs shows that the multienzyme complex is coded by two genetically unlinked gene loci, *fas 1* and *fas 2* (*30, 36*). Furthermore, the following assignment of enzyme activities to each gene locus can be made: *fas 1* (molecular weight 180,000)—acetyl transacylase, malonyl (palmitoyl) transacylase, dehydratase, and enoyl reductase; *fas 2* (molecular weight 185,000)—4′-phosphopantetheine binding region, β-ketoacyl synthase, and β-ketoacyl reductase. The malonyl and palmitoyl transacylase are coded by the same structural gene indicating a single enzyme catalyzes both reactions (*36*). Whether or not the structures of the yeast and animal complexes are similar remains to be shown.

A mechanism has been proposed for the yeast and avian liver enzymes based on studies of acetyl, malonyl, and palmityl binding to the multienzyme complex (*30, 32, 37*). Acetyl binding sites were studied by isolating peptides containing radioactively labeled acetyl groups after incubation of the enzyme with [^{14}C]acetyl-CoA. Three different protein binding sties were found, a 4′-phosphopantetheine sulfhydryl, a cysteine sulfhydryl, and a serine or threonine hydroxyl. The malonyl binding sites found in similar investigations were the 4′-phosphopantetheine sulfhydryl and a serine or threonine hydroxyl. Three different palmityl binding sites were found, analogous to the acetyl binding sites. On the basis of these results and studies of intermediate steps with model substrates, the mechanism can be depicted schematically as in Fig. 10-5. In this mechanism, the reaction is primed by loading an acetyl group on the hydroxyl site. This is then passed to the 4′-phosphopantetheine and in turn to the cysteine. A malonyl moiety is then loaded on the hydroxyl group and passed to the 4′-phosphopantetheine. The condensation, reduction, dehydration, and reduction [Eqs. (10-15)–(10-18)] all occur with the intermediate linked to the 4′-phosphopantetheine. However, investigations with model substrates indicate that the last three reactions can occur without this linkage. The function of the 4′-phosphopantetheine is to serve as a rotating arm, carrying the intermediates from site to site. As previously mentioned the span of this coenzyme arm is about 40 Å. The cycle can be continued by transferring the buturyl group to the cysteine and loading another malonyl group on to the serine or threonine hydroxyl. When the correct chain length is reached, the fatty acid is removed by a deacylase (or transferred to CoA by a transacylase in yeast).

While the proposed mechanism is attractive, essentially no information is available concerning the proximities of the many catalytic sites and the rates of the individual reactions in the sequence. Such information is essential if a molecular reaction mechanism is to be established for the processing of the reaction intermediates. In fatty acid synthase, the sequestering and channeling of intermediates and the determination of the fatty acid chain length are important functions of the multienzyme complex. The specific mechanisms for control also are obscure. Nutritional factors have been shown to

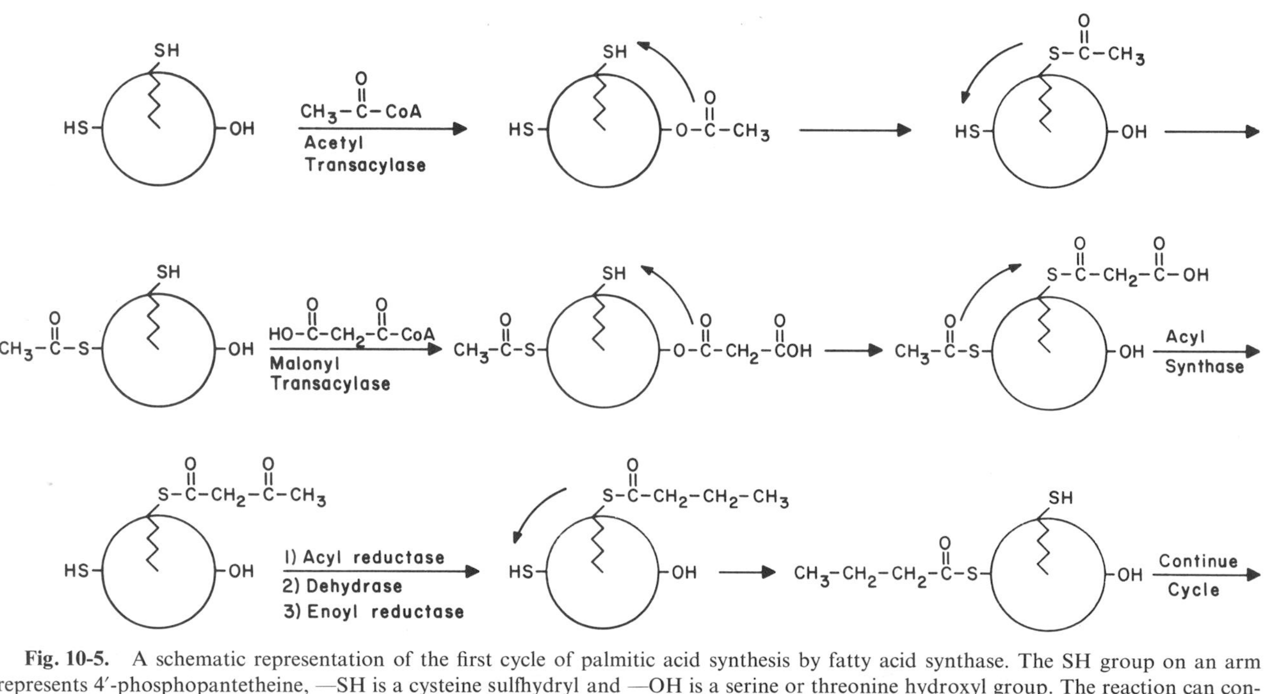

Fig. 10-5. A schematic representation of the first cycle of palmitic acid synthesis by fatty acid synthase. The SH group on an arm represents 4′-phosphopantetheine, —SH is a cysteine sulfhydryl and —OH is a serine or threonine hydroxyl group. The reaction can continue by transfer of malonic acid from malonyl-CoA to the —OH loading site and repetition of the cycle. When the fatty acid chain becomes sufficiently long, free palmitic acid is released from the 4′-phosphopantetheine in mammalian systems or palmityl–CoA is released in the yeast system. [From G. G. Hammes, *Biochem. Soc. Symp.* **46**, 73 (1981).]

be of great importance in regulating the synthesis of the enzyme, but attempts to establish the occurrence of allosteric regulation have been inconclusive (*30*).

CONCLUSION

Obviously the molecular mechanisms underlying the function of multienzyme complexes are not yet well understood. However, many well-defined systems now are available for study with the chemical, physical, and structural methods that have been successfully utilized in elucidating simpler systems.

REFERENCES

1. K. Kirschner and H. Bisswanger, *Annu. Rev. Biochem.* **45**, 143 (1976).
2. A. Ginsburg and E. R. Stadtman, *Annu. Rev. Biochem.* **39**, 429 (1970).
3. L. J. Reed and D. J. Cox, *in* "The Enzymes" (P. Boyer, ed.), Vol. 1, p. 213. Academic Press, New York, 1970.
4. L. J. Reed, *Acc. Chem. Res.* **7**, 40 (1974).
5. G. G. Hammes, *Biochem. Soc. Symp.* **46**, 73 (1981).
6. R. E. MacKenzie and C. M. Baugh, *Biochim. Biophys. Acta* **611**, 187 (1980).
7. C. Yanofsky and I. P. Crawford, *in* "The Enzymes" (P. Boyer, ed.) Vol. 7, p. 1. Academic Press, New York, 1972.
8. E. W. Miles, *Adv. Enzymol.* **49**, 127 (1979).
9. J. A. DeMoss, *Biochim. Biophys. Acta* **62**, 279 (1962).
10. E. J. Faeder and G. G. Hammes, *Biochemistry* **9**, 4043 (1970).
11. E. J. Faeder and G. G. Hammes, *Biochemistry* **10**, 1041 (1971).
12. K. Kirschner, W. Weischet, and R. L. Wishoril, *in* "Protein-Ligand Interactions" (H. Sund and G. Blavier, eds.) p. 2. de Gryter, Berlin. (1975).
13. J. Tschopp and K. Kirschner, *Biochemistry* **19**, 4521 (1980).
14. T. E. Creighton, *Eur. J. Biochem.* **13**, 1 (1970).
15. W. H. Matchett, *J. Biol. Chem.* **249**, 4041 (1974).
16. K. J. Angelides, S. K. Akiyama and G. G. Hammes, *Proc. Natl. Acad. Sci. U.S.A.* **76**, 3279 (1979).
17. M. J. Danson and R. N. Perham, *Biochem. J.* **159**, 677 (1976).
18. D. C. Speckhard, B. H. Ikeda, S. S. Wong, and P. A. Frey, *Biochem. Biophys. Res. Commun.* **77**, 708 (1977).
19. G. B. Shepherd and G. G. Hammes, *Biochemistry* **16**, 5234 (1977).
20. J. H. Collins and L. J. Reed, *Proc. Natl. Acad. Sci. U.S.A.* **74**, 4223 (1977).
21. S. K. Akiyama and G. G. Hammes, *Biochemistry* **19**, 4208 (1980).
22. S. K. Akiyama and G. G. Hammes, *Biochemistry* **20**, 1491 (1981).
23. K. J. Angelides and G. G. Hammes, *Biochemistry* **18**, 1223 (1979).
24. D. J. Bates, M. J. Danson, G. Hale, E. A. Hooper, and R. N. Perham, *Nature (London)* **268**, 313 (1977).
25. K. J. Angelides and G. G. Hammes, *Proc. Natl. Acad. Sci. U.S.A.* **75**, 4877 (1978).

26. M. C. Ambrose-Griffin, M. J. Danson, W. G. Griffin, G. Hale, and R. N. Perham, *Biochem. J.* **187**, 393 (1980).
27. P. A. Frey, B. H. Ikeda, G. R. Gavino, D. C. Speckhard, and S. S. Wong, *J. Biol. Chem.* **253**, 7234 (1978).
28. M. C. Ambrose and R. N. Perham, *Biochem. J.* **155**, 429 (1976).
29. K. J. Angelides and G. G. Hammes, *Biochemistry* **18**, 5531 (1979).
30. J. J. Volpe and P. R. Vagelos, *Annu. Rev. Biochem.* **42**, 21 (1973).
31. K. Bloch and D. Vance, *Annu. Rev. Biochem.* **46**, 263 (1977).
32. F. Lynen, *Eur. J. Biochem.* **112**, 431 (1980).
33. M. J. Arslanian, J. K. Stoops, Y. A. Oh, and S. J. Wakil, *J. Biol. Chem.* **251**, 3194 (1976).
34. F. A. Lornitzo. A. A. Quershi, and J. W. Porter, *J. Biol. Chem.* **250**, 4520 (1975).
35. D. Oesterhelt, H. Bauer, and F. Lynen, *Proc. Natl. Acad. Sci. U.S.A.* **63**, 1377 (1969).
36. A. Knobling, D. Schiffman, H.-D. Sickinger, and E. Schweizer, *Eur. J. Biochem.* **56**, 359 (1975).
37. G. T. Phillips, J. E. Nixon, J. A. Dorsey, P. H. Butterworth, C. J. Chesteron, and J. W. Porter, *Arch. Biochem. Biophys.* **138**, 380 (1970).
38. O. A. Moe, Jr., D. A. Lerner, and G. G. Hammes, *Biochemistry* **13**, 2552 (1974).

11

Membrane-Bound Enzymes

INTRODUCTION

Thus far only soluble enzyme systems have been considered. However, a large number of important enzymes are attached to or imbedded in membranes. Only an introduction to this topic is presented here; an entire book could well be devoted to this subject. How does the characterization of membrane-bound enzymes differ from the characterization of soluble enzymes? The most obvious difference is that membranous systems are not soluble; the best that can be done is to obtain a uniform suspension. To purify a membrane-bound enzyme, it must be removed from the membrane by some type of solubilization procedure, generally utilizing detergents. The trick is to solubilize the enzyme of interest sufficiently well so that traditional methods of protein purification, such as ion exchange and molecular sieve chromatography, salt precipitations, density-gradient centrifugation, etc., can be utilized. Unfortunately excessive solubilization inevitably leads to inactivation of the enzyme. Because the membrane environment is very different from that of aqueous solution, solubilized enzymes often are very unstable. The assay of a membrane-bound enzyme can be difficult. In some cases, the membrane environment is required for activity and/or the function of the enzyme may require a closed membrane structure (e.g., transport). The assay then must involve reconsitution of the solubilized enzyme into a membranelike environment. This environment could be a biological membrane or a mixture of lipids, usually phospholipids. Reconstitution can be exceedingly difficult! The physical and chemical characterization of membrane-bound enzymes also is very difficult. Even if the solubilized enzyme can be assayed and is active, the physical and chemical properties of the solubilized enzyme may not be the same as those of the native enzyme. At the very least, many detergent molecules are tightly bound to the enzyme. Characterization of the membrane-bound enzyme *in situ* is even more difficult. Traditional physical methods do not cope well with heterogeneous solutions, and chemical reagents must be specially designed to work in the hydrophobic environment of membranes. The most successful method for obtaining three-dimensional structural information thus far has been electron microscopy combined with image reconstruction techniques (*1*). Fluorescence resonance energy transfer and nuclear magnetic resonance techniques also are applicable. This chapter presents a brief introduction to membrane structure, followed by discussions of a few specific membrane-bound enzymes that illustrate the diversity in the characteristics of membrane-bound enzymes found in physiological systems.

MEMBRANE STRUCTURE AND FUNCTION

The basic structure of a membrane is a phospholipid bilayer with a mosaic of protein molecules (cf. *2*). This structure is shown schematically in Fig. 11-1. The phospholipids, by virtue of having polar head groups and long chain fatty acids esterified to the glycerol framework, form a structure in which the

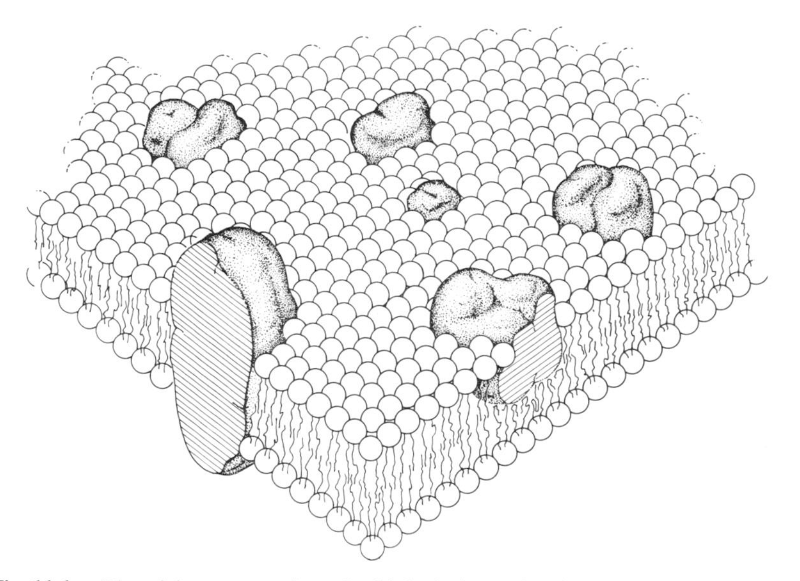

Fig. 11-1. Pictorial representation of a biological membrane. The lipid bilayer is represented by the small balls and wavy lines. The solid bodies with lined surfaces represent proteins. [From S. J. Singer and G. L. Nicolson, *Science* **175**, 720 (1972). Copyright (1972) American Association for the Advancement of Science.]

polar head groups are on the surfaces of the bilayer, and the long chain hydrocarbons form a very hydrophobic interior. When phospholipids are suspended in water and sonicated, closed phospholipid vesicles are formed (Fig. 11-2). Such vesicles often are used to reconstitute solubilized enzymes (cf. *3*). Membrane-bound proteins can be roughly classified into three categories: (1) those that are bound to the polar surface without penetrating into the hydrophobic part of the bilayer; (2) those that are firmly imbedded in both the polar and hydrophobic parts of the membrane, but do not span the membrane; and (3) those that span the entire membrane and are accessible to molecules both inside and outside the membrane. The structure of membranes is not static (*4*). Most of the lipids diffuse freely on the membrane surface, as do many of the proteins. In fact, this free diffusion appears to be a necessity for the operation of many physiological processes. Some proteins do not diffuse appreciably; they appear to be anchored, ultimately through some type of interaction with other proteins inside the cell.

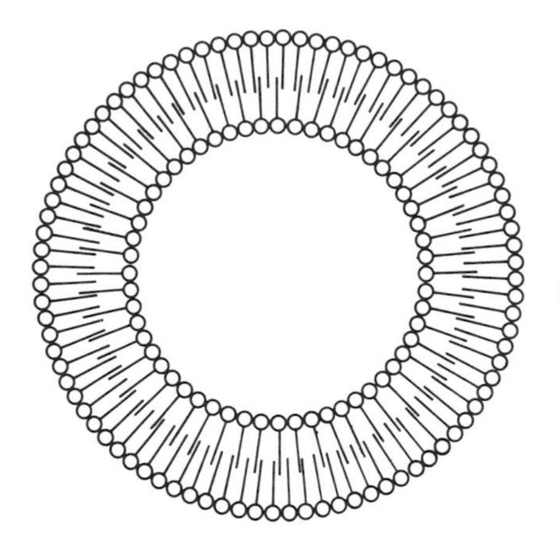

Fig. 11-2. Pictorial representation of a phospholipid vesicle.

In simplistic terms, the function of a membrane is to provide a permeability barrier, and it does so with a closed asymmetric surface. This provides a method for segregating cellular components from each other. The placement of enzymes in membranes has at least four important physiological implications. First, the enzymes are localized to where they are needed. A classic example of this is the electron transport chain in mitochondria where several enzymes are confined to a region in which they can interact with each other and their substrates. Second, the membrane forms a lipophilic environment quite different from that of water. This environment can specifically activate an enzyme, and thus restrict its activity to the membrane. This is the case for β-hydroxybutyrate dehydrogenase, which is discussed. More generally, a lipophilic environment can provide a mechanism for restricting the location of an enzyme. Third, the membrane provides a means for the coupling of enzyme activity and transport across the membrane. For example, the hydrolysis of ATP is responsible for the transport of Na^+, K^+, and Ca^{2+}, and the establishment of a pH gradient across the membrane is necessary for ATP synthesis. Finally membrane-bound enzymes can provide a mechanism for communication between the inside and outside of a cell. The obvious illustration of this is adenylate cyclase: hormones on the outside of the cell regulate the synthesis of cyclic-AMP on the inside of the cell, and cyclic-AMP is an allosteric effector for specific enzymes. The modulation of enzymic activity by interaction of the enzyme with other membrane components provides another mechanism for the regulation of enzymatic processes. We now turn to several specific membrane-bound enzymes that illustrate the general principles just discussed.

β-HYDROXYBUTYRATE DEHYDROGENASE

The enzyme D(−)β-hydroxybutyrate dehydrogenase catalyzes the reaction

$$\text{Acetoacetate} + \text{NADH} + \text{H}^+ \rightleftharpoons \text{D-}\beta\text{-hydroxybutyrate} + \text{NAD}^+ \qquad (11\text{-}1)$$

The enzyme is tightly bound to the mitochondrial membrane, but can be solubilized by the use of phospholipase A or detergents (*5*). The purified solubilized enzyme has one type of polypeptide chain of molecular weight 32,000 (*6*). It is totally inactive in the absence of the phospholipid lecithin (phosphatidylcholine) or lecithin analogs. Lecithin has the structure

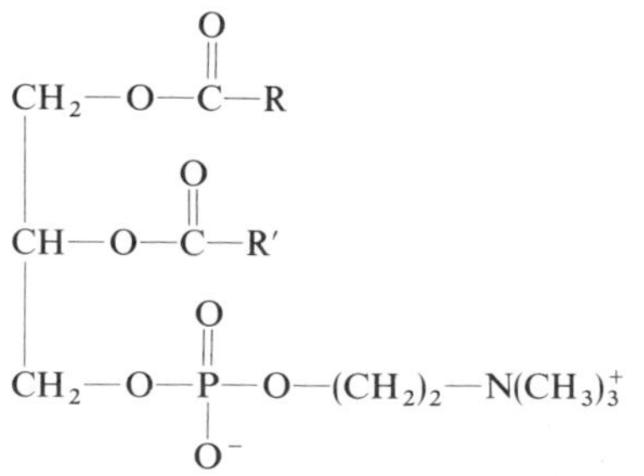

$$\begin{array}{l} CH_2\text{—O—}\overset{\overset{\displaystyle O}{\|}}{C}\text{—R} \\ | \\ CH\text{—O—}\overset{\overset{\displaystyle O}{\|}}{C}\text{—R}' \\ | \\ CH_2\text{—O—}\underset{\underset{\displaystyle O^-}{|}}{\overset{\overset{\displaystyle O}{\|}}{P}}\text{—O—}(CH_2)_2\text{—}N(CH_3)_3^+ \end{array}$$

where R and R′ are long chain hydrocarbons. Enzyme activity can be reconstituted by incubating the purified enzyme with lecithin or its analogs.

The specificity of the enzyme–lecithin interaction has been probed by utilizing lecithins with different fatty acid side chains and related compounds for reconstitution (*7, 8*). As an example, the dependence of the catalytic rate on the chain length of saturated fatty acids in lecithin is shown in Fig. 11-3. A minimum chain length (four carbons) is required for activity, but the rate does

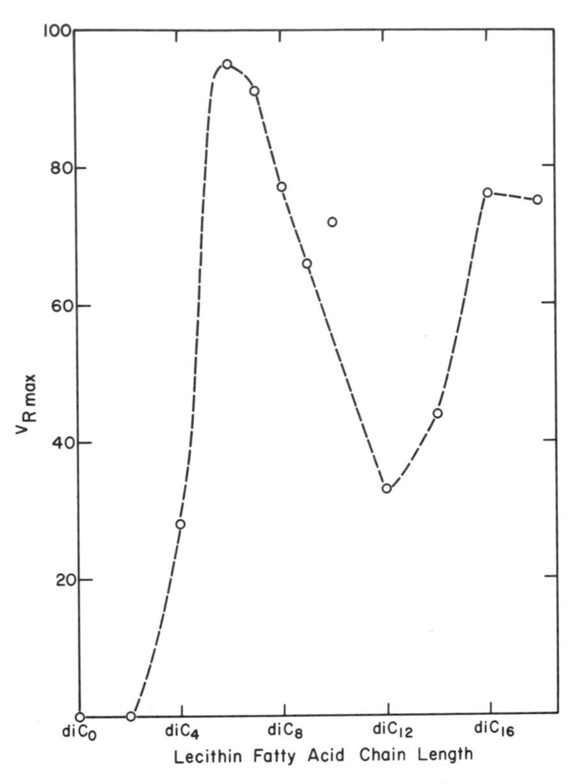

Fig. 11-3. A plot of the maximal relative activity, v_{Rmax}, of β-hydroxybutyrate dehydrogenase–lecithin mixtures versus the number of carbon atoms in the saturated fatty acid side chains of the lecithins. [From A. K. Grover, A. J. Slotboom, G. H. de Haas, and G. G. Hammes, *J. Biol. Chem.* **250**, 31 (1975).]

not depend greatly on chain length above this minimum. This is the case even though the lecithins with short chains are soluble in water (less than eight carbons), whereas lecithins with longer chains form vesicular structures. However, while short-chain lecithins are good activators, an enzyme–lecithin complex that is stable over long time periods only forms with lecithins that are capable of forming lamellar vesicular structures. The degree of saturation of the fatty acid chain also influences the activation process. However, activation does not require a specific structure in the hydrophobic region of lecithin; all that is required is a hydrophobic chain. Thus D and L stereoisomers of lecithin, lysolecithin, dialkoxyphosphatidylcholines, and stearyl phosphorylcholine all activate the enzyme. Whereas the acyl linkage is not crucial for activation, the phosphorylcholine moiety is. Some modification is possible: substitution of butyl or propyl for the ethyl group between the two charged groups does not appreciably alter the activation, but no activation is seen with an isopropyl group substitution. A quaternary ammonium salt is essential, although an ethyl group can be substituted for one of the methyl groups. Thus, the activator site on the enzyme has three specific binding regions: one for a hydrophobic moiety, one for a phosphoryl moiety (negative charge), and one for a quaternary ammonium salt (positive charge). The interaction of the enzyme with lecithin also has been studied with ^{13}C nuclear magnetic resonance (*9*). The lecithin was labeled with ^{13}C either in the polar or hydrophobic moiety. The spin lattice relaxation times and spectral line shapes were measured as a function of the enzyme to phospholipid ratio. When enzyme is added to lecithin, the rotational motion of the head groups is constrained, whereas the motion in the hydrophobic region is increased. The latter effect is probably due to disorder in the bilayer caused by the specific interaction of the hydrophobic chains with the enzyme. Line shape analyses suggest the lateral motion of the lecithin is more constrained in the presence of the enzyme than in its absencc. In spite of the specific interaction of the enzyme with the mitochondrial membrane, its amino acid composition is very similar to that of typical soluble enzymes; the specificity should be apparent in the amino acid sequence, however.

Steady-state kinetic studies of the enzyme indicate the mechanism involves ternary complex formation and an ordered addition of substrates, with NADH binding first (*10*). In the absence of lipid activators the enzyme will not bind NADH. The ability to bind NADH also can be destroyed by modification of a single sulfhydryl residue (*11*). A second nonessential thiol can be modified in the presence of NADH. An arginine and a histidine residue also have been implicated in the catalytic activity by chemical modification studies (*12*).

The lipid specificity of β-hydroxybutyrate dehydrogenase has the important function of restricting the enzymatic activity to the membrane surface. However, the physiological function of the enzyme is not well understood; it has been implicated in an electron shuttle mechanism and as part of a control system for regulating the phosphate potential. In any event, it serves as an excellent model system for studying enzyme–lipid interactions.

$(Na^+ + K^+)$-ACTIVATED ATPase

In bisoynthetic pathways, the hydrolysis of ATP often is coupled to a reaction that is thermodynamically unfavorable. Because of the large negative standard free energy associated with ATP hydrolysis, this coupling shifts the overall equilibrium in such a manner that the desired product is obtained. Membrane-bound ATPases are coupled to ion transport; in this case the coupling permits ion gradients across the membrane to be created and maintained. The $(Na^+ + K^+)$-activated ATPase is an ubiquitous enzyme that transports K^+ into the cell and Na^+ out of the cell, both against concentration gradients. The free energy change, ΔG, associated with a concentration gradient is

$$\Delta G = RT \ln([c_2]/[c_1]) \tag{11-2}$$

where z is the charge on the ion and $\mathscr{F}$ is the Faraday. This total free energy ions, a gradient of electrical charge also can exist, resulting in a membrane potential, $\Delta\psi$. Therefore, the total free energy change for the transport of a charged ion is

$$\Delta G = RT \ln([c_2]/[c_1]) + z\mathscr{F}\Delta\psi \tag{11-3}$$

where z is the charge on the ion and $\mathscr{F}$ is the Faraday. This total free energy change is called the *electrochemical gradient*; it is often expressed in electrical units (volts or millivolts) as $\Delta\mu = \Delta G/[z\mathscr{F}]$. The unfavorable (positive) free energy change associated with ion transport can be compensated for by the favorable standard free energy change associated with ATP hydrolysis.

The $(Na^+ + K^+)$-activated ATPase has been purified from many different sources including the eel electrical organ, kidney, and brain. Many comprehensive reviews of the system are available (*13–15*). The first studies of the enzyme utilized homogenized crab nerve (*16*). An ATPase activity was found that required Na^+, K^+, and Mg^{2+} for maximum activity. However, the first convincing demonstration that the transport of Na^+ and K^+ was coupled to ATP hydrolysis occurred with the enzyme in erythrocytes where defined internal concentrations of salts can be obtained (*17*). If the Na^+ concentration is high on the inside and the K^+ concentration high on the outside, then K^+ moves in and Na^+ out as internal ATP is hydrolyzed. The stoichiometry of the overall reaction is

$$\begin{aligned} 3Na^+_{inside} + 2K^+_{outside} + ATP^{-4} + H_2O &\xrightarrow{Mg^{2+}} \\ &3Na^+_{outside} + 2K^+_{inside} + ADP^{-3} + P_i^{-2} + H^+ \end{aligned} \tag{11-4}$$

The purified protein contains two types of polypeptide chains, α and β, probably in a 1:1 stoichiometry. However, the active enzyme may be either $\alpha\beta$ or $(\alpha\beta)_2$, with the latter seeming most probable (*15*). Considerable controversy exists with regard to the molecular weights of the polypeptide chains because different methods of assessing molecular weight give different results and corrections must be made for bound detergent which is used to purify the enzyme. The α polypeptide appears to have a molecular weight of about 105,000 and the β polypeptide a molecular weight of about 40,000 (*18*), but reported values range from 85,000 to 120,000 and from 37,000 to

50,000, respectively (*15*). The β polypeptide is covalently coupled to carbohydrate ($\sim 10\%$ by weight). The purified enzyme has been reconstituted into phospholipid vesicles, and the reconstituted enzyme pumps ions (*19*). The enzyme is randomly oriented in the vesicle, but since ATP is added externally only enzyme with the ATP hydrolysis site on the outside is functional.

Several important binding sites have been identified on the enzyme. The glycoside ouabain

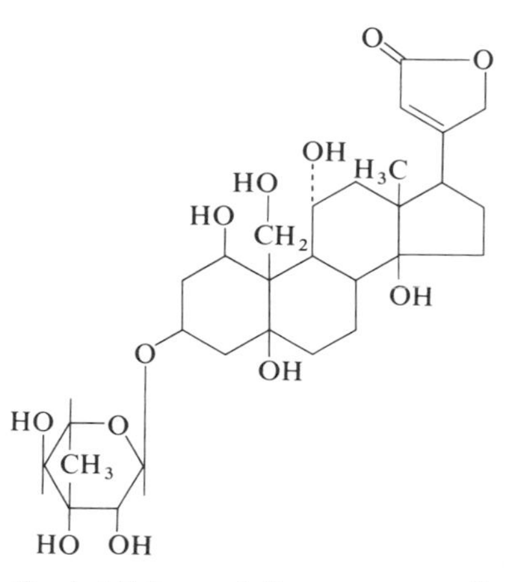

is a potent and specific inhibitor of the enzyme; the photoaffinity label 2-nitro-5-azidobenzoyl-ouabain can be covalently linked to the α polypeptide (*20*). The catalytic site is readily labeled because the enzyme is phosphorylated by ATP during catalysis, and the phosphorylated protein can be isolated. An aspartate residue on the α polypeptide is phosphorylated (*21*). Vanadate ion is a specific and very strong inhibitor of the enzyme: the dissociation constant is about 4 nM (*22*). This ion has been postulated to be a *transition state analog* because it resembles the transition state expected for the hydrolysis of the phosphate–aspartate anhydride bond, namely, a trigonal bipyramid. Ouabain and vanadate binding and ATP phosphorylation all have the same stoichiometry, but it is still not clear whether the stoichiometry is one per $\alpha\beta$ or per $(\alpha\beta)_2$. Obviously the number of monovalent cation sites per functional unit is of considerable importance. Equilibrium binding studies have suggested about three sodium sites per phosphorylation site (*23*) and two K^+ sites per ouabain binding site (*24*). However, the specificity of these sites is not clearly established, and whether or not Na^+ and K^+ compete for the same binding sites also is unclear. A single set of monovalent cation binding sites appears to be consistent with existing data. Nuclear magnetic resonance has been used to study metal binding to the enzyme (*25*). Experiments with Mn^{2+} as a paramagnetic probe for Mg^{2+} suggest that ATP is close to, but not directly coordinated to, Mn^{2+}. Measurement of the nuclear magnetic relaxation times of $^{205}Tl^{+}$ and $^{7}Li^{+}$ in the presence of Mn^{2+} or Cr(III)ATP indicate that one or more monovalent cation sites are close to both the Mn^{2+} site (5–7 Å) and the Cr(III)ATP site (4–7 Å). However, the functional role of these monovalent cation sites has not been elucidated. ATP protectable modification of the enzyme (α polypeptide) has been found with sulfhydryl, arginine, and tyrosine reagents, and these groups may be near the catalytic site. The function of the β polypeptide is unknown, but it is necessary for catalytic activity and is always present at a concentration similar to that of the α polypeptide.

The orientation of the enzyme in the membrane has not been firmly established; however, photoaffinity labeling of the cardiac glycoside binding site indicates this site is on the external side of the plasma membrane, whereas ATP phosphorylation occurs on the cytoplasmic side of the membrane (*26*). Therefore, the α polypeptide spans the membrane. Furthermore, antibodies specific for the cytoplasmic side of the enzyme do not bind when added to the external side, indicating movement of the protein through the bilayer does not occur during transport (*27*). Electron microscopy studies suggest the α chain is extended away from both sides of the membrane, but the β chain extends appreciably only on the exterior of the membrane (*28*). Studies of the proteolytic susceptibility of the membrane-bound protein suggest the amino terminal domain of the α polypeptide interacts strongly with the interior membrane. Delipidated enzyme is completely inactive. A pictorial representation of the enzyme in the membrane is shown in Fig. 11-4.

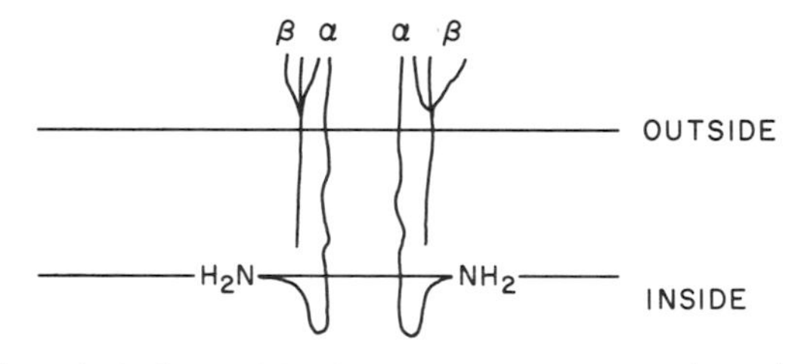

Fig. 11-4. A hypothetical model for the orientation of the $(Na^{+} + K^{+})$-activated ATPase in a membrane.

Many kinetic studies of the $(Na^{+} + K^{+})$-activated ATPase have been carried out, and the results obtained are very complex (cf. *14*, *15*). If the ATPase activity is studied in the presence of Na^{+} and K^{+}, a "biphasic" Lineweaver–Burke plot is obtained; this can be interpreted in terms of an apparent Michaelis constant in the micromolar range and another in the the 0.1–1 m*M* range. Several different mechanisms have been used to explain these results. For example, a high affinity and low affinity catalytic site might exist or negative cooperativity might occur. Alternatively, a "flip–flop" mechanism has been proposed in which the release of products at one site requires low affinity substrate binding at a second site. The binding at the second site then becomes very tight, until low affinity binding at the first site causes relase of product at the second site, etc. More recent results of transient kinetics and other steady-state kinetics suggest an alternative mechanism. If the enzyme is mixed with Na^+ and MgATP, phosphorylation of the enzyme occurs. The reaction can be reversed by ADP, but maximum net hydrolysis requires the addition of K^+. The phosphorylation is dependent only on saturation of a high affinity MgATP site, whereas the dephosphorylation is inhibited by MgATP binding at a low affinity site. This suggests two quite different conformations of the enzyme exist, E_1 and E_2; the overall reaction can be written as

$$
\begin{aligned}
E_1 + Na^+ + \text{MgATP} &\rightleftharpoons E_1 - P \cdot Mg + Na^+ + \text{ADP} \\
E_1 - P \cdot Mg &\rightleftharpoons E_2 - P \cdot Mg \\
E_2 - P \cdot Mg + K^+ &\longrightarrow E_2 + P_i + K^+ + Mg^{2+}
\end{aligned}
\tag{11-5}
$$

Detailed studies of these reactions have led to the postulation of various intermediates and mechanisms for coupling transport and the enzymatic reaction. One of these mechanisms is depicted in Fig. 11-5 (*15, 29*). The phosphorylation ($A \rightleftharpoons B \rightleftharpoons C$) occurs on the inside of the membrane. Conversion to the E_2 conformation ($C \rightleftharpoons D$) results in the extrusion of Na^+. The K^+-stimulated hydrolysis ($D \rightleftharpoons E \rightleftharpoons F$) utilizes K^+ on the outside of the cell. At low concentrations of MgATP the conversion of F to A results

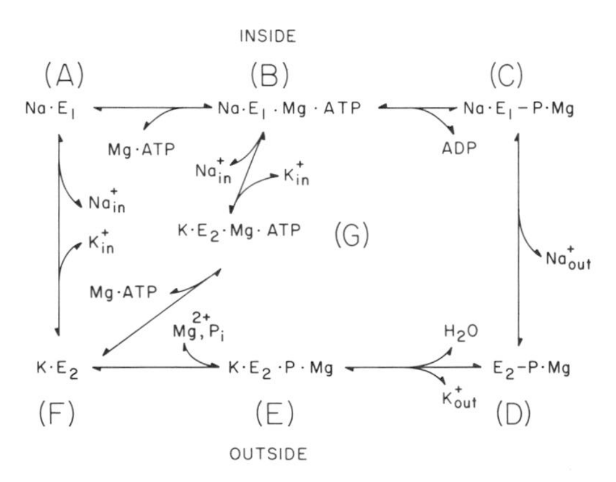

Fig. 11-5. A possible kinetic mechanism for ATP-driven transport of Na^+ and K^+. E_1 and E_2 designate conformations of the transport sites that favor the binding of cytoplasmic Na^+ and extracellular K^+, respectively. The stoichiometries of Na^+ and K^+ are omitted. [From L. C. Cantley, *Curr. Top. Bioenerg.* **11**, 201 (1981).]

in the transport of K^+ to the inside, whereas at high concentrations of MgATP, transport is accomplished through the pathway $F \rightleftharpoons G \rightleftharpoons B$. The Na^+ and K^+ stoichiometries have been omitted for the sake of simplicity and only two different conformational states of the enzyme are shown. Very likely all of the states shown represent different conformations. If MgATP is assumed to bind tightly to E_1 and weakly to E_2, a biphasic Lineweaver–Burke plot is obtained. Thus, this mechanism appears to account for all of the experimental observations.

The mechanism in Fig. 11-5 is a formal representation of the reaction, but what is happening on a molecular scale? The answer to this question is not known, but some general mechanistic features can be outlined. Clearly the binding of Na^+ must be strong when the monovalent cation sites are exposed to the inside and weak when exposed to the outside. The inverse situation must prevail for K^+. This change in metal ion specificity undoubtedly involves conformational changes. The mechanism by which the monovalent cation sites are exposed to the inside and outside also could be small conformational changes that block access of the sites to the outside and inside, respectively. The next step is to determine what these conformational changes are in terms of the three-dimensional structure of the enzyme.

ATP SYNTHESIS

The enzyme just discussed utilizes the free energy obtained from ATP hydrolysis to pump ions across the membrane. Obviously enzymes that hydrolyze ATP are capable of synthesizing ATP: an enzyme catalyzes reactions equally well in both the forward and reverse directions. A mechanism for making ATP is to simply run an ion pump backward, that is impose an ion gradient across the membrane that is large enough so that the enzyme synthesizes ATP instead of hydrolyzes it. This, in fact, is exactly what is done physiologically except that a special enzyme, usually called a coupling factor complex, has been developed for this role. The structure and function of this enzyme are quite ubiquitous in nature: a similar enzyme is found in bacteria, yeast, green plants, and mitochondria. While the molecular mechanism for ATP synthesis is not yet established, the chemiosmotic hypothesis provides a description of the overall mechanism (*30–32*). In this hypothesis, photon excitation or oxidation of organic molecules (depending on the organism) induces a flow of electrons within the membrane, which is accompanied by a unidirectional movement of protons across the membrane; this creates an electric potential and a proton gradient across the membrane. The reverse flow of protons provides the energy for the coupling factor to synthesize ATP. The reverse reaction, ATP hydrolysis, is accompanied by proton movement across the membrane in the opposite direction. A schematic representation of the chemiosmotic mechanism is presented in Fig. 11-6. If only a proton gradient is present, the overall process can be represented as

$$\mathrm{ATP} + n\mathrm{H}_o^+ \rightleftharpoons \mathrm{ADP} + \mathrm{P}_i + n\mathrm{H}_i^+ \tag{11-6}$$

where the subscripts o and i represent outside and inside the vesicle in Fig. 11-6, respectively. The equilibrium constant for this reaction is

$$K_{\mathrm{eq}} = \frac{[\mathrm{ADP}][\mathrm{P}_i]}{[\mathrm{ATP}]}\left[\frac{\mathrm{H}_i^+}{\mathrm{H}_o^+}\right]^n \tag{11-7}$$

A more complete formulation would include the divalent metal ion required in the enzymatic process and the proton released when ATP is hydrolyzed. Since the metal ion concentration and outside pH are usually constant, these factors can be included in the value of K_{eq}. The value of n is still controversial, but probably is 2 or 3. Note that the contribution of the proton gradient to the standard free energy is $-nRT\ln([\mathrm{H}_i^+]/[\mathrm{H}_o^+])$; the customary

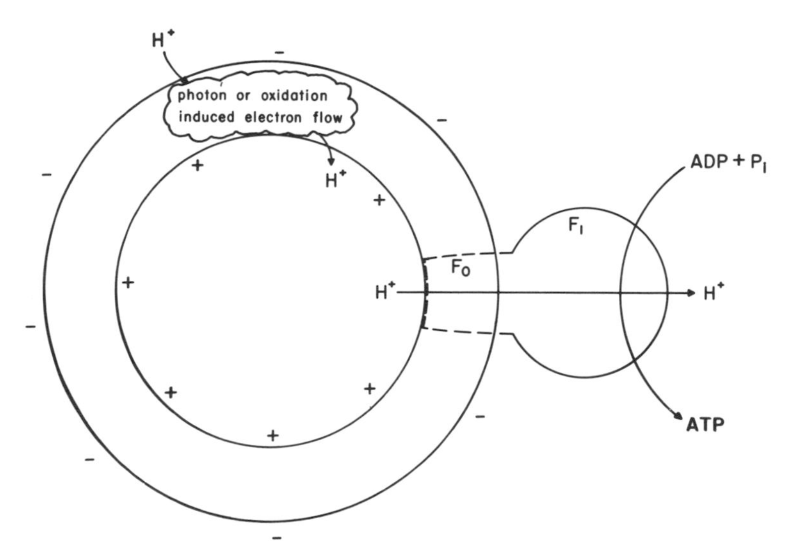

Fig. 11-6. Schematic representation of the coupling factor complex ($F_1 \cdot F_0$) and the chemiosmotic mechanism for chloroplasts, sonicated mitochondria and bacteria, and reconstituted systems. Outward proton flow through the coupling factor complex is accompanied by ATP synthesis. In the reverse reaction, ATP hydrolysis by the complex is accompanied by inward proton movement. The initial pH gradient is established by electron transport. In native mitochondria and bacteria, protons are ejected during electron flow and the position of the coupling factor is reversed with F_1 being on the inner membrane surface. [From B. A. Baird and G. G. Hammes, *Biochim. Biophys. Acta* **549**, 31 (1979).]

notation is to convert this to volts per proton by dividing the standard free energy by $n\mathscr{F}$ where $\mathscr{F}$ is the Faraday. If a membrane potential, $\Delta\psi$, is present, the total electrochemical potential is

$$\Delta\mu_{H^+} = \Delta\psi - 2.303(RT/\mathscr{F})\Delta pH \tag{11-8}$$

Thus, both the membrane potential and pH gradient contribute to the free energy balance necessary for ATP synthesis.

Coupling factor complexes from a variety of sources are composed of two multisubunit parts. One portion, coupling factor 1 (F_1) is peripheral to the membrane; it can be easily and reversibly dissociated from the membrane without disruption of the membrane structure. The solubilized F_1 hydrolyzes ATP, but can no longer synthesize ATP because a pH gradient is not possible. The remaining portion of the complex, F_0, is an integral membrane protein and mediates proton transport. Membranes depleted of F_1 are devoid of ATP synthetic activity even if some F_1 molecules remain. This is because the membrane can no longer maintain a proton gradient. The coupling factor F_1, therefore, has two important functions: it prevents dissipation of the pH gradient and catalyzes ATP production. We will concentrate on the coupling factor complex from one particular source, chloroplasts. Many reviews of coupling factor complexes are available (cf. *32–37*). Two purified preparations of the chloroplast coupling factor complex have been extensively studied (*36,37*): the F_1 portion, which can be dissociated from the membrane with ethylene-diaminetetraacetic acid, and detergent solubilized $F_1 \cdot F_0$. The $F_1 \cdot F_0$ complex can be reconstituted into phospholipid vesicles, and ATP can be synthesized if a pH gradient is created across the phospholipid bilayer (*38*).

Polyacrylamide gel electrophoresis in the presence of sodium dodecyl sulfate reveals that soluble F_1 from chloroplasts (CF_1) contains five different types of subunits designated α, β, γ, δ, and ε. The subunit molecular weights are 59,000, 56,000, 37,000, 17,500, and 13,000, respectively. The molecular weight of CF_1 is 325,000, and it is an approximately spherical molecule with a diameter of about 110 Å. The stoichiometry of subunits in CF_1 is still controversial, but several independent measurements have suggested it is $\alpha_2\beta_2\gamma\delta\varepsilon_2$. The functions of the individual subunits are only known in general terms. The catalytic and nucleotide binding sites are on α and β; in fact, the $\alpha_2\beta_2$ complex alone has ATPase activity. The γ subunit appears to be involved in controlling the proton permeability; the δ subunit is required for CF_1 to bind to the membrane portion of the complex (CF_0); and the ε subunit may be involved in the regulation of catalytic activity. The spatial relationships of the subunits in the intact CF_1 molecule have been investigated with cleavable chemical cross-linking reagents and two-dimensional sodium dodecyl sulfate polyacylamide gel electrophoresis (*39*). With this technique, the subunits adjacent to each other are cross-linked, and the compositions of the cross-linked aggregates are determined by gel electrophoresis. If cross-linking is observed, two polypeptides must be close to each other; however, if cross-linking is not observed two polypeptides are not adjacent *or* reactive groups on the two polypeptides are not close. Therefore, a negative finding cannot be interpreted unambiguously. The tentative arrangement of subunits found from this work is shown in Fig. 11-7. Fluorescence resonance energy transfer measurements have been used to determine the distances between five functionally important sites on the enzyme (*36*).

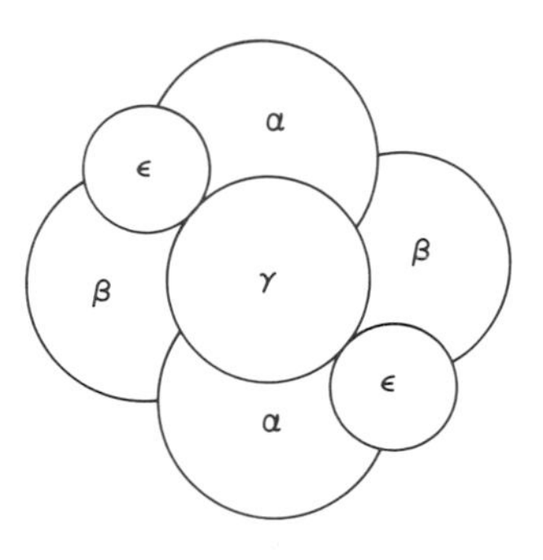

Fig. 11-7. A pictorial representation of the subunit arrangement in CF_1 derived from chemical cross-linking experiments. [From B. A. Baird and G. G. Hammes, *Biochim. Biophys. Acta* **549**, 31 (1979).]

The $CF_1 \cdot CF_0$ complex has a molecular weight of about 405,000. In addition to the five different types of polypeptide chains found in CF_1, four polypeptides of molecular weight 17,500, 15,500, 13,500, and 8000 are found. The polypeptide of molecular weight 8000 is a proteolipid and appears to mediate passive proton transfer when reconstituted into phospholipid vesicles. The stoichiometry of polypeptide chains in CF_0 is not known. The stalk and ball structure for the coupling factor complex shown in Fig. 11-6 was derived from electron microscopy and also is consistent with fluorescence resonance energy transfer measurements in which individual subunits were labeled with specific fluorescent antibodies (the energy donor) and the phospholipid vesicle was labeled with energy acceptors (*36*).

A complicating aspect of the mechanism of ATP synthesis is that adenine nucleotides are both substrates and effectors. Three distinct nucleotide binding sites have been found on CF_1 (*40*). One site binds ADP tightly; long-term dialysis will not dissociate ADP from this site. However, ADP can be exchanged into this site, providing ADP is bound at other sites which regulate the exchange process. The exchange is relatively slow; it has a half-time of about 30 min under optimal conditions. When radioactively labeled ATP is incubated with the enzyme, radioactively labeled ADP is incorporated into the site so that the exchange process is associated with a weak ATPase activity. A second nucleotide site binds MgATP very tightly;

the dissociation process requires days; ADP (and ATP without Mg^{2+}) binds reversibly to this site with a dissociation constant of about 1 μM. A third site binds ADP and ATP reversibly with a dissociation constant of about 1 μM. The subunit location of these three sites has been probed with nucleotide photoaffinity labels. Primarily the β subunit is labeled when the photoaffinity label is on the site that binds MgATP tightly, whereas α and β subunits are labeled equally when the label is at the site that reversibly binds nucleotide. Thus far labeling of the site containing tightly bound ADP has not been possible although this site definitely is on the $\alpha_2\beta_2$ complex.

The CF_1 molecule is a latent ATPase that requires Ca^{2+} as a cofactor. Activation is most easily done by heating the enzyme solution in the presence of dithiothreitol. The molecular mechanism of activation is not known, but gross structural changes are not seen, and a disulfide linkage on the γ subunit apparently is reduced. Steady-state kinetic studies of ATP hydrolysis give normal Michaelis–Menten kinetics. However, ADP is an allosteric inhibitor with a dissociation constant of about 1 μM. Nucleotide turnover at the sites binding ADP and MgATP tightly is too slow to be of catalytic significance. Thus we are left with the dilemma that only one nucleotide binding site qualifies as the catalytic site, in spite of a probable subunit stoichiometry of $\alpha_2\beta_2$. The concept of a catalytic site shared by two polypeptide chains is not without precedent: see the discussion of phosphofructokinase in Chapter 9. Consistent with this idea is the finding that modification of a single tyrosine residue on a β polypeptide (per enzyme molecule) completely inactivates the enzyme (*41*). The possibility exists, of course, that yet undetected nucleotide binding site(s) are the catalytic site(s).

The reconstituted $CF_1 \cdot CF_0$ enzyme has nucleotide binding characteristics very similar to CF_1 although the data are not as clear-cut (*42*). This enzyme requires Mg^{2+} for ATP hydrolysis and synthesis. The simplest model system for studying the mechanism of ATP synthesis and hydrolysis and proton pumping is one in which $CF_1 \cdot CF_0$ and bacteriorhodopsin are incorporated into phospholipid vesicles (*43, 44*). When light shines on these vesicles, bacteriorhodopsin pumps protons into the vesicles thereby creating a pH gradient. The magnitude of the pH gradient can be varied by changing the light intensity. If ADP and P_i are added to the system while the light is on, ATP is synthesized. The kinetics of both ATP hydrolysis and synthesis can be studied with different pH gradients and external pH values. The data obtained thus far indicate that only the maximum velocities depend on ΔpH; therefore, the proton pumping occurs after the substrates are bound. The detailed mechanism for synthesis and hydrolysis is not known although many speculations have been made. A "flip–flop" mechanism has been proposed in which one tightly bound nucleotide is always bound to the enzyme (*32*). This mechanism is difficult to reconcile with the nucleotide binding characteristics and turnover on CF_1 but cannot be excluded. Stereochemical experiments indicate that ATP hydrolysis occurs with inversion at the γ phosphorus; therefore, a direct, in line, transfer between substrates is probable without an intermediate transfer of the phosphoryl group to the enzyme (*45*). The molecular linkage between the proton pumping and catalytic mechanism also remains to be established. However, an obvious type of mechanism is the occurrence of conformational changes during the reaction so that specific ionizable groups have high pK values when facing the outside of the vesicle and low pK values when facing the inside. Some type of proton channel seems very likely, but again direct proof is lacking. Clearly the catalytic, transport and regulatory mechanisms for this sytem still pose fascinating problems.

ADENYLATE CYCLASE

Hormone-sensitive adenylate cyclase is found in most animal cells (cf. *46–49*). The interaction of specific hormones with the outside of the cellular membrane raises and lowers the rate of production of cyclic-AMP by adenylate cyclase on the inside of the cell. Cyclic-AMP controls diverse metabolic processes, one of the most important being the phosphorylation–dephosphorylation cascade discussed in Chapter 8. Table 11-1 contains a listing of some of the hormone effectors of adenylate cyclase and the tissues

Table 11-1

Some Hormone Effectors of Adenylate Cyclase (*47, 48*)

Hormone	Source of material
Catecholamines	Muscle, liver, spleen, kidney, brain pineal gland, adipose, pancreatic islets, erythrocytes, leukocytes, thymocytes, fat cells lung fibroblasts, parotid gland, endothelium, peritoneal macrophages
ACTH	Fat cell
Glucagon	Liver, fat cell, pancreatic islets
Secretin	Fat cell
Thyrotropin	Thyroid
Prostaglandin	Thyroid, kidney, platelets
Oxytocin	Bladder
Corticotropin	Adrenal, fat cell
Calcitonin	Kidney
Parathyroid	Kidney

involved. The hormonal stimulation also requires the presence of a guanine (or related purine) nucleotide, and these compounds stimulate enzymatic activity even in the absence of a hormone. Fluoride ion is another ubiquitous activator, but the physiological relevance of this finding is unknown. The primary difficulty in studying the adenylate cyclase system is the small amount of material present in membranes: the maximum abundance is about 10 pmoles/mg membrane proteins, and the amount in many membranes is even less. In addition to this, the usual difficulties associated with integral membrane proteins exist: detergent solubilization must be used and the proteins are labile when removed from the membrane. In spite of these difficulties, considerable progress is being made in elucidating the molecular mechanism underlying the operation of the hormone-sensitive adenylate cyclase system.

The system contains three distinct parts: the hormone receptor, the catalytic protein, and the guanine nucleotide binding protein. The hormone receptor binding site is located on the outside of the cell, and each type of hormone has its own receptor so that a target cell may contain several different types of receptors. The most direct demonstration that adenylate cyclase and hormone receptors are distinct proteins was from cell fusion experiments (*50*). A type of cell was found that has adenylate cyclase but no receptors. When this type of cell was fused with cells that had receptor and inactive adenylate cyclase, a hormone-stimulated adenylate cyclase activity was observed. Many similar experiments now have been carried out, and they indicate that the receptor and enzyme diffuse freely and interact in the plane of the membrane. Several hormone receptor proteins have been purified, and their ligand binding properties are quite similar to those of the membrane-bound proteins (*49*). Typical molecular weights appear to be about 100,000. Adenylate cyclase has been purified from a variety of sources, and in a few cases has been resolved into catalytic and regulatory proteins (*49*). The catalytic protein has a molecular weight of about 150,000–200,000, and its subunit composition is unknown. Both MgATP and MnATP are substrates, but the latter is 10-fold more active. The regulatory protein binds guanine nucleotides and fluoride. It can be purified by affinity chromatography with GTP–agarose and can be covalently labeled with guanine derivatives (*49,51*). A polypeptide chain of approximately molecular weight 45,000 has been isolated, but this may not be the entire regulatory protein. The regulatory protein also appears to increase the ability of the catalytic subunit to use MgATP as substrate. While the regulatory and catalytic proteins can be resolved, whether or not they exist as a complex in the membrane or as separate entities is not known. Membranes deficient in the regulatory protein can be reconstituted by the addition of detergent solubilized regulatory protein to give hormone-stimulated activity (*52*). However, the complete reconstitution of the individual proteins into depleted membranes or phospholipid vesicles has proved elusive. A reconstitution of partially resolved soluble proteins with a membrane extract–phospholipid mixture has been reported (*53*).

The question of how the three components of the adenylate cyclase system interact in the membrane is largely a matter of speculation. The fusion and reconstitution experiments already mentioned suggest that all three components have some independent motion. Additional evidence suggests the hormone receptor proteins diffuse freely. In the adipocyte the alterations in catalytic rate caused by mixtures of hormones indicate that different receptors compete for a pool of adenylate cyclase (*47*). In the turkey erythrocyte, *cis*-vaccenic acid was found to increase the rate of activation by a GTP analog plus epinephrine (*54*). This increase in rate was inversely related to the "microviscosity" of the phospholipid bilayer. (The "microviscosity" is not a well-defined quantity, but gives some measure of the restriction of motion in the membrane.) A model consistent with these data is that the hormone receptor and enzyme diffuse freely and the rate of activation is controlled by their collision rate. A more detailed mechanism for the hormone receptor–enzyme interaction is presented below. A schematic representation of the hormone-sensitive adenylate cyclase system is presented in Fig. 11-8. The regulatory and catalytic proteins are shown as separate entities in the inactive state although this may not actually be the case.

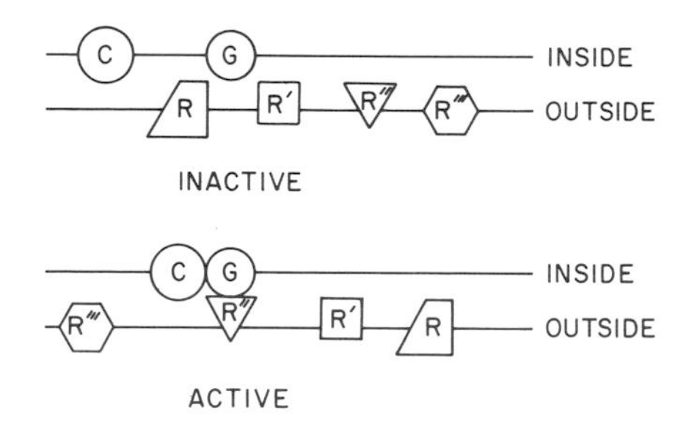

Fig. 11-8. Schematic representation of an active and inactive adenylate cyclase system. Here C is the adenylate cyclase, G is the regulatory protein which binds GTP, and R, R′, R″, R‴ are different hormone receptors.

The regulation of GTP by adenylate cyclase merits further attention. The most important and general observation is that a guanine (or related purine) nucleotide is an absolute requirement for hormonal stimulation of enzyme activity (*55*). Moreover, these purine nucleotides decrease the affinity of the receptor for the hormone (*56*). This is attributed to the association of the hormone receptor with the regulatory protein. Further insight into the regulatory process was obtained with the discovery of a catecholamine-stimulated GTPase activity in turkey erythrocyte (*57*). This led to the postulate that GTP-liganded adenylate cyclase is the active species and that hydrolysis of GTP to GDP is the primary mechanism of inactivation. Additional studies have led to the hypothetical model for a GTPase regulatory cycle shown in Fig. 11-9 (*49*). The active species, designated by a star, is the complex between the catalytic and regulatory proteins with GTP bound. When GTP hydrolyzes, an inactive species is formed. In the absence of hormone, the rate limiting step in reactivation is the dissociation of GDP. However, when the hormone–receptor complex interacts with GDP–regulatory protein complex, the dissociation of GDP is enhanced, as required by the reciprocal effect of nucleotide on hormone binding. This permits GTP to bind and activate adenylate cyclase. The catalytic protein is shown as binding only to the regulatory protein–GTP complex, although information as to whether or not it binds to other states is lacking. This model is compatible with the free diffusion of all three protein components of the system.

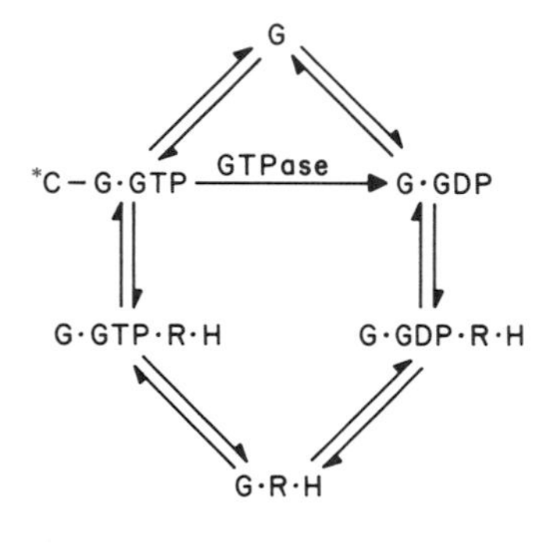

Fig. 11-9. A model for the regulatory GTPase cycle in which GDP displacement is enhanced by the hormone–receptor complex. In the diagram G is the regulatory protein, C the catalytic protein, R the hormone receptor, and H the hormone. [Adapted with permission from E. M. Ross and A. G. Gilman, *Annu. Rev. Biochem.* **49**, 533. © 1980 by Annual Reviews Inc.]

Obviously much is missing from this picture of how the hormone-sensitive adenylate cyclase system works. Further purification of proteins and reconstitution studies are needed, and quantitative studies of the thermodynamics and kinetics of the interactions of ligands and proteins remain to be done.

CONCLUSION

A quantitative understanding of membrane-bound enzymes and their coupling with transport and receptor proteins clearly is in its infancy. The brief discussion presented is intended to whet the appetite of the reader for what remains to be done.

REFERENCES

1. R. Henderson, *Annu. Rev. Biophys. Bioeng.* **6**, 87 (1977).
2. A. L. Lehninger, "Biochemistry," p. 304. Worth, New York, 1975.
3. E. Racker, *in* "Methods in Enzymology" (S. Fleischer and L. Packer, eds.), Vol. **55**, p. 699 Academic Press, New York, 1979.
4. W. W. Webb, L. S. Barak, D. W. Tank and E.-S. Wu, *Biochem. Soci. Symp.* **46**, 191 (1981).
5. H. G. O. Bock and S. Fleischer, *in* "Methods in Enzymology" (S. Fleischer and L. Packer, eds.), Vol. 32, p. 374. Academic Press, New York, 1974.
6. H. M. Menzel and G. G. Hammes, *J. Biol. Chem.* **248**, 4885 (1973).
7. A. K. Grover, A. J. Slotboom, G. H. de Haas and G. G. Hammes, *J. Biol. Chem.* **250**, 31 (1975).
8. Y. A. Isaacson, P. W. Deroo. A. F. Rosenthal, R. Bittman, J. O. McIntyre, H. G. Bock, P. Gazzotti, and S. Fleischer, *J. Biol. Chem.* **254**, 117 (1979).
9. S. Fleischer, J. O. McIntyre, W. Stoffel, and P. D. Tunggal, *Biochemistry* **18**, 2420 (1979).
10. N. C. Nielson and S. Fleischer, *J. Biol. Chem.* **248**, 2549 (1973).
11. N. Latruffe, S. C. Brenner, and S. Fleischer, *Biochemistry* **19**, 5285 (1980).
12. M. S. El Kebbaj, N. Latruffe, and Y. Gaudemer, *Biochem. Biophys. Res. Commun.* **96**, 1569 (1980).
13. J. C. Skou, *Q. Rev. Biophys.* **1**, 401 (1975).
14. J. D. Robinson and M. S. Flashner, *Biochim. Biophys. Acta* **549**, 145 (1979).
15. L. C. Cantley, *Curr. Top. Bioenerg.* **11**, 201 (1981).
16. J. C. Skou, *Biochim. Biophys. Acta* **23**, 394 (1957).
17. R. Whittam, *Biochem. J.* **84**, 110 (1962).
18. D. F. Hastings and J. A. Reynolds, *Biochemistry* **18**, 817 (1979).
19. S. M. Goldin, *J. Biol. Chem.* **252**, 5630 (1977).
20. B. Forbush, III, and J. F. Hoffman, *Biochemistry* **18**, 2308 (1979).
21. R. L. Post and S. Kume, *J. Biol. Chem.* **248**, 6993 (1973).
22. L. C. Cantley Jr., L. G. Cantley, and L. Josephson, *J. Biol. Chem.* **253**, 7361 (1978).
23. C. Kaniike, E. Lindenmeyer, E. Wallick, L. Lane, and A. Schwartz, *J. Biol. Chem.* **251**, 4794 (1976).
24. M. Matsui, Y. Hayashi, H. Homareda, and M. Kimimura, *Biochem. Biophys. Res. Commun.*, **75**, 373 (1977).
25. C. M. Grisham, *Adv. Inorg. Biochem.* **1**, 193 (1979).
26. A. Ruoho and J. Kyte, *Proc. Natl. Acad. Sci. U.S.A.* **71**, 2352 (1974).
27. J. Kyte, *J. Biol. Chem.* **249**, 3652 (1974).
28. F. Vogel, H. W. Meyer, R. Gresse, and K. R. H. Repke, *Biochim. Biophys. Acta* **470**, 497 (1977).
29. R. L. Post, C. Hegyvary, and S. Kume, *J. Biol. Chem.* **247**, 6530 (1972).
30. P. Mitchell, *Nature (London)* **191**, 144 (1961).
31. E. Racker, "A New Look at Mechanisms in Bioenergetics." Academic Press, New York, 1976.
32. P. D. Boyer, B. Chance, L. Ernster, P. Mitchell, E. Racker, and E. C. Slater, *Annu. Rev. Biochem.* **46**, 955 (1977).
33. H. S. Penefsky, *Adv. Enzymol.* **49**, 223 (1979).
34. R. H. Fillingame, *Annu. Rev. Biochem.* **49**, 1079 (1980).
35. Y. Kagawa, *Biochim. Biophys. Acta* **505**, 45 (1978).
36. B. A Baird and G. G. Hammes, *Biochim. Biophys. Acta* **549**, 31 (1979).
37. N. Shavit, *Annu. Rev. Biochem.* **49**, 111 (1980).
38. U. Pick and E. Racker, *J. Biol. Chem.* **254**, 2793 (1979).

39. B. A. Baird and G. G. Hammes, *J. Biol. Chem.* **251**, 6953 (1976).
40. M. F. Bruist and G. G. Hammes, *Biochemistry* **20**, 6298 (1981).
41. D. W. Deters, N. Nelson, H. Nelson, and E. Racker, *J. Biol. Chem.* **250**, 1041 (1975).
42. R. A. Cerione and G. G. Hammes, *Biochemistry* **20**, 3359 (1981).
43. E. Racker and W. Stoeckenius, *J. Biol. Chem.* **249**, 662 (1974).
44. T. G. Dewey and G. G. Hammes, *J. Biol. Chem.* **256**, 8941 (1981).
45. M. R. Webb, C. Grubmeyer, H. S. Penefsky, and A. R. Trentham, *J. Biol. Chem.* **255**, 11637 (1980).
46. J. P. Perkins, *Adv. Cyclic Nucleotide Res.* **3**, 1 (1973).
47. M. Rodbell, M. C. Lin, Y. Solomon, C. Londos, J. P. Harwood, B. R. Martin, M. Rendell, and M. Berman, *Adv. Cyclic Nucleotide Res.* **5**, 3 (1975).
48. R. J. Lefkowitz, L. E. Limbird, C. Mukherjee, and M. G. Caron, *Biochim. Biophys. Acta* **457**, 1 (1976).
49. E. M. Ross and A. G. Gilman, *Annu. Rev. Biochem.* **49**, 533 (1980).
50. J. Orly and M. Schramm, *Proc. Natl. Acad. Sci. U.S.A.* **76**, 1174 (1979).
51. T. Pfeuffer, *J. Biol. Chem.* **252**, 7224 (1977).
52. E. M. Ross and A. G. Gilman, *Proc. Natl. Acad. Sci. U.S.A.* **74**, 3715 (1977).
53. F. M. Hoffman, *J. Biol. Chem.* **254**, 255 (1979).
54. E. Hanski, G. Rimon, and A. Levitzki, *Biochemistry* **18**, 846 (1979).
55. M. Rodbell, L. Birnbaumer, S. L. Pohl, and H. M. J. Krans, *J. Biol. Chem.* **246**, 1877 (1971).
56. M. Rodbell, H. M. J. Krans, S. L. Pohl, and L. Birnbaumer, *J. Biol. Chem.* **246**, 1872 (1971).
57. D. Cassel and Z. Selinger, *Biochim. Biophys. Acta* **452**, 538 (1976).

Appendix

PRACTICE PROBLEMS

1. Consider the following fluorescence resonance energy transfer experiment.
   a. An energy donor is bound at a single site on an enzyme; it has a fluorescence lifetime of 5 nsec and a quantum yield of 0.5. An energy acceptor is bound at a single site on the enzyme and the fluorescence lifetime is reduced to 3 nsec and the quantum yield to 0.3. Calculate the distance between the donor and acceptor assuming a value for $R_{\circ}$ of 30 Å.
   b. Assume that *two acceptor* molecules are bound per enzyme molecule and that they are equidistant from the single donor. Calculate the distance between the donor and acceptors using the same data as above.
   c. Assume that *two donor* molecules are bound per enzyme molecule and that they are equidistant from the single acceptor. Calculate the distance between the donors and the acceptor using the same data as in a.
2. The following data were obtained for the measurement of T_1 of the protons of succinate bound to aspartate transcarbamoylase in the presence of covalently bound Zn^{2+} and in the presence of covalently bound Mn^{2+} [S. Fan, L. W. Harrison, and G. G. Hammes, *Biochemistry* **14**, 2219 (1975)].

Metal	Frequency (MHz)	$1/T_{1M}$ (sec^{-1})
Zn	100	22.8
Mn	100	62.8
Zn	220	9.8
Mn	220	19.4

Calculate the distance between the Mn^{2+} and the bound succinate protons. Assume τ_c is frequency independent and that fast exchange occurs. Hint: Using Eq. (2-25), the paramagnetic contribution to the relaxation time can be written as

$$\frac{1}{T_{1M}} = \left(\frac{815}{r}\right)^6 \frac{3\tau_c}{1 + \omega^2\tau_c^2}$$

where r is in Ångstroms, τ_c is in seconds, and $\omega = 2\pi$ (frequency). For Mn^{2+}, the relaxation time is determined by both paramagnetic and diamagnetic effects, whereas for Zn^{2+} only diamagnetic relaxation occurs.

3. The hydration of CO_2

$$CO_2 + H_2O \longrightarrow HCO_3^- + H^+$$

is catalyzed by the enzyme carbonic anhydrase. A stopped flow apparatus has been used to study the steady-state kinetics of the forward and reverse reactions at pH 7.1, 0.5° C, and 2×10^{-3} M phosphate buffer with bovine carbonic anhydrase. Some typical data obtained are given in the table below [H. DeVoe and G. B. Kistiakowsky, *J. Am. Chem. Soc.* **83**, 274 (1961)].

Hydration		Dehydration	
$10^{-3}/v$ $(M^{-1}\ sec^{-1})$[a]	$10^3[CO_2]$ (M)	$10^{-3}/v$ $(M^{-1}\ sec^{-1})$[a]	$10^3[HCO_3^-]$ (M)
36	1.25	95	2
20	2.5	45	5
12	5	29	10
6	20	25	15

[a] Reciprocal initial velocity with a total enzyme concentration of 2.8×10^{-9} M.

a. Calculate the steady-state kinetic parameters for the forward and reverse reactions. Calculate the turnover numbers for the forward and reverse reactions.
b. Calculate the equilibrium constant for the overall reaction.
c. Calculate lower bonds of the rate constants for a mechanism with an arbitrary number of isomeric reaction intermediates.

4. Derive the steady-state rate equation for the following mechanism

$$E + S \rightleftharpoons X_1 \rightleftharpoons X_2 \rightleftharpoons X_3 \longrightarrow E + P$$

5. The following initial velocity data have been obtained for the inhibition of the ribonuclease catalyzed hydrolysis of cytidine 2′,3′-cyclic phosphate by orthophosphate. Decide what type of inhibition is occurring and calculate the inhibition constant.

$1/v$ (arbitrary units)	10^4 [S] (M)	10^3 [I] (M)
1.3	11.3	0
1.6	11.3	5
2.05	11.3	10
2.4	11.3	15
2.7	11.3	20
1.7	6.26	0
2.3	6.26	5
2.9	6.26	10
3.55	6.26	15
4.0	6.26	20
2.45	3.76	0
3.3	3.76	5
4.35	3.76	10
5.3	3.76	15
6.2	3.76	20

6. Consider the three two substrate–two product reactions discussed in Chapter 3.

I. $$\mathrm{A + E \underset{k_{-1}}{\overset{k_1}{\rightleftharpoons}} EA}$$

$$\mathrm{EA + B \underset{k_{-2}}{\overset{k_2}{\rightleftharpoons}} X \underset{k_{-3}}{\overset{k_3}{\rightleftharpoons}} ED + C}$$

$$\mathrm{ED \underset{k_{-4}}{\overset{k_4}{\rightleftharpoons}} E + D}$$

II. $$\mathrm{A + E \underset{k_{-1}}{\overset{k_1}{\rightleftharpoons}} X_1 \underset{k_{-2}}{\overset{k_2}{\rightleftharpoons}} X_2 + C}$$

$$\mathrm{B + X_2 \underset{k_{-3}}{\overset{k_3}{\rightleftharpoons}} X_3 \underset{k_{-4}}{\overset{k_4}{\rightleftharpoons}} E + D}$$

III. $$\mathrm{E + A \overset{K_1}{\rightleftharpoons} EA}$$

$$\mathrm{E + B \overset{K_2}{\rightleftharpoons} EB}$$

$$\mathrm{EA + B \overset{K_2}{\rightleftharpoons} X_1 \underset{k'}{\overset{k}{\rightleftharpoons}} X_2 \rightleftharpoons EC + D \text{ (or ED + C)}}$$

$$\mathrm{EB + A \overset{K_1}{\rightleftharpoons} X_1 \underset{k'}{\overset{k}{\rightleftharpoons}} X_2 \rightleftharpoons EC + D \text{ (or ED + C)}}$$

$$\mathrm{EC \overset{K_{-2}}{\rightleftharpoons} E + C}$$

$$\mathrm{ED \overset{K_{-1}}{\rightleftharpoons} E + D}$$

Assume rapid establishment of equilibria for biomolecular steps.

a. Derive the expressions for the initial velocities of the forward and reverse reactions and put into the forms given in Chapter 3.
b. Derive the "Haldane" relationships for these mechanisms, that is obtain an expression for the overall equilibrium constants in terms of the ϕs associated with the forward and reverse reactions.

7. The reaction catalyzed by the enzyme hexokinase is

$$\text{MgATP} + \text{G} \rightleftharpoons \text{MgADP} + \text{G6P}$$

where G is glucose and G6P is glucose 6-phosphate. What mechanisms are consistent with the initial-rate steady-state data given below for yeast hexokinase at pH 8, 25°C, and 0.3 *M* $(CH_3)_4NCl$ [G. G. Hammes and D. Kochavi, *J. Am. Chem. Soc.* **84**, 2069, 2073, 2076 (1962)].

10^4 [MgATP] (*M*)	10^4 [G] (*M*)	10^3 $[E_0]/v$ (sec)	10^4 [MgATP] (*M*)	10^3 [G] (*M*)	10^3 $[E_0]/v$ (sec)
4.73	10	2.55	19.4	10	1.83
	5	2.90		5	2.17
	2	4.54		2	3.17
	1	7.48		1	4.87
9.20	10	2.21	40.0	10	1.72
	5	2.51		5	2.03
	2	3.90		2	2.93
	1	6.10		1	4.44

8. A simplified mechanism for the chymotrypsin catalyzed hydrolysis of esters can be written as

$$E + S \underset{k_{-1}}{\overset{k_1}{\rightleftharpoons}} ES \underset{k_{-2}}{\overset{k_2}{\rightleftharpoons}} EP_2 + P_1 \underset{k_{-3}}{\overset{k_3}{\rightleftharpoons}} E + P_1 + P_2$$

In this mechanism the products, P_1 and P_2, are released in consecutive steps.
a. Use the King–Altman method to derive the steady-state rate equation for the initial velocity ($[P_1] = [P_2] = 0$).
b. If you wished to solve the transient rate equations, how many *independent* rate equations are needed to describe this system?

9. Consider the simple Michaelis–Menten mechanism

$$E + S \underset{k_{-1}}{\overset{k_1}{\rightleftharpoons}} X \underset{k_{-2}}{\overset{k_2}{\rightleftharpoons}} E + P$$

Derive the relaxation times associated with this mechanism.

10. Assume an enzyme has two ionizable groups that are involved in catalysis. The diprotonated species can be written as HEH where the left side of E represents one ionizable group and the right side the other ionizable group.

a. Assume EH is the active species and calculate its concentration in terms of the total enzyme concentration, the hydrogen ion concentration, and any ionization constants you may wish to define.
b. Assume HE is the active species and carry out the same calculation as in a.
c. Is it possible to distinguish whether EH or HE is the active species from the pH dependence of the enzymatic reaction? Explain your answer.
Hint: The ionization of the enzyme can be written as

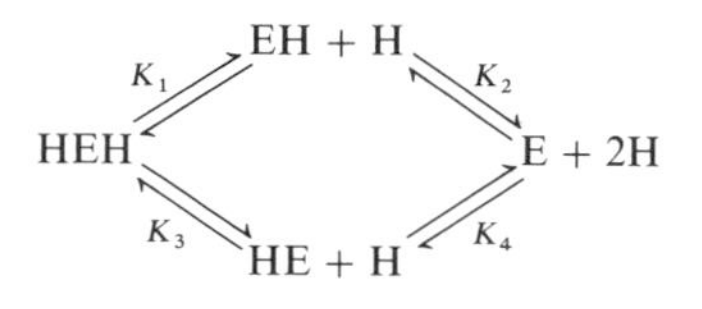

The principle of detailed balance requires that $K_1K_2 = K_3K_4$. These Ks are the *microscopic* dissociation constants for the ionizable groups. As you will see if you work this problem correctly the measured dissociation constants are functions of the microscopic dissociation constants.

11. The following data have been obtained for the hydrolysis of the mixed anhydride acetyl phosphate [G. DiSabato and W. P. Jencks, *J. Am. Chem. Soc.* **83**, 4400 (1961)].

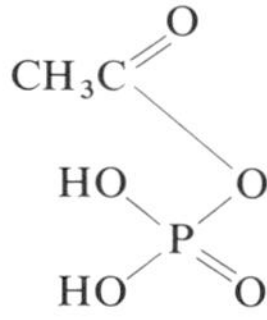

pH	$10^{-3}\ k_{obs}$ (min^{-1})	pH	$10^{-3}\ k_{obs}$ (min^{-1})
0.2	65.0	6.0	4.0
0.5	40.0	7.0	4.0
1.2	20.0	8.0	4.0
2.0	11.0	9.0	4.0
3.0	10.0	10.0	4.5
4.0	9.0	11.0	9.0
4.5	7.0	11.5	25.0
5.0	5.0	12.0	80.0

Assume only the mono and di-ionized forms exist over the pH range 0.2–12.0 and write down a rate law that fits the data. Evaluate the constants in your rate law.

[This need not be done quantitatively, which would require a computer fitting, but indicate how you estimate the constants.]

12. A hypothetical mechanism for the base catalyzed iodination of acetone can be written as

$$B + HCH_2\overset{\overset{O}{\|}}{C}CH_3 \underset{}{\overset{k}{\underset{\text{slow}}{\rightleftharpoons}}} BH + CH_2\overset{\overset{O}{\|}}{C}CH_3 \xrightarrow[\text{fast}]{I_2} ICH_2\overset{\overset{O}{\|}}{C}CH_3 + I^- + BH$$

   a. If B is the basic form of the buffer and K_A is the ionization constant of the buffer, derive an expression for the pH dependence of the ratio of the pseudo first order rate constant to the total buffer concentration, $k_{obs}/[B_T]$.
   b. If $pK_A = 8$ and the pK of the methyl group hydrogen is assumed to be 17, calculate an approximate value of the rate constant k based on your knowledge of the rates of proton transfer reactions. Clearly indicate any assumptions you make.

13. The hydrolysis of an ester has been found to be catalyzed by imidazole and the observed pseudo first order rate constant can be described by the expression

$$k_{obs} = k_1[Im][ImH^+] + k_2[Im]$$

   where Im is the free base form of imidazole and ImH^+ the protonated form.

   a. What would be the shape of a plot of k_{obs} versus the concentration of free base, at constant buffer ratio and ionic strength?
   b. How would you calculate k_1 and k_2 for this system?

14. The following data were obtained for the binding of ADP to a chloroplast ATPase. How many binding sites are present per mole of enzyme and what is the intrinsic binding constant? In the table below, r is the ratio of

r	[ADP] (μM)
0.500	0.719
0.694	1.23
1.06	2.36
1.22	3.30
1.32	4.55
1.63	8.81
1.69	16.32
1.91	27.15
2.21	40.62
2.15	79.75
2.40	115.5

the moles of ADP bound to the total protein concentration and [ADP] is the concentration of free ADP. [Note that nonspecific binding, i.e., low affinity binding occurs; you will have to decide how to handle this problem.]

15. The following initial velocities, v, were measured for the carbamoylation of aspartic acid by carbamoyl phosphate as catalyzed by the enzyme aspartate transcarbamoylase. The concentration of carbamoyl phosphate was 1 mM, and the concentration of aspartic acid was varied.

v (arbitary units)	Aspartate (mM)
0.90	2
1.60	3
2.25	4
3.20	5
3.65	6
4.70	8
5.05	10
5.25	12
5.80	15
6.00	17
6.05	20

a. Construct a Hill plot of the data and estimate the "minimum" number of aspartate binding sites.
b. In fact six aspartate binding sites are present per mole of protein. Using this fact and the initial velocity data, make a table of r versus aspartate and a plot of r/[aspartate] versus r. Indicate what assumptions you made in doing this.
c. Indicate which models would be appropriate to describe these data; please indicate appropriate equations. Unfortunately, a computer is required for data fitting (unless you have plenty of free time).

16. Consider an enzyme which has two different conformations, A and B, both of which can bind ligands but with different binding constants. The equilibrium constant, [B]/[A], in the absence of ligands is K_0.
a. Assume that two ligands L1 and L2 bind to conformations A and B with *intrinsic* binding constants K_{A1}, K_{A2}, K_{B1}, and K_{B2} and that two equivalent independent binding sites exist for each ligand. Derive an expression for the ratio $\sum B/\sum A$ in the presence of L1 where the sums indicate all species, liganded and unliganded.
b. Derive an expression for the ratio $\sum B/\sum A$ in the presence of both L1 and L2.
c. A little thought and proper arrangement of the expressions you have

derived should allow you to generalize the results of a and b to n equivalent binding sites. Please do so.

d. The result of c can be further generalized to any number of ligands. Write down an expression for m ligands.

[Using these techniques, for example, a more general Monod–Wyman–Changeux model can be generated.]

17. The binding of a regulatory ligand to an allosteric enzyme was studied with equilibrium and kinetic methods. Three relaxation times were observed. The equilibrium and kinetic data obtained are summarized in the table below. The kinetic data were obtained at an enzyme concentration of 1 μM, whereas the binding data were obtained at an enzyme concentration of 10 μM–1 mM; [L] is the *free* ligand concentration. Determine a binding mechanism that is consistent with all of the data. (Use the allosteric models discussed in Chapter 8 as the basis of your considerations.) Determine all possible equilibrium and rate constants from the data. Construct plots of the relaxation times and r versus [L] from the data, and show that the theoretical curves calculated with the constants you have determined are consistent with the data.

	Data			
[L]	$1/\tau_1$ (sec^{-1})	$1/\tau_2$ (sec^{-1})	$1/\tau_3$ (sec^{-1})	r
20 μM	102	1000	10.6	0.239
50 μM	105	1010	11.2	0.461
100 μM	200	1100	11.6	0.679
200 μM	300	1200	11.7	0.913
400 μM	500	1400	11.1	1.15
1 mM	1100	2000	9.5	1.45
2 mM	2100	3000	8.1	1.65
5 mM	5100	6000	6.6	1.83
(∞, extrapolated)	—	—	5.0	—

18. The following binding, initial velocity and fast reaction data have been obtained for an enzyme. The turnover number of the enzyme is sufficiently small so that direct measurements of substrate binding could be made. On the basis of these data, derive a simple regulatory mechanism. Use the data to determine all parameters associated with this mechanism: equilibrium constants, rate constants, etc. Show that your mechanism and parameters are consistent with the data by calculation of theoretical curves of r versus [S], $v/[E_0]$ versus [S], and $1/\tau_1$, $1/\tau_2$, $1/\tau_3$ versus [S_0]. Show graphs of the calculated and experimental curves. [Note: This problem has been designed to give manageable calculations and simple

numerical parameters; if you generate complex and strange things, you probably have gone wrong somewhere. However, the use of a pocket calculator is recommended.]

Binding and Initial Velocity Data

[S] (μM)	r	$10^3 v/[E_0]$ (sec^{-1})
2	0.144	0.72
4	0.361	1.81
8	0.816	4.08
10	1.0	5.00
20	1.52	7.60
40	1.81	9.05
100	1.94	9.70

Fast Reaction Data
$[E_0] = 0.7\ \mu M$

$[S_0]$ (μM)	$10^{-3}/\tau_1$ (sec^{-1})	$10^{-3}/\tau_2$ (sec^{-1})	$1/\tau_3$ (sec^{-1})
[0, extrapolated	10	0.01	1.0]
10	11	0.11	2.0
40	14	0.41	9.58
80	18	0.81	21.3
100	20	1.01	26.5
200	30	2.01	45.9
500	60	5.01	70.7
[∞, extrapolated	—	—	101.0]

In the tables, [S] is the free substrate concentration; r is the moles of ligand bound/mole of protein; v is the initial velocity; $[E_0]$ is the total enzyme concentration; $[S_0]$ is the total substrate concentration; and the τ_i are relaxation times.

Index

F

H

I

K

L

M

N

X

Molecular Biology

An International Series of Monographs and Textbooks

HAROLD A. SCHERAGA. Protein Structure. 1961

STUART A. RICE AND MITSURU NAGASAWA. Polyelectrolyte Solutions: A Theoretical Introduction, *with a contribution by Herbert Morawetz*. 1961

SIDNEY UDENFRIEND. Fluorescence Assay in Biology and Medicine. Volume I—1962. Volume II—1969

J. HERBERT TAYLOR (Editor). Molecular Genetics. Part I—1963. Part II—1967. Part III—Chromosome Structure—1979

ARTHUR VEIS. The Macromolecular Chemistry of Gelatin. 1964

M. JOLY. A Physico-chemical Approach to the Denaturation of Proteins. 1965

SYDNEY J. LEACH (Editor). Physical Principles and Techniques of Protein Chemistry. Part A—1969. Part B—1970. Part C—1973

KENDRIC C. SMITH AND PHILIP C. HANAWALT. Molecular Photobiology: Inactivation and Recovery. 1969

RONALD BENTLEY. Molecular Asymmetry in Biology. Volume I—1969. Volume II—1970

JACINTO STEINHARDT AND JACQUELINE A. REYNOLDS. Multiple Equilibria in Protein. 1969

DOUGLAS POLAND AND HAROLD A. SCHERAGA. Theory of Helix-Coil Transitions in Biopolymers. 1970

JOHN R. CANN. Interacting Macromolecules: The Theory and Practice of Their Electrophoresis, Ultracentrifugation, and Chromatography. 1970

WALTER W. WAINIO. The Mammalian Mitochondrial Respiratory Chain. 1970

LAWRENCE I. ROTHFIELD (Editor). Structure and Function of Biological Membranes. 1971

ALAN G. WALTON AND JOHN BLACKWELL. Biopolymers. 1973

WALTER LOVENBERG (Editor). Iron-Sulfur Proteins. Volume I, Biological Properties—1973. Volume II, Molecular Properties—1973. Volume III, Structure and Metabolic Mechanisms—1977

A. J. HOPFINGER. Conformational Properties of Macromolecules. 1973

R. D. B. FRASER AND T. P. MACRAE. Conformation in Fibrous Proteins. 1973

OSAMU HAYAISHI (Editor). Molecular Mechanisms of Oxygen Activation. 1974

FUMIO OOSAWA AND SHO ASAKURA. Thermodynamics of the Polymerization of Protein. 1975

LAWRENCE J. BERLINER (Editor). Spin Labeling: Theory and Applications. Volume I, 1976. Volume II, 1978

T. BLUNDELL AND L. JOHNSON. Protein Crystallography. 1976

HERBERT WEISSBACH AND SIDNEY PESTKA (Editors). Molecular Mechanisms of Protein Biosynthesis. 1977

TERRANCE LEIGHTON AND WILLIAM F. LOOMIS, JR. (Editors). The Molecular Genetics of Development: An Introduction to Recent Research on Experimental Systems. 1980

ROBERT B. FREEDMAN AND HILARY C. HAWKINS (Editors). The Enzymology of Post-Translational Modification of Proteins, Volume 1. 1980

WAI YIU CHEUNG (Editor). Calcium and Cell Function, Volume I: Calmodulin. 1980

OLEG JARDETZKY and G. C. K. ROBERTS. NMR in Molecular Biology. 1981

DAVID A. DUBNAU (Editor). The Molecular Biology of the Bacilli, Volume I: *Bacillus subtilis*. 1982

GORDON G. HAMMES. Enzyme Catalysis and Regulation. 1982

In preparation

CHARIS GHELIS and JEANNINE YON. Protein Folding. 1982

WAI YIU CHEUNG (Editor). Calcium and Cell Function, Volume II. 1982

P. R. CAREY. Biochemical Applications of Raman and Resonance Raman Spectroscopies. 1982

GUNTER KAHL and JOSEF S. SCHELL (Editors). Molecular Biology of Plant Tumors. 1982